Topics in Mining, Metallurgy and Materials Engineering

Series Editor

Carlos P. Bergmann, Federal University of Rio Grande do Sul, Porto Alegre, Rio Grande do Sul, Brazil

"Topics in Mining, Metallurgy and Materials Engineering" welcomes manuscripts in these three main focus areas: Extractive Metallurgy/Mineral Technology; Manufacturing Processes, and Materials Science and Technology. Manuscripts should present scientific solutions for technological problems. The three focus areas have a vertically lined multidisciplinarity, starting from mineral assets, their extraction and processing, their transformation into materials useful for the society, and their interaction with the environment.

**** Indexed by Scopus (2020) ****

More information about this series at http://www.springer.com/series/11054

Mostafa Mohamed Ali Elbeblawi ·
Hassan Ali Abdelhak Elsaghier ·
Mostafa Tantawy Mohamed Amin ·
Wael Rashad Elrawy Abdellah

Surface Mining Technology

Springer

Mostafa Mohamed Ali Elbeblawi
Department of Mining and Metallurgical Engineering
Faculty of Engineering
Assiut University
Assiut, Egypt

Hassan Ali Abdelhak Elsaghier
Department of Mining and Metallurgical Engineering
Faculty of Engineering
Assiut University
Assiut, Egypt

Mostafa Tantawy Mohamed Amin
Department of Mining and Metallurgical Engineering
Faculty of Engineering
Assiut University
Assiut, Egypt

Wael Rashad Elrawy Abdellah
Department of Mining and Metallurgical Engineering
Faculty of Engineering
Assiut University
Assiut, Egypt

ISSN 2364-3293 ISSN 2364-3307 (electronic)
Topics in Mining, Metallurgy and Materials Engineering
ISBN 978-981-16-3567-0 ISBN 978-981-16-3568-7 (eBook)
https://doi.org/10.1007/978-981-16-3568-7

This Springer imprint is published by the registered company Springer Nature Singapore Pte Ltd.
The registered company address is: 152 Beach Road, #21-01/04 Gateway East, Singapore 189721, Singapore

Preface

Surface mining methods account for the vast majority of mineral production worldwide. It is currently used to extract almost all non-metallic minerals (e.g. 95%), most metallic minerals (e.g. 90%), and a significant portion of coal (e.g. 60%). Alternatively, surface mining methods account for mining about 25 billion tonnes of ore and waste materials out of 30 billion tonnes each year.

This book introduces a variety of surface mining methods to both undergraduate mining students and mining engineers. Drilling, blasting, loading, and hauling are common operations in most methods. Because of technological advancements over the years, surface methods can now be applied to deeper and leaner deposits. A great deal of effort was expended in this book to demonstrate procedures for exploitation of open surface mine through its closure (e.g. land reclamation).

The content of this book is interestingly organized to cover a wide range of topics, beginning with the definition of mining, surface mining terminology, the choice between surface and underground mining, surface mining types, mining plan options, slope stability, the concept of cut-off, surface mine field ratios, stripping ratio considerations, equipment optimization and selection, mining work delineation or development, volume of trenches, surface mining methods, and systems and safety in surface mining. As a result, students can learn how to apply what they have learned in class to real-world engineering projects. The book, in particular, assists students in preparing for jobs in surface mining (e.g. open pit mine and quarries).

This book aims to teach students how to use maps to locate the target (e.g. ore), how to apply for permits for access, prospecting, mining, mine closure, and land reclamation, how to use data collected in the field to determine mine feasibility, and how to choose the best method for locating ore bodies (e.g. geochemical, geological, or geophysical sampling) and to list the factors that influence the feasibility of a prospective mine, such as the geological layout of the mineral to be mined, market economics, labour costs, access to mine equipment, and environmental factors.

Chapter 1 presents an overview of introduction to mining. Chapter 2 demonstrates the principles of surface mining of mineral deposits. Chapter 3 presents the slope stability. Chapter 4 introduces the prevention of slides and falls in surface mines. Chapter 5 describes the surface mine development and closes up special topics (e.g.

the concept of stripping ratios and pit limits). Chapter 6 presents the essential equipment used in surface mining. Chapter 7 describes rock extraction with scrapers, bulldozers, and loaders. Chapter 8 introduces the surface mining methods and systems. The book back matter gives a detailed glossary of surface mining terms.

Authors are indebted to reviewers for their comments which enabled us to correct errors and clarify the presentation.

Authors would like to thank the staff at Springer, in particular Ramesh Nath Premnath, Boopalan Renu, and N. S. Pandian, for their help and support.

Assiut, Egypt

Mostafa Mohamed Ali Elbeblawi
Hassan Ali Abdelhak Elsaghier
Mostafa Tantawy Mohamed Amin
Wael Rashad Elrawy Abdellah

Contents

List of Figures

List of Tables

Chapter 1
Introduction to Mining

1.1 Advancements in Mining Technology

Many metals occur in their native state or in readily accessible ores. Thus, the extraction and working of metals dates much further back in time than does the mining industry. Some of the earliest known mines were those developed by the Greeks in the sixth century B.C. As were mines for many centuries thereafter, the workers in these mines were slaves and prisoners of war. By the time, the Roman Empire reached its peak; it had established mines throughout the European continent, in the British Isles, and in parts of North Africa.

The first scientific description of mining operation was the book De Re Metallica by the Saxon physician Georgius Agricola (1494–1555). De Re Metallica, which remained an authoritative reference for nearly 200 years, was translated from Latin to English by mining engineer and former United States president Herbert Hoover (1874–1964) and his wife Lou Henry Hoover (1874–1944). It is not possible to chronicle all of the developments that made mining what it is today. A more complete chronology of the important events is outlined in Table 1.1.

1.2 Introductory to Mining

Mining is the branch of industry involving the exploration and removal the rocks or ores or minerals from the earth. Mining is one of the oldest and most important endeavors of humankind, because it provides the raw ingredients for most of the material world around us (see Table 1.2) and, like agriculture, is the lifeblood of civilization. The main objective of any type of mining is to remove the valuable material economically and safely with minimum damage to the surrounding environment. Surface mine method is considered the first stage in the history of the development of mining technology and more than 80% of the total output of material are produced by this method. This method is applied only when rocks, ores, and minerals occurring

M. M. Ali Elbeblawi et al., *Surface Mining Technology*, Topics in Mining, Metallurgy and Materials Engineering, https://doi.org/10.1007/978-981-16-3568-7_1

Table 1.1 Chronological development of mining technology

Date	Event
450,000 B.C.E	First mining (at surface), by Paleolithic humans for stone implements
40,000	Surface mining progresses underground, in Swaziland, Africa
30,000	Fired clay pots used in Czechoslovakia
18,000	Possible use of gold and copper in native form
5000	Fire setting, used by Egyptians to break rock
4000	Early use of fabricated metals; start of Bronze Age
3400	First recorded mining, of turquoise by Egyptians in Sinai
3000	Probable first smelting, of copper with coal by Chinese; first use of iron Implements by Egyptians
2000	Earliest known gold artifacts in New World, in Peru
1000	Steel used by Greeks
100 C.E	Thriving Roman mining industry
122	Coal used by Romans in present-day United Kingdom
1185	Edict by bishop of Trent gives rights to miners
1524	First recorded mining in New World, by Spaniards in Cuba
1550	First use of lift pump, at Joachimstal, Czechoslovakia
1556	First mining technical work, De Re Metallica, published in Germany by Georgius Agricola
1585	Discovery of iron ore in North America, in North Carolina
1600s	Mining commences in eastern United States (iron, coal, lead and gold)
1627	Explosives first used in European mines, in Hungary (possible prior use in China)
1646	First blast furnace installed in North America, in Massachusetts
1716	First school of mines established, at Joachimstal, Czechoslovakia
1780	Beginning of Industrial Revolution; pumps are first modern machines used in mines
1800s	Mining progresses in United States; gold rushes help open the West
1815	Sir Humphrey Davy invents miner's safety lamp in England
1855	Bessemer steel process first used, in England
1867	Dynamite invented by Nobel, applied to mining
1903	Era of mechanization and mass production opens in U.S. mining with development of first low-grade copper porphyry, in Utah; although the first modern mine was an open pit, subsequent operations were underground as well
1940	First continuous miner initiates the era of mining without explosives
1945	Tungsten carbide bits developed by McKenna Metals Company (now Kennametal)

Table 1.2 Commonly mined materials and end uses

Mined material	End uses
Coal	Generating electricity, making iron and steel, manufacturing chemicals and other products
Sand and Gravel	Building roads, homes, schools, offices, and factories
Limestone	As dimension stone, the chief raw ingredient in cement, a fertilizer and soil conditioner, as a flux in the melting of iron
Iron ore	Steel products (automobiles, ships, buildings)
Aluminum ore (Bauxite)	Military aircrafts, naval vessels, pots and pans, beverage cans
Copper ore	Electrical motors, generators, communications equipment, wiring
Silver ore	Electric and electronic circuitry, coins, jewelry, photographic film
Gold ore	Jewelry, satellites, sophisticated electronic circuits
Zinc	Die casting, galvanizing brass and bronze, protective coating on steel, chemical compounds in rubber and paints
Lead	Batteries, solder, electronic components
Clay	Bricks, paper, paint, glass, pottery, linoleum, concrete, wallboard, spackling, pencils, microwavable containers, vegetable oil
Gypsum	Concrete, wallboard, spackling, caulking, potting soil
Phosphate	Plant fertilizers
Salt (Halite)	Cooking, drinking water, plastics, de-icing detergents
Asbestos	Fireproof fabrics, yarn, cloth, paper, paint filler, gaskets, roofing composition
Antimony	Hardening alloy for lead, sheet and pipes, semiconductor technology, collapsible tubes and foil
Barium	Used as a heavy additive in oil-well-drilling mud, in the paper and rubber industries, in radiography
Feldspar	Glass and ceramic industries, pottery and enamelware, soaps, abrasives; bond for abrasive wheels, fertilizer
Beryllium	Nuclear industry and in light, very strong alloys used in the aircraft industry
Chromite	Chemical and metallurgical industries use
Cobalt	Super alloys for jet engines, chemicals (paint driers, catalysts, magnetic coatings)
Nickel	An alloy to stainless steel plays key role in the chemical and aerospace industries
Lithium	Ceramics and glass, in primary aluminum production, in the manufacture of lubricants and greases, rocket propellants
Mica	Electronic insulators (mainly in vacuum tubes), in paint, as joint cement, as a dusting agent, in plastics
Molybdenum	In alloy steels (47% of all uses)
Quartz	Used as a semiprecious gem stone pressure gauges, oscillators, resonators, and wave stabilizers

(continued)

Table 1.2 (continued)

Mined material	End uses
Titanium	In jet engines, airframes, and space and missile applications
Tungsten	Filament in light bulbs, as carbide in drilling equipment, in heat and radiation shielding textile dyes, enamels, paints
Silica	In manufacture of glass and refractory materials, ceramic, abrasives, water filtration
Vanadium	In metal alloys, in the production of aerospace titanium alloys
Sulfur	In the manufacture of sulfuric acid, fertilizers, chemicals, explosives, dyestuffs, petroleum refining, rubber, fungicides
Manganese	Essential to iron and steel production

on or close to the surface or at shallow deposits, but nowadays much deeper deposits are worked by surface mine method. For further reading: Darling [1] and Hartman [2].

The technology and integrated mechanization of surface mining are considered as the science of the laws of organization and accomplishment of surface mining work on the basis of integrated mechanization at all stages of existence of mining enterprises. Surface mining machinery has shown a marked tendency to increase in size to enable the deposits to be mined economically at high outputs, and surface mining is tending to become a material handling problem on a massive scale. The main reasons for the large increase in surface mining are economic. Usually where a deposit is accessible for surface mining; it is a case of determining "cut-off" limits when surface mining operations become uneconomic and underground mining must be adopted (it may also be uneconomic) This is often the case with stock work type deposits which continue in depth, when surface mining may be followed by underground mining, using systems such as block caving. For further reading: Freeman [3] and Fung [4].

There are many terms and expressions unique to mining that characterize the field and identify the user of such terms as a "mining person." The student of mining is thus advised to become familiar with all the terms used in mining, particularly those that are peculiar to either mines or minerals. Some terms distinguish various types of mined minerals. Geologically, one can distinguish the following mineral categories:

1.2.1 Mineral

Naturally occurs as an inorganic element or compound having an orderly internal structure and a characteristic chemical composition, crystal form, and physical properties.

1.2.2 *Rock*

Any naturally formed aggregate of one or more types of mineral particles. Economic differences in the nature of mineral deposits are evident in the following terms.

1.2.3 *Ore*

It is defined as: a mineral deposit that has sufficient utility and value to be mined at a profit.

1.2.3.1 Gangue

It is known as: the valueless mineral particles within an ore deposit that must be discarded.

1.2.3.2 Waste

It is defined as: the material which associated with an ore deposit that must be mined to get at the ore and must then be discarded. Gangue is a particular type of waste.

A further subdivision of the types of minerals mined by humankind is also common. These terms are often used in the industry to differentiate between the fuels, metals, and nonmetallic minerals. The following are the most common terms used in this differentiation.

1.2.4 *Metallic Ores*

They are defined as: those ores of the ferrous metals (iron, manganese, molybdenum, and tungsten), the base metals (copper, lead, zinc, tin, and etc.), the precious metals (gold, silver, the platinum group metals), and the radioactive minerals (uranium, thorium, and radium).

1.2.5 *Nonmetallic Minerals (Industrial Minerals)*

They are defined as: the nonfuel mineral ores those are not associated with the production of metals. These include phosphate, potash, halite, trona, sand, gravel, limestone, sulfur, and many others. Fossil fuels (also known as mineral fuels): the organic mineral substances that can be utilized as fuels, such as coal, petroleum, natural gas, coalbed methane, gilsonite, oily shale, and tar sands.

1.3 Surface Mine Terminology

The design of the main engineering parameters of an open pit mine depends on geotechnical and operational considerations. Figure 1.1 illustrates the surface mine parameters: Bench-is a ledge that forms a single level of operation above which mineral or waste materials are mined back to a bench face, see WestOne [5]. Bench height (digging height)-is a vertical distance between the highest point (crest) of a bench and the lowest (toe) of the bench. Bench slope—is the angle, measured

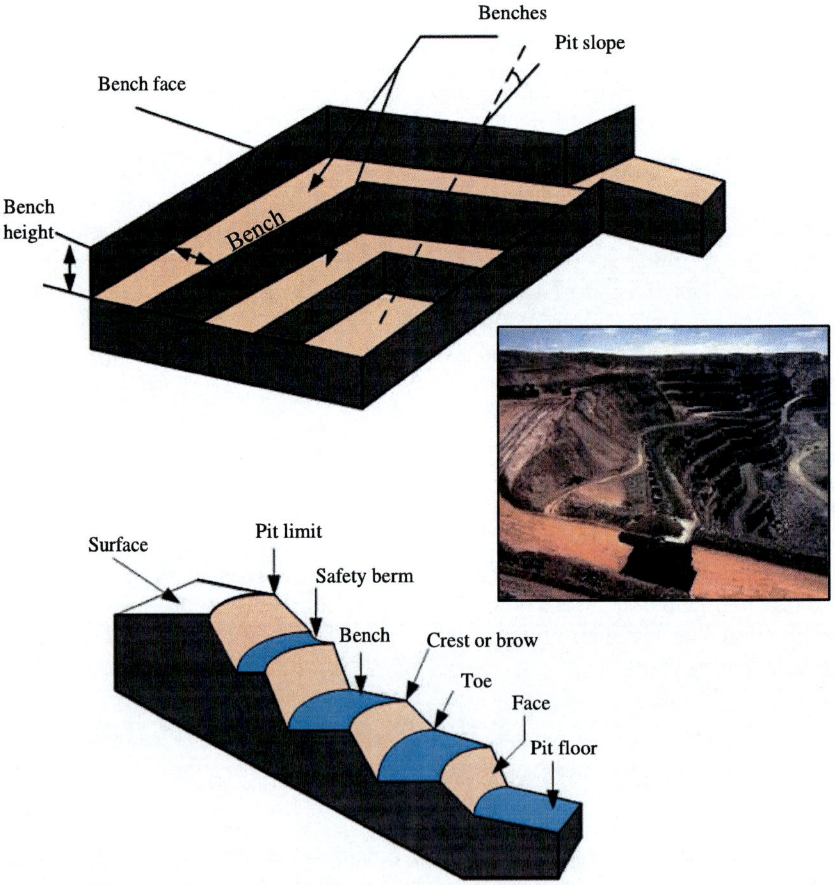

Fig. 1.1 Open-pit and bench terminology

in degrees between the horizontal and an imaginary line joining the crest and the toe of the bench. Pit slope—is the angle at which the wall of an open pit stands as measured along an imaginary plane extended from the slope crest to its toe. Berm—is a horizontal shelf within the pit wall slope that is left for stability. The berm slope angle and width are determined geotechnical configuration.

Pit limit—is the vertical and lateral extent to which the mining of a mineral deposit by open pitting may be economically carried out. Bench face (bank)—is a steep sloping mass of any earthy or rock material rising above the digging level from which the soil or rock is to be dug. Toe—is the base of a bank (bench) or slope in a quarry ore open-pit mine.

1.4 The Choice Between Surface and Underground Mining

In practice there is seldom scope for competition between two systems. Where deposit is accessible to surface mining it is often a case of determining the "cut-off limit" when surface operations are uneconomic and underground methods must be adopted i.e. stock works, pipes, steeply dipping stratified deposits, Fig. 1.2, see Atkinson [6].

The following geological settings are obviously most economically mined by surface methods, Figs. 1.3 and 1.4. While the next geological setting is obviously

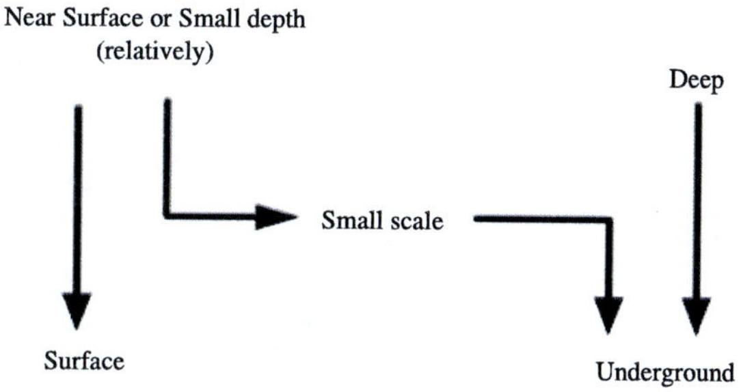

Fig. 1.2 Choice between surface or underground mining methods

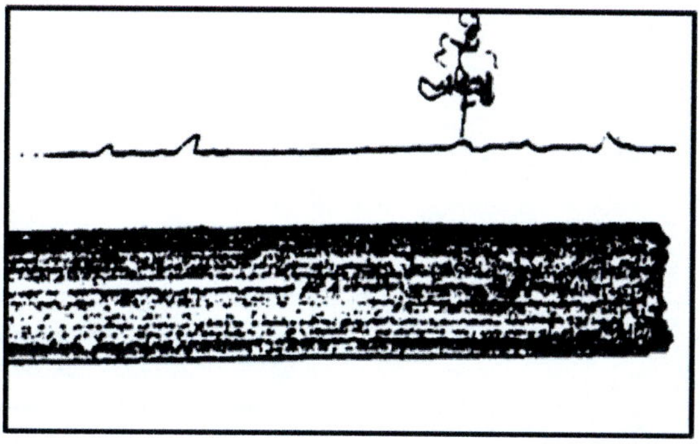

Fig. 1.3 Orebody near to surface

Fig. 1.4 Orebody with gently dipping

most economically mined by underground methods, Figs. 1.5 and 1.6, see Atkinson [6, 7].

As shown in Fig. 1.5, the choice is self-evident, see Atkinson [6], but in the following geological setting, Fig. 1.6, it is not clear. However, the choice must be an economic decision.

Fig. 1.5 Orebody at great depth below surface

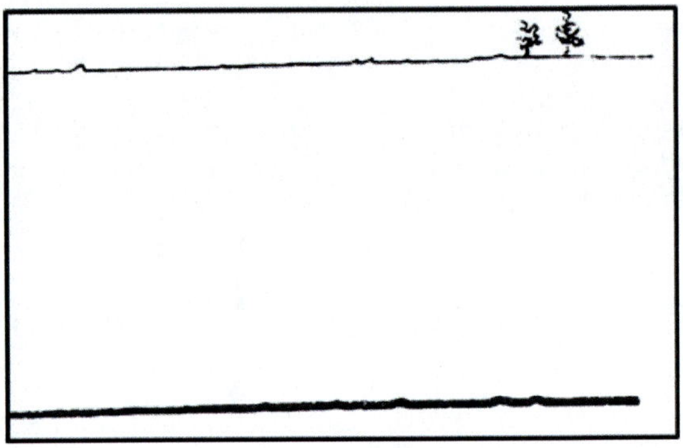

Fig. 1.6 Fat orebody at great depth

1.5 Surface Mining

When a mineral occurs fairly close to the surface in a massive or wide tabular body, or where the mineral itself is part of the surface soil or rock, it is generally more economical to mine it by means of surface mining methods. Strip mining, open-pit, opencast mining and quarrying are the most common mining methods that start from the earth's surface and maintain exposure to the surface throughout the extraction period. For both access and safety, the excavation usually has stepped or benched side slopes and can reach depths exceeding 600 m. Methods of surface mining can be subdivided into various classes and subclasses as in Table 1.3, see Bohnit [8], Freeman [3], Fung [4], Osanloo and Ataei [9].

1.5.1 Open Pit Mining

It is a method of extraction used where the overburden is limited and easily stripped, but where waste has to be transported to external dumps. It is generally used where deposits are limited laterally but are thicker than in open-cast mining. Open pit is a term properly applied to a surface mining method in which reclamation is deferred until all, or nearly all, of the deposit is removed within economic limits. The concept of open pit mining is simple, but planning for development is complex and costly. It may be necessary to blend different ore types to maintain character and grade of the mill feed, or it may be necessary to ship different ore types separately—oxide must be treated separately from sulfide ores, and low-grade ores may go to leach dumps, or gold-bearing oxide capping to special leach pads. Grade and tonnage of material available will determine pit limits and how much waste rock can be stripped. The ultimate limit to the pit is determined by the economics of removing overburden (Stripping ratio: Ore/Waste). Deposits generally decline in grade outward, so cut-off and pit limits may vary greatly with economic parameters. Slight variations in cost/value may have a great influence upon ore resources.

Table 1.3 Types of surface mining methods

Class	Subclass	Method
Mechanical		Open pit Open cast Glory holing Quarrying Strip mining Auger mining
Aqueous	Placer	Dredging Hydraulic mining
	Solution	Surface techniques In situ leaching

1.5.2 Open Cast Mining

It is mining in which large strips of land are excavated in order to extract materials. Open cast mining is used where the resource is near the surface and widespread. It is used mainly for coal which resembles open pit mining but differs in one unique respect: the overburden is not transported to waste dumps for disposal but cast or hauled directly into adjacent mined—out panels. Open pit mining and open cast mining rank as the most important surface mining methods among all mining methods, because of the highest productivity and the lowest cost of any of the broadly used mining methods.

1.5.3 Glory Holing

Glory holing involves a mine opening at the surface from which ore is removed by gravity through raises connected to adits haulage ways beneath, and transport the ore to the surface. It is suited to mining on a hillside, and irregular deposits can be mined without dilution by waste wall rock. Mining can be quite selective and little waste rock accumulates on the surface. However, reclamation is difficult.

1.5.4 Quarrying or Quarry Mining

Quarrying is the open, or surface excavation of rock to be used for various purposes, including construction, ornamentation, road building or as an industrial raw material. Quarrying methods depend mainly on the desired size and shape of the stone and its physical characteristics. It is usually restricted to mining dimension stone prismatic blocks of marble, granite, limestone, sandstone, slate, etc. that are used for primary construction of buildings or decorative facing materials for exterior and interior portions of buildings. Quarries generally have benches with vertical faces from a few feet to 200 feet in height. For building or dimension stone, where the rock needs to be extracted in large homogeneous rectangular blocks, controlled blasting such as smooth blasting can be used. Several methods are used to break out the blocks, including splitting, diamond saws and diamond wire cutting.

1.5.5 Strip Mining

It is a surface mining in which reclamation is contemporaneous with extraction. Area Mining or strip mining is generally carried out on a large scale, and consequently is low-cost. It essentially involves removing the overlying strata or overburden and extracting the valuable mineral deposit. It is applicable to shallow, flat-lying deposits of coal, oil-shale, clay, sand, gravel, and some uranium, phosphate and placer deposits. As the overburden is removed from one portion of a mineral deposit, it is used to fill in the trench left by the previous removal. Deposition of waste is, thus, much less of a problem and reclamation can readily be accomplished.

1.5.6 Auger Mining

Auger mining refers to a method of removing coal, clay, phosphate, oil-shale, etc. from thin seams exposed in deep trenches or high-walls in strip mines. The auger consists of two principal pieces. The first is a cutting head, generally from 1.5 to 8 feet in diameter. It may be single or multiple. The second is a prime mover, usually a skid mounted carriage, providing a mounting for the engine, drive head, and controls. As coal arrives at the surface it is transported via a conveyor belt or a front-end loader to a waiting truck. Operations are usually low-cost and highly productive, but recovery ranges from 40 to 60%. It can be implemented with relatively low capital costs.

1.5.7 Placer Mining or Alluvial Mining

Placer mining is a method for the recovery of heavy minerals using water to excavate, transport, and/or concentrate the mineral. Placers are deposits of detrital material containing valuable mineral liberated as discrete grains through weathering and erosion processes, usually occurring as unconsolidated sediments. Rich placers usually result from several cycles of erosion and re-concentration in one place. Ore bodies can be very large and low-grade, but low-cost. Most high-grade surficial placer deposits which historically supporting the small prospector have been exhausted.

Placer mining affects large surface areas for the volume of material mined, is highly visible and has serious environmental problems with surface disturbance and stream pollution. A variety of placer deposit types exist Residual Placers are composed of mineralized rock weathered in place, and are common in tropical countries. Hillside Slope Placers form a transition between source and stream, with less liberation of mineral grains than in stream placers. Stream or fluvial placers (creek, river, bench, terrace, gravel plain or swamp, and delta) are formed by running water which carries away lighter minerals and concentrates heavy minerals in areas where current is reduced or on steam rock bottoms or on top of clay seams. Dry or Bajada

Placers are formed in arid climates as a result of violent storm and wind action. There is less sizing and liberation of mineral grains than in stream placers.

Glacial Till or Glacio-Fluvial Placers are usually poorly sorted with poor liberation of grains. They are difficult to evaluate because of lack of uniformity and lack of continuity of values unless subjected to stream action. Beach Placers are formed by bottom currents and/or beach wave action on pre-existing placers, deltaic deposits, and coastal mineralized bedrock. Economically, the three most important placer types have been fluvial, beach, and off-shore marine placers. Mining Methods vary greatly as a consequence of the great variability in size and characteristics of placer deposits. They consist of the following:

1.5.7.1 Panning and Sluicing

The traditional prospectors gold pan is an efficient device for washing and separating the heavy minerals in placer deposits and is commonly used as a prospecting and testing tool for evaluating placer deposits. However, as a production device it is slow, and even in the hands of a skilled operator only a small volume of material can be processed. Most surface deposits rich enough to be economically mined and concentrated by panning have long since been mined. However, it is still used as a recreational tool. In sluicing the placer gravel is shoveled, along with a stream of water, into the head of an inclined elongated sluice box with riffles positioned across the bottom. These trap the heavy minerals and the lighter minerals are washed over the top and out as relatively barren waste. Sometimes fine gold is trapped as an amalgam when mercury is placed within the riffles or on a copper plate at the exit of the sluice box. The gold in the amalgam is recovered by retorting off the mercury.

1.5.7.2 Hydraulic Mining

It involves directing a high-pressure stream of water, via a monitor or nozzle, against the base of the placer bank. The water caves the bank, disintegrates the ground and washes the material to and through sluice boxes, and/or jigs, and/or tables situated down-slope. Hydraulic mining totally disturbs large areas and puts much debris into the drainage system. Presently, hydrolyzing is used primarily in Third World countries. It is closely controlled or prohibited in the U.S.

1.5.7.3 Dredging

It involves floating washing plants capable of excavating gravel, processing it and stacking the tailings away from the dredge pond. Several types of excavation methods are in use: Draglines and Backhoe Plants. Dragline use in placer mining with washing plants is limited to shallow digging depths. Its bucket is less controllable on the bottom than the backhoe, and it is less able to dig into the bottom to clean up all the ore that may be there. However, it has the advantage of a longer reach. The digging reach of the backhoe extends to as much as 70 feet below the surface, see Krause and Dwire [10].

It has the advantage of relatively low first cost, excellent mobility, and an ability to excavate hard material. Bucket Wheel Hydraulic Dredges are becoming more popular for underwater excavation, except where a high content of soft clay exists or where excessive oversize material occurs. It is dependent upon flooded pump openings that convey the material mined to the washing plant, and therefore it cannot work above water level. Placement of the pump suction is critical. Suction Cutter Dredges are similar to the Bucket Wheel Dredge except the digging device consists of a series of cutting arms rotating in a basket about a suction intake. The rotating arms break up the bank material, slurring it so it can be drawn into the dredge suction. It has proven to be successful in mining unconsolidated beach sands and offshore placers. Bucket-line Dredges are capable of continuous excavation and are very efficient.

They mine, process, and discard tailings to waste in one continuous stream. However, no storage opportunities exist, and the stream moves through the system by the force of gravity. Buckets, supported by a ladder, dig the mine face. Material moves up the ladder and dumps into a hopper that feeds the washing plant. They are capable of high excavation rates. Various methods are used to position the dredge—anchored by wire ropes or piling (SPUDS) at the rear of the dredge. Boulders can cause serious problems.

1.5.7.4 In Situ Leaching

In situ leaching is an alternative to mechanical mining that is growing in popularity because of low capital and labor costs, short preproduction time, and low surface environmental impact. Sub-surface groundwater contamination can pose a problem. For example, promising tests carried out at Miami, Arizona by Occidental Minerals were discontinued on the basis of being a threat to Miami's water supply. It is applicable to a wide variety of commodities that are soluble in water or an aqueous lixiviant.

Three general area of application include, first, extraction of water soluble salts, such as potash, trona, and common salt (NaCl). Second, in situ leaching is used in the Frasch Process, in extraction of sulfur from salt domes using hot water injection to supply heat to melt the sulfur and allow it to be pumped to the surface. Finally, hard rock in situ mining uses a lixiviant. It is applicable to extraction of uranium, copper and gold. Permeability, innate or induced, within the rock is critical, as is the distribution of metal values relative to the flow channels. Various methods are used, or proposed, to enhance permeability, including hydrocracking, use of explosives, undercutting and caving.

The pattern and spacing of injection and production wells is critical and varies with rock conditions. Amenability of a deposit to in situ leaching is a function of: (1) The pattern and character of value distribution (depth, shape, grade, mineral type and distribution, and structural and/or stratigraphic features; (2) fluid flow characteristics of the rock (permeability, porosity, natural groundwater flow, fracture character, frequency and orientation; (3) solvent effectiveness (rate of mineral dissolution, reactions with host and gangue minerals and the effects of reactions upon permeability) and (4) recovery of values from the leach solutions. Evaluation requires both qualitative and quantitative determinations, with particular attention directed to controls to avoid groundwater pollution. Deposits in hard rock that favor in situ leach fall generally into six categories: (1) stratiform sandstone deposits, (2) stockwork deposits, (3) breccia bodies, (4) fault zones, (5) shattered irregular bodies, and (6) surficial deposits.

1.6 Underground Mining

Under certain circumstances surface mining can become prohibitively expensive and underground mining may be considered. A major factor in the decision to operate by means of underground mining rather than surface mining is the strip ratio or the number of units of waste material in a surface mine that must be removed in order to extract one unit of ore. Once this ratio becomes large, surface mining is no longer attractive. The objective of underground mining is to extract the ore below the surface of the earth safely, economically, and with the removal of as little waste as possible. These cost need to be weighed against the extraction of the ore. In open pit mine up to 90–95% of the ore body can be removed. In underground mining generally more ore has to be left behind as it is used to support the mine roof. The entry from the surface to an underground mine may be through an Adit, or horizontal tunnel, a shaft or a declined shaft, Fig. 1.7, see Gauteng [11]. A typical underground mine has a number of roughly horizontal levels at various depths below the surface and these spread out from the access to the surface. Ore is mined in stopes, or rooms. Material left in place to support the ceiling is called a pillar and can sometimes be recovered afterward. A vertical internal connection between two levels of a mine is called a winze if it was made by driving downward and a raise if it was made by driving upward.

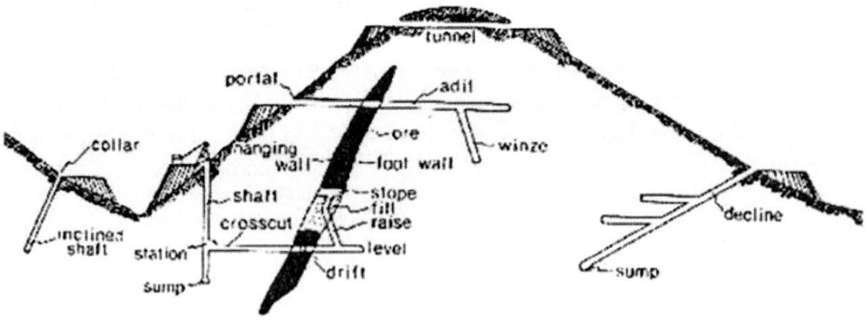

Fig. 1.7 Idealized cross-section of underground mine

A modern underground mine is a highly mechanized operation requiring little work with pick and shovel. Rubber-tired vehicles, rail haulage and multiple drill units are commonplace. In order to protect miners and their equipment much attention is paid to mine safety. Mine ventilation provides fresh air underground and at the same time removes noxious gases as well as dangerous dusts that might cause lung disease, e.g. silicosis. Roof support is accomplished with timber, concrete or steel supports or, most commonly, with roof bolts, which are long steel rods used to bind the exposed roof surface to the rock behind it. Shafts, which are generally vertical, but may be inclined depending on the orientation of the ore body, can be distinguished from adits and tunnels, which are horizontal. Shafts and adits are the main access routes through which men; supplies, ore and waste are transported. They are the chief service openings during the development and operation of a mine, and provide space for compressed-air pipes or electric cables. By law at least two access points to the ore body are required for adequate ventilation and safety concerns (see GDACE Mining and Environmental Impact Guide Chap. 6).

1.7 Preparation of Open Pit Field for Mining

The essential operations of surface mining are:

- Choice of the site of the mine according to nearest presence of materials and then solving the transportation problem.
- All natural obstacles, debris and structures within the limits of the open-pit field and in the approach zone of transport lines are removed or moved to other places and then dressing of the surface. Natural obstacles include forests, shrub, streams, rivers, lakes, marshes in flat-land regions, and rock overhangs in mountainous regions. Structures within open-pit fields may include existing automobile and railway roads passing within the technical boundaries of the quarry and old structures.
- Primary drainage for underground water with the mine field. Water-drainage channels are constructed with the run-off towards lower portions of the terrain relief. The drainage system of a deposit must provide normal conditions for permanent mining and mining-exploitation work in the open-pit.
- Development schemes and mining method by construction of roads, trenches and benches.
- Excavation of ore and removal of overburden according to the development schemes and mining method and systems.
- Loading and transportation to places of dressing or industrial plants.

The development of a successful surface mine design should consider the following:

- Topography, climate and other physical factors.
- Geological structure and geotechnical considerations for pit slope geometry.

- Groundwater inflow, dam locations and other infrastructure requiring engineering design.
- Rock types from overburden and ore to determine density, hardness, degree of weathering, crushing characteristics (ore), drillability and suitability for road base.
- Mine and crusher's location, haul road design (amount of cut and/or fill) and distances.
- Waste dump locations, haul distances and profiles, and amenability of low grade dump leaching in the future.
- Operating cost estimation, administration and other ancillary costs.
- Metal price and discount rate selection based on the degree of risk, current spot prices and price forecasting.
- Initial pit optimization and cut-off grade determination.
- Cut-off grade optimization, detailed design, scheduling and financial analysis.

The factors favoring surface mining are:

- Higher productivity.
- Greater concentration of all operations and simplified management of men and machines.
- Greater output per mine.
- Lower capital cost per annual ton mined.
- Lower operating cost per ton.
- Possibility of moving a higher ratio of waste, the exploitation of lower grade reserves.
- Greater geological certainty.
- Less limitation on size and weight of machines.
- Simpler auxiliary operations and services.
- Increased recovery at mineral and less dilution.
- Greater reserves available for mining
- Simplified planning and control.
- Increased safety.

The main disadvantages of surface mining are due to adverse climatic conditions and the effect on the environment:

- Rainfall, snow, fog and severe heat or cold result in a drop in efficiency of labor and machines.
- Large open pit mines require large areas of land tar mining and spoil heaps which are lost, at least temporarily, to agriculture.
- For night working large areas must be illuminated.
- Where operations start from the outcrop the weathered zone may yield a low grade product e.g. coking coal may be oxidized.

1.8 Stages in the Life of a Mine

The overall sequence of activities in modern mining is often compared with the five stages in the life of a mine: prospecting, exploration, development, exploitation, and reclamation. Prospecting and exploration, precursors to actual mining, are linked and sometimes combined. Geologists and mining engineers often share responsibility for these two stages—geologists more involved with the former, mining engineers more with the latter. Likewise, development and exploitation are closely related stages; they are usually considered to constitute mining proper and are the main province of the mining engineer. Reclamation of the mine site has become a necessary part of the mine life cycle because of the demands of society for a cleaner environment and stricter laws regulating the abandonment of a mine. The five stages in the life of a mine are summarized in Table 1.4. Scope of mining activities from exploration to marketing is illustrated in schematic diagram, Fig. 1.8.

1.8.1 Prospecting

Prospecting, the first stage in the utilization of a mineral deposit, is the search for ores or other valuable minerals (coal or nonmetallic). Because mineral deposits may be located either at or below the surface of the earth, both direct and indirect prospecting techniques are employed. The direct method of discovery, normally limited to surface deposits, consists of visual examination of either the exposure (outcrop) of the deposit or the loose fragments (float) that have weathered away from the outcrop. Geologic studies of the entire area augment this simple, direct technique.

By means of aerial photography, geologic maps, and structural assessment of an area, the geologist gathers evidence by direct methods to locate mineral deposits. Precise mapping and structural analysis plus microscopic studies of samples also enable the geologist to locate the hidden as well as surface mineralization. The most valuable scientific tool employed in the indirect search for hidden mineral deposits is geophysics, the science of detecting anomalies using physical measurements of gravitational, seismic, magnetic, electrical, electromagnetic, and radiometric variables of the earth. The methods are applied from the air, using aircraft and satellites; on the surface of the earth; and beneath the earth, using methods that probe below the topography. Geochemistry, the quantitative analysis of soil, rock, and water samples, and geobotany, the analysis of plant growth patterns, can also be employed as prospecting tools.

Table 1.4 Stages in the life of a mine

Stage	Procedure	Time
Prospecting (Mineral deposit)	Precursors to Mining (Search for ore) • Prospecting methods • Direct: physical geologic • Indirect: geophysical, geochemical • Locate favorable loci (maps, literature, old mines) • Air: aerial photography, airborne geophysics, satellite • Surface: ground geophysics, geology • Spot anomaly, analyze, evaluate	1–3 years
Exploration (Ore body)	Defining extent and value of ore (examination/evaluation) • Sample (drilling or excavation), assay, test • Estimate tonnage and grade • Valuate deposit (Hoskold formula or discount method): present value = income − cost • Feasibility study: make decision to abandon or develop	2–5 years
Development (Prospect)	Mining Proper (Opening up ore deposit for Production) • Acquire mining rights (purchase or lease), if not done in stage 2 • File environmental impact statement, technology assessment, permit • Construct access roads, transport system • Locate surface plant, construct facilities • Excavate deposit (strip or sink shaft)	2–5 years
Exploitation (Mine)	Large-scale production of ore • Factors in choice of method: geologic, geographic, economic, environmental, societal safety • Types of mining methods surface: open pit, open cast, etc. underground: room and pillar, block caving, etc • Monitor costs and economic payback (3–10 years)	10–30 years
Reclamation (Real estate)	Post-mining (Restoration of site) • Removal of plant and buildings • Reclamation of waste and tailings dumps • Monitoring of discharges	1–10 years

1.8.2 Exploration

The second stage in the life of a mine, exploration, determines as accurately as possible the size and value of a mineral deposit, utilizing techniques similar to but more refined than those used in prospecting. The line of demarcation between

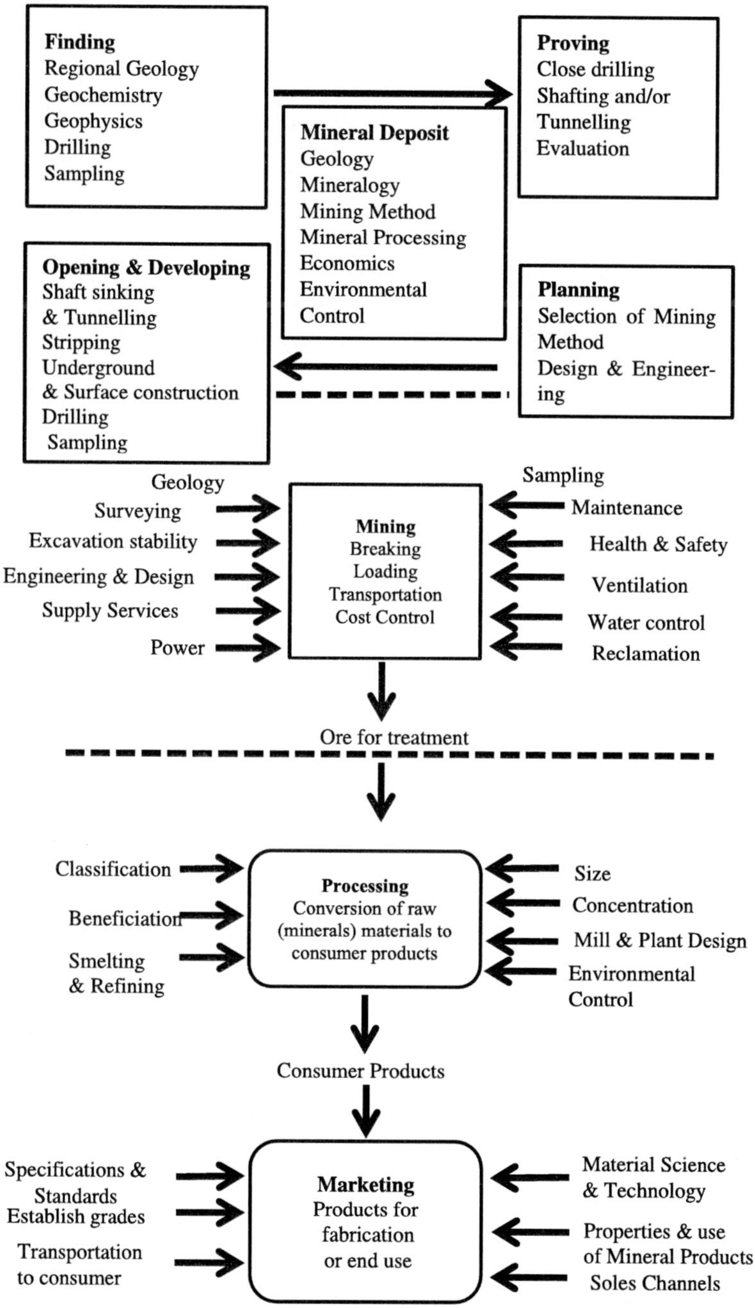

Fig. 1.8 Schematic diagram of scope of mining activities

prospecting and exploration is not sharp; in fact, a distinction may not be possible in some cases.

Exploration generally shifts to surface and subsurface locations, using a variety of measurements to obtain a more positive picture of the extent and grade of the ore body. Representative samples may be subjected to chemical, metallurgical, X ray, spectrographic or radiometric evaluation techniques that are meant to enhance the investigator's knowledge of the mineral deposit. Samples are obtained by chipping outcrops, trenching, tunneling, and drilling; in addition, borehole logs may be provided to study the geologic and structural makeup of the deposit. Rotary, percussion, or diamond drills can be used for exploration purposes. However, diamond drills are favored because the cores they yield provide knowledge of the geologic structure.

The core is normally split along its axis; one half is analyzed, and the other half is retained intact for further geologic study. An evaluation of the samples enables the geologist or mining engineer to calculate the tonnage and grade, or richness, of the mineral deposit. He or she estimates the mining costs, evaluates the recovery of the valuable minerals, determines the environmental costs, and assesses other foreseeable factors in an effort to reach a conclusion about the profitability of the mineral deposit. The crux of the analysis is the question of whether the property is just another mineral deposit or an ore body. For an ore deposit, the overall process is called reserve estimation, that is, the examination and valuation of the ore body. At the conclusion of this stage, the project is developed, traded to another party, or abandoned.

1.8.3 Development

In the third stage, development, the work of opening a mineral deposit for exploitation is performed. With it begins the actual mining of the deposit, now called the ore. Access to the deposit must be gained either (1) by stripping the overburden, which is the soil and/or rock covering the deposit, to expose the near-surface ore for mining or (2) by excavating openings from the surface to access more deeply buried deposits to prepare for underground mining. In either case, certain preliminary development work, such as acquiring water and mineral rights, buying surface lands, arranging for financing, and preparing permit applications and an environmental impact statement (EIS), will generally be required before any development takes place.

When these steps have been achieved, the provision of a number of requirements—access roads, power sources, mineral transportation systems, mineral processing facilities, waste disposal areas, offices, and other support facilities—must precede actual mining in most cases. Stripping of the overburden will then proceed if the minerals are to be mined at the surface. Economic considerations determine the stripping ratio, the ratio of waste removed to ore recovered; it may range from as high as 45 yd^3/ton (38 m^3/tonne) for coal mines to as low as 1.0 yd^3/ton (0.8 m^3/tonne) in metal mines. Some nonmetallic mines have no overburden to remove; the mineral is simply excavated at the surface, see Dirk [12], Gregory and Winterfeldt [13].

Development for underground mining is generally more complex and expensive. It requires careful planning and layout of access openings for efficient mining, safety, and permanence. The principal openings may be shafts, slopes, or adits; each must be planned to allow passage of workers, machines, ore, waste, air, water, and utilities. Many metal mines are located along steeply dipping deposits and thus are opened from shafts, while drifts, winzes, and raises serve the production areas. Many coal and nonmetallic mines are found in nearly horizontal deposits. Their primary openings may be drifts or entries, which may be distinctly different from those of metal mines.

1.8.4 Exploitation

Exploitation, the fourth stage of mining, is associated with the actual recovery of minerals from the earth in quantity. Although development may continue, the emphasis in the production stage is on production. Usually only enough development is done prior to exploitation to ensure that production. It is once started, can continue uninterrupted throughout the life of the mine. The mining method selected for exploitation is determined mainly by the characteristics of the mineral deposit and the limits imposed by safety, technology, environmental concerns, and economics.

Geological conditions, such as the dip, shape, and strength of the ore and the surrounding rock, play a key role in selecting the method. Traditional exploitation methods fall into two broad categories based on locale: surface or underground. Surface mining includes mechanical excavation methods such as open pit and open cast (strip mining), and aqueous methods such as placer and solution mining. Underground mining is usually classified in three categories of methods: unsupported, supported, and caving.

1.8.5 Reclamation

The final stage in the operation of most mines is reclamation, the process of closing a mine and re-contouring, re-vegetating, and restoring the water and land values. The best time to begin the reclamation process of a mine is before the first excavations are initiated. In other words, mine planning engineers should plan the mine so that the reclamation process is considered and the overall cost of mining plus reclamation is minimized, not just the cost of mining itself.

The new philosophy in the mining industry is sustainability, that is, the meeting of economic and environmental needs of the present while enhancing the ability of future generations to meet their own needs. In planning for the reclamation of any given mine, there are many concerns that must be addressed. The first of these is the safety of the mine site, particularly if the area is open to the general public. The removal of office buildings, processing facilities, transportation equipment, utilities, and other surface structures must generally be accomplished. The mining company is then required to seal all mine shafts, adits, and other openings that may present physical hazards. Any existing highwalls or other geologic structures may require mitigation to prevent injuries or death due to geologic failures. The second major issue to be addressed during reclamation of a mine site is restoration of the land surface, the water quality, and the waste disposal areas so that long-term water pollution, soil erosion, dust generation, or vegetation problems do not occur. The restoration of native plants is often a very important part of this process, as the plants help build a stable soil structure and naturalize the area. It may be necessary to carefully place any rock or tailings with acid-producing properties in locations where rainfall has little effect on the material and acid production is minimized. The same may be true of certain of the heavy metals that pollute streams. Planning of the waste dumps, tailings ponds, and other disturbed areas will help prevent pollution problems, but remediation work may also be necessary to complete the reclamation stage of mining and satisfy the regulatory agencies. The final concern of the mine planning engineer may be the subsequent use of the land after mining is completed. Old mine sites have been converted to wildlife refuges, shopping malls, golf courses, airports, lakes, underground storage facilities, real estate developments, solid waste disposal areas, and other uses that can benefit society. By planning the mine for a subsequent development, mine planners can enhance the value of the mined land and help convert it to a use that the public will consider favorable. The successful completion of the reclamation of a mine will enhance public opinion of the mining industry and keep the mining company in the good graces of the regulatory agencies. The fifth stage of the mine is thus of paramount importance and should be planned at the earliest possible time in the life of the mine.

1.9 Unit Operations of Mining

During the development and exploitation stages of mining when natural materials are extracted from the earth, remarkably similar unit operations are normally employed. The unit operations of mining are the basic steps used to produce mineral from the deposit, and the auxiliary operations that are used to support them. The steps contributing directly to mineral extraction are production operations, which constitute the production cycle of operations. The ancillary steps that support the production cycle are termed auxiliary operations. The production cycle employs unit operations that are normally grouped into rock breakage and materials handling. Breakage generally consists of drilling and blasting, and materials handling encompasses loading or excavation and haulage (horizontal transport) and sometimes hoisting (vertical or inclined transport).

References

1. Darling P (2011) SME mining engineering handbook. Society for Mining Metallurgy and Exploration, Inc.
2. Hartman HL (2002) Introductory mining engineering. Wiley, Hoboken
3. Freeman C (1983) Surface mining—a review of progress. In: Proceedings, surface mining and quarrying. Institution of Mining and Metallurgy, London, pp 265–276
4. Fung R (ed) (1981) Surface coal mining technology. Noyes Data Corp., Park Ridge, NJ, pp 7–12, 49–115, 186–229
5. WestOne (1984) "Quarring1", WestOne prospect place west Perth WA 6005, Australia
6. Atkinson T (1985) Surface mining course notes: mining engineering department. University of Nottingham, UK
7. Atkinson T (1992a) Selection and sizing of mining equipment, Sec. 13.3 in SME mining engineering handbook, 2nd edn. In: Hartman HL (ed). Society for Mining, Metallurgy and Exploration, Littleton, CO, pp 1311–1333
8. Bohnet EL, Kunze L (1990) Waste disposal planning and environmental protection aspects. In: Kennedy RA (ed) Surface mining, 2nd edn. SME
9. Osanloo M, Ataei M (2003) Factors affecting the selection of site for arrangement of pit rock-dumps. J Min Sci 39:148–153. https://doi.org/10.1023/B:JOMI.0000008460.62695.44
10. Krause AL, Dwire DL (1999) Site selection approach for mine tailings. In: Proceedings of the 6th international conference on tailing and mine waste '99. Balkema, Rotterdam, Brookfield
11. Gauteng (2008) Mining and environmental impact guide, mining method. Department of Agriculture, Johannesburg (Chap 6)
12. van Zyl D (1993) Mine waste disposal. In: Daniel DE (ed) Geotechnical practice for waste disposal. Chapman and Hall
13. Gregory R, Keeney R, Von Winterfeldt D (1992) Adapting the environmental impact statement process to inform decision makers. J Policy Anal Manag 11(1)

Chapter 2
Principles of Surface Mining of Mineral Deposits

2.1 Mine Layout

A classification of mineral deposits is illustrated in Fig. 2.1. This indicates the influence of the shape of a deposit on the design work. In general, stratified deposits are in softer, younger sedimentary rocks and non-stratified are in harder, older, igneous rocks. The shape of a mineral deposit fundamentally determines the shape of the pit; the physical characteristics of the mineral and waste determine the choice of machines, see Mario [11]. In summary, the factors affecting layout are:

2.1.1 The Shape and Depth of the Deposit

This is of particular importance since the selection of the transport system is restricted by the distance and vertical height through which overburden and ore must be moved, that is, routes and gradients.

2.1.2 The Properties of the Ore and Overburden

1. Slope -angle limitations, both of working faces, sidewalls and dumps.
2. Degree of hardness and abrasiveness, which influence the selection of excavation and transport equipment.
3. Grade variations which may require selective mining to be practiced and affect the shape of the pit due to the cut-off grade limits.
4. Competence of the various rocks to support the equipment.

M. M. Ali Elbeblawi et al., *Surface Mining Technology*, Topics in Mining, Metallurgy and Materials Engineering, https://doi.org/10.1007/978-981-16-3568-7_2

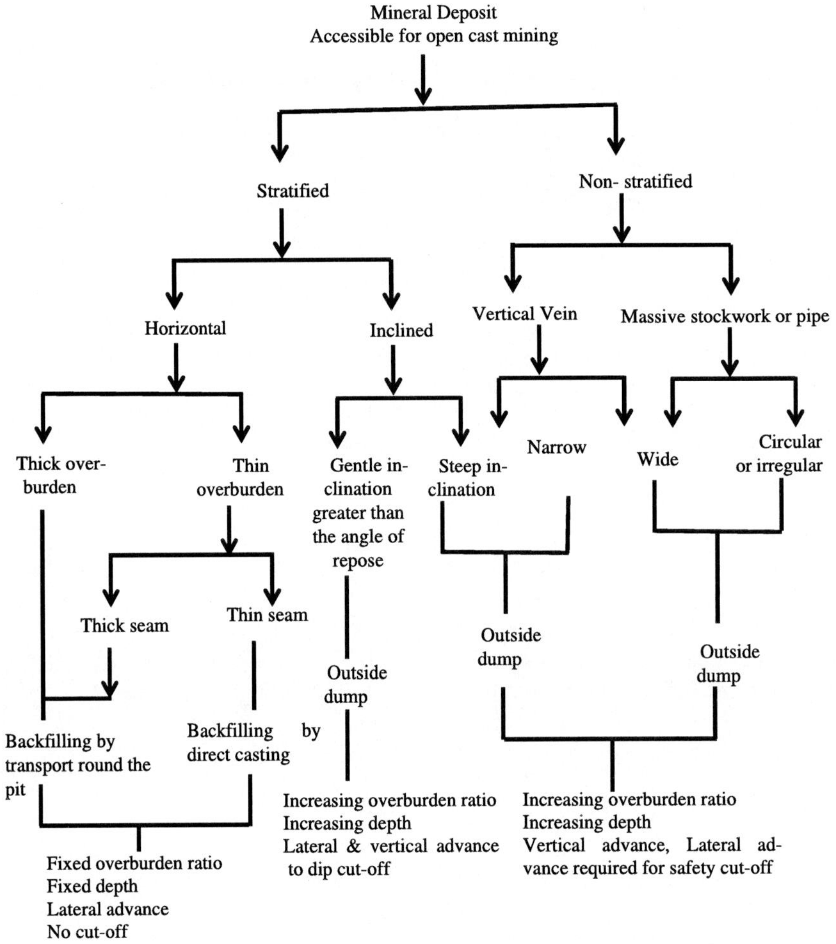

Fig. 2.1 Classification of deposits for open pit mining

2.1.3 Geometry of Excavating Equipment

The drainage requirements, particularly the maintenance of drainage slopes to sumps so as to safeguard faces, benches, dumps and access roads at all times during the life of the mine from the influence of ground water. Saturated slopes are more prone to collapse than well-drained slopes.

The dimensions, of open pit, are determined by the dimensions of the deposit itself. The objective is to plan for optional dimensions of the pit so the total capital outlays per 1 tonne of ore mined should be minimal, account being taken of the period over which the initial investments are made.

In general the extent across the strike is smaller than the length. However, circular type pits occur where length and width approach one another. Pit types range from quarry-type operations of relatively limited life and yearly output (these may be hill-side type operations) to large pit operations (in general below-ground pits) with long life and multimillion tonne annual production rates.

2.2 Types of Surface Mining Deposits

Minerals extracted in surface mining are used in various industries and accordingly all mineral deposits can be divided into coal deposits, ore deposits, deposits of building construction rock, deposits of minerals for cement manufacture, deposits of minerals for chemical processing. Exploited deposits of minerals may be in very different natural conditions. Deposits can be classed into various types, in the first place, by their typical geometric characteristics, Fig. 2.1, see Bullivant [4].

2.2.1 Deposit Shape

- Isometric, i.e. more or less extending in all directions (massive deposits, bosses, nests), Fig. 2.2c, h.
- Bedded (sheet) deposits, which are stretched preferably in two directions and have a relatively low capacity (beds and blanket deposits), Fig. 2.2a, b, d and g.
- Pipe-like and columnar (shoot) deposits extended in one direction only, Fig. 2.2e.
- Intermediate and transient forms between the indicated ones (lenses, veins, saddle-shaped deposits, folds, bends, tectonically dislocated suits of strata, Fig. 2.2a, f.

2.2.2 The Relief of Deposit Surface

The surface relief of a deposit determines the method of mining and the applicable mechanization means.

- Flat, Fig. 2.2a.
- As a slope of a hill, Fig. 2.2b.
- A hill proper, Fig. 2.2c.
- Hilly, Fig. 2.2d.
- The surface of a deposit may be under water.

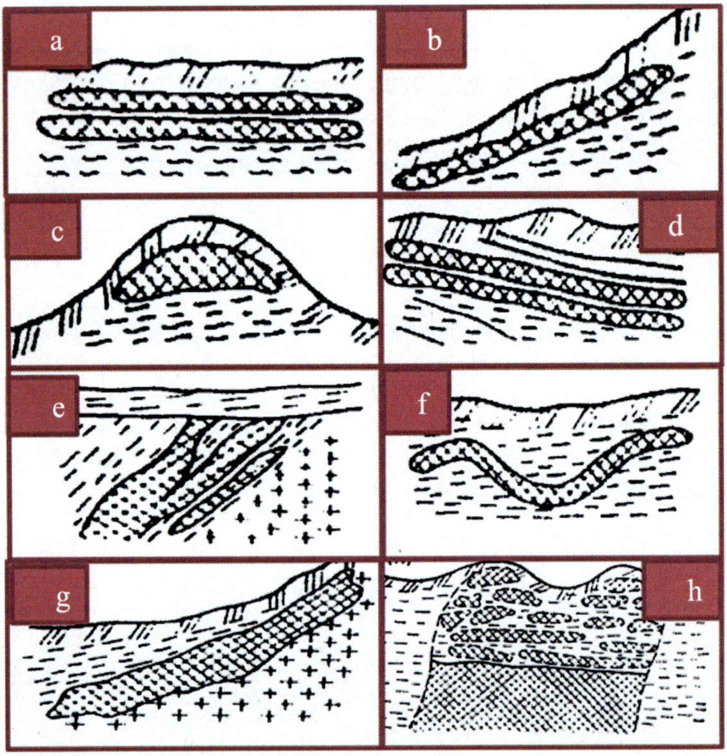

Fig. 2.2 **a–h** The shape of deposits predetermines the shape of quarries and open-pit fields

2.2.3 Position of Deposit

The depth of occurrence of deposit is relative to the prevailing surface level. The depth of occurrence of the deposits is distinguished as follows:

- Surface-type deposits which open immediately onto the surface or are covered by a thin depth of overburden (up to 20–30 m), Fig. 2.2a.
- Deep deposits located much lower than the prevailing surface level, with the overburden depth being from 40 to 250 m, Fig. 2.2e, f. Such deposits can be extracted by either surface or underground mining depending on which the methods turn out to be more favorable economically;
- High-type deposits which are located above the prevailing surface level, Fig. 2.2b, c; they also may be extracted by either surface or underground mining;
- High-deep deposits, i.e. deposits located partially above and partially below the prevailing surface level, Fig. 2.2g.

The position (bedding) of a deposit may be conformable (regular) or unconformable (irregular) relative to the surface relief; a deposit may occupy a whole hill or only a portion of it (slope). The position of a deposit relative to the earth surface determines the size of a quarry (along the depth and in plan) and the applicable engineering facilities, especially haulage means, see Rzhevsky [13].

2.2.4 Deposit Dip Angle

- Gently dipping which are characterized by a low angle of dip (up to 80°) and wavy occurrence of the main portion of the deposit, Fig. 2.2a, d, horizontal deposits are a particular case of this type.
- Inclined, or dipping, with the angles of dip from 8°–10° to 25°–30°, Fig. 2.2b.
- Steeply dipping, with the angles of dip more than 30° Fig. 2.2g.
- Steep, with the angles of dip 56°–90°.
- Deposits of complex bedding which are typical of anticline and syncline folds, Fig. 2.2f, and sharp geological dislocations; they have a variable angle of dip.

This classification of deposits has been done from the standpoint of surface mining technology. For instance, when mining horizontal and gently dipping deposits, Fig. 2.3a, it is possible to locate waste dumps in the mind-out area of a quarry; this is sometimes possible in the mining of dipping and steeply dipping stretched deposits. In mining dipping deposits, the stability of the final flanks of a quarry and the location of stripping workings do not necessitate removal of enclosing rock from the lying wall of the deposit, Fig. 2.3b. With a steep bedding of a deposit, it is required to mine the enclosing rock in the hanging and lying wall of the deposit, Fig. 2.3c.

2.2.5 Depth of Deposits

It may be very low, low, medium, high, and very high. This division is adopted since the capacity of a deposit is usually associated with the number of working benches that can be mined simultaneously. The conditions and order of mining of horizontal and dipping (and steeply dipping) deposits are different, because of which the indices of the same capacity classes of these deposits are numerically different.

Simple deposits, Fig. 2.2b, g, are characterized by a homogeneous structure without substantial strata or inclusions; in this case all minerals of a deposit are extracted jointly (total mining).

Complex deposits, Fig. 2.2a, d may contain, together with high-quality mineral, its grades of poorer quality and streaks or inclusions of gangue with clearly distinguishable contacts; in such cases separate (selective) mining of the main high-quality mineral and off-grade minerals and gangue is resorted to.

Disseminated deposits, Fig. 2.2h, may have a complicated structure, with high- and poorer-quality minerals and gangue being distributed irregularly and without

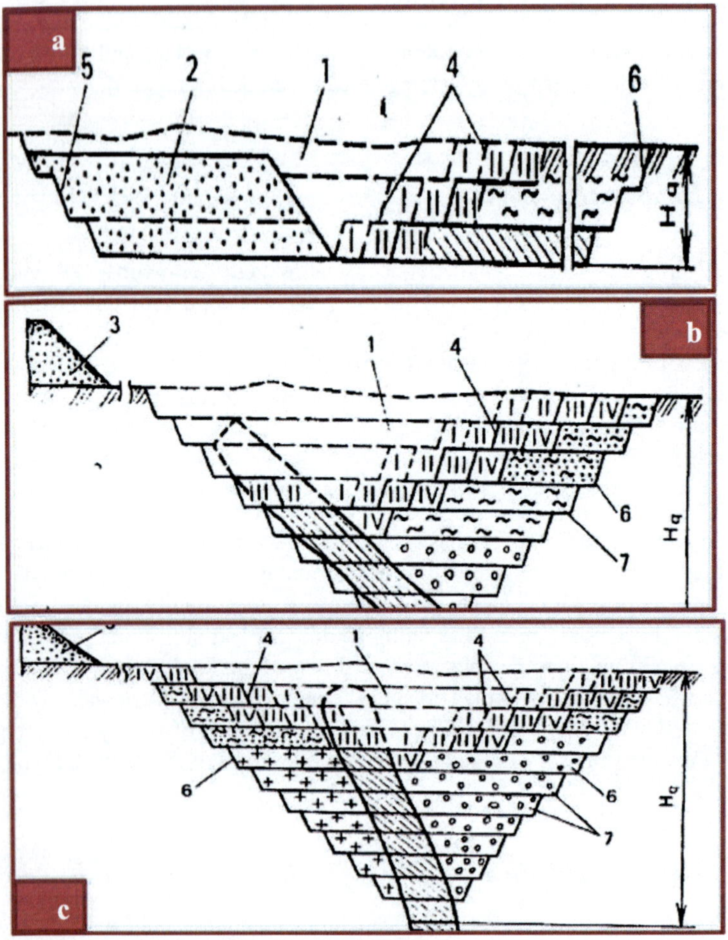

1. Represents mined-out space.
2. Internal dumps.
3. External dump.
4. Working flank.
5. Non-working flank.
6. Final contour of quarry.
7. Berms.
8. I-.IV: Sequence of working on benches

Fig. 2.3 Mining works of ore deposits based on their dip angle

clear contacts in the earth's crust; the choice between the total or selective mining is then clone after a detailed exploitation prospecting.

2.2.6 *Mineral Quality*

- Either uniformly when the quality of the mineral as specified by consumer's requirements is essentially the same within the limits of the deposits; in that case extraction (either total or selective) can be carried out in various districts of the deposit independently and without blending, see Abdel Sabour et al. [1].
- Or non-uniformly, when quality is distributed unevenly across the depth and over the plan area of the deposit; in that case it is preferable to plan simultaneous extraction from various districts of the deposit, provide a number of working extraction sections and blend the extracted mineral.

2.2.7 *Rock Type*

- The overburden and mineral are both compact igneous or metamorphic rock.
- The overburden consists of inhomogeneous rock and the mineral and enclosing rock are of compact or weathered igneous or metamorphic type; in that case the thick capping layer of overburden may be represented by alternating soft, dense, weathered and compact igneous and metamorphic rock.
- Soft and dense rock of the overburden, with the mineral and enclosing rock being of the compact or weathered igneous or metamorphic type.
- Weathered igneous or metamorphic rock of the overburden and the mineral of the weathered igneous or metamorphic type or rather dense type.
- Soft overburden rock and inhomogeneous (diverse) mine.
- Soft overburden rock and soft or dense mineral.

These factors are decisive for the selection of suitable engineering facilities, the order and possibilities of surface mining work.

2.3 Kinds of Surface Mining

The principal kinds of surface mining are classified according to the position of deposits relative to the surface, Fig. 2.4, as follows:

- Surface-kind mining. This kind is applicable with most types of placer deposits, natural building construction rock (stone), an appreciable portion of ore fields and a small portion of ore deposits of horizontal or gently dipping type. The quarry in surface mining has a rather low (up to 40–60 m) and essentially constant depth. The overburden and mineral are diverse, more often soft or weathered igneous and metamorphic type, Fig. 2.4a.
- Deep-kind mining. It is employed on many ore deposits and partially for mining coal from dipping and steep deposits. In this case the quarry is deepened gradually

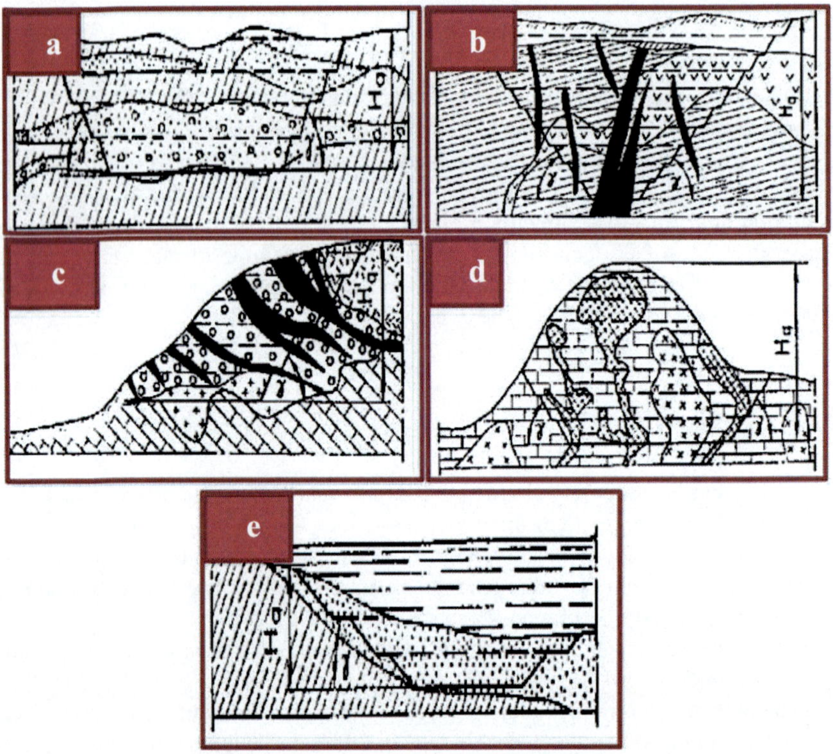

Fig. 2.4 Kind of surface mining

and its ultimate depth may be as large as 800 m. All kinds of rock can be mined in deep-mining quarries, Fig. 2.4b.

- On-slope mining. This kind of mining mainly relates to surface mining of various ores, raw materials for chemical processing, building construction rock and sometimes of coal. The deposit is located much higher than the prevailing surface level; the number of working benches and the plan dimensions of the quarry may be diverse. The mineral and overburden are mostly compact igneous and metamorphic rock, Fig. 2.4c.

- On-slope-deep mining. This kind is used for surface mining of various ores, chemical raw materials, building construction rock and coal where the surface of the open-pit field has an intricate relief. The mineral and overburden are contact or weathered igneous and metamorphic rock, sometimes diverse, Fig. 2.4d.
- Underwater mining. The deposit is located under water and the overburden usually has a relatively small depth. This kind is employed. In particular, for mining in river flood-lands and from the bottom of lakes and seas. The rock may be soft, dense, weathered igneous and metamorphic or diverse, Fig. 2.4e.

Each of these kinds of surface mining differs from the other by some specifics of preparation of the deposit for exploitation, the order of working, stripping of mining levels, location of waste dumps and accordingly by the kind of integrated mechanization of mining work. Workings of the first kind are most efficient economically. In that case extraction of the mineral is done to the full capacity and the overburden rock is unloaded into the mined-out space.

Stripping and mining of mineral deposit by deep-kind method are done layer wise (in slices) from the top downwards. As a rule, the mined rock (overburden and mineral) is moved upwards onto the surface and the overburden is stored in external waste dumps. Before mining each new layer, development work is carried out to ensure proper stripping of mining levels. The depth of the quarry gradually increases to the limits determined by the boundaries of the open-pit field. The movement of the overburden, enclosing rock and mineral characterizes the on-slope method of surface mining by haulage means downwards to waste dumps and processing plant. The on-slope-deep method of surface mining combines the characteristic features of the 2nd and 3rd kind. The kinds of deposits (with flat-land relief) suitable for surface mining, as applied to rounded, elongated and stretched forms and I the total angle of dip, and the respective letter abbreviations are shown (for teaching purposes) in Fig. 2.5 (part I) and Fig. 2.6 (part II).

2.4 Kinds and Sizes of Open-Pit Fields

A deposit or part of it that is surface mined from one quarry is called an open-pit field. An open-field pit is a three-dimensional figure specified by its dimensions, depth, and the slope angles of flanks. An open-pit field is a part of the lot assigned to the quarry within the boundaries of which the waste dumps, the pay-ore area and other industrial structures are located. Table 2.1 gives the shape of ore deposits.

The dimensions of an open-pit field determine the total scope of mining work and the probable production utilization (capacity) of a quarry. The ultimate depth (Hq) of quarries depends on the natural conditions. For surface-type surface mining, it remains essentially the same during the entire working period. For deep, on-slope and combined-type mining, the ultimate depth is established at the stage of quarry design.

The ultimate depth of modern quarries or pits measures from a few meters up to 450 m. Projects envisage the possibility of surface mining up to 800 m deep.

The dimensions of the quarry floor (lf and bf) are established by contouring the mined section of the deposit at the elevation of the ultimate quarry depth, Fig. 2.7. The minimum dimensions of the quarry floor are specified by the conditions of safe extraction and loading of the extracted rock on the lower bench (b1 not less than 30 m and lf not less than 100 m). The slope angles of open-pit flanks (pit slope angles), γ, are determined by the stability of the massif adjoining the slope and the arrangement of haulage lines: transport berms and inclined trenches. The tendency

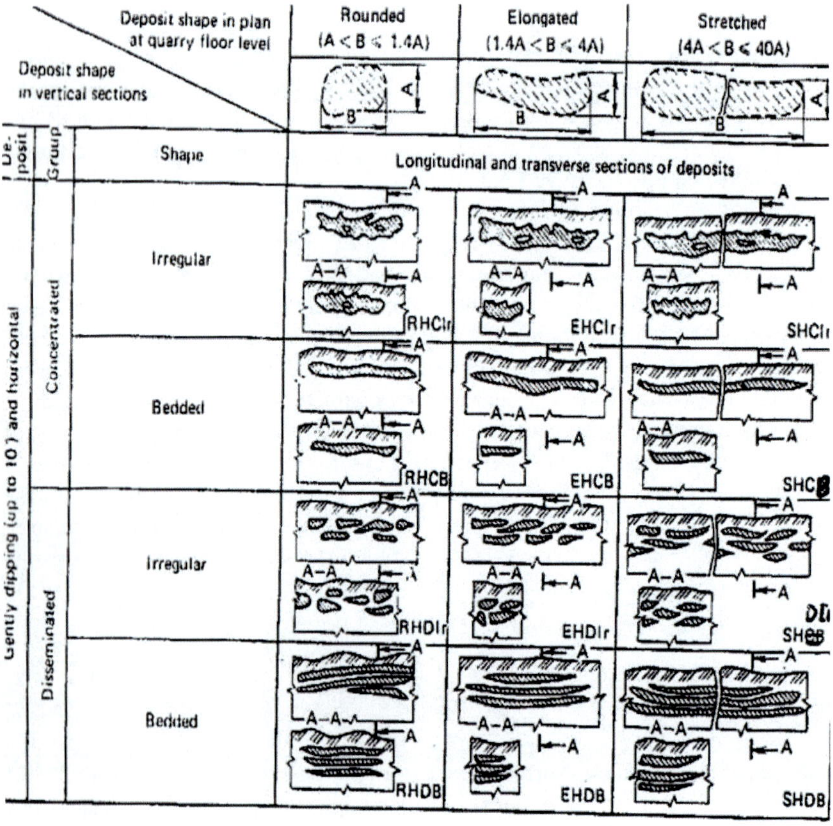

Fig. 2.5 Shapes of horizontal and gently dipping deposits (part I)

is to have steeper slopes so as to minimize the total volume of the overburden to be removed, see Abdel Sabour et al. [1].

The dimensions of a quarry along and across the surface strike of the deposit, Lq and Bq, as shown in Fig. 2.7 are determined by the size of the deposit, the size of the quarry floor, the pit slope angles, and by certain topographic and hydrographic conditions. As a rule, the dimensions of a quarry are established graphically. With very large deposits, L_q and B_q are determined from the conditions of opening of mining levels and the division of a deposit into individual open-pit fields. The strike of an elongated open-pit field should be matched with the specified quarry capacity, i.e. the total length of the required excavation lines.

If only one side of a quarry is worked, Eq. 2.1:

$$\frac{L_q - L_f}{2} = \frac{f\,N_e\,L_{ei}}{n_{wb}} \tag{2.1}$$

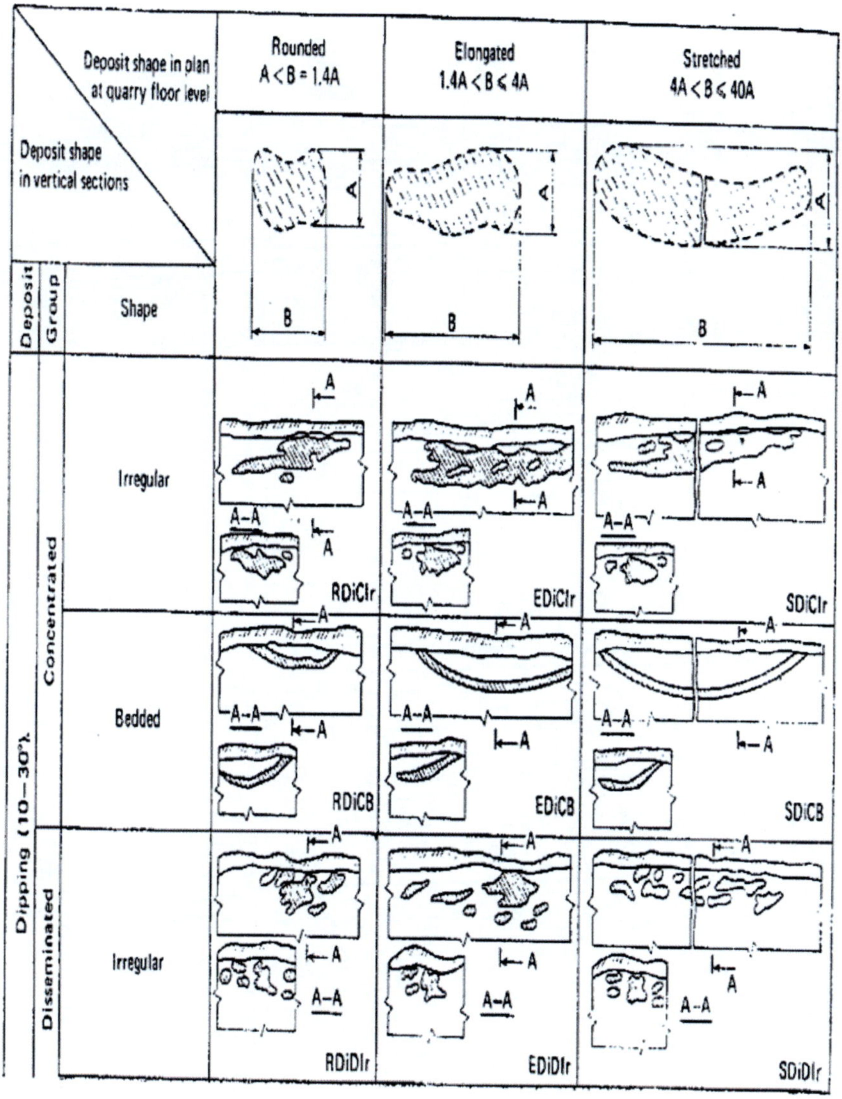

Fig. 2.5 (continued)

where:

f is the coefficient of face reserve,
N_e is the number of excavators in operation,
L_{ei} is the length of a single excavation line, m, and
n_{wb} is the number of working benches.

An increase of the length of an open-pit field can offer the following advantages:

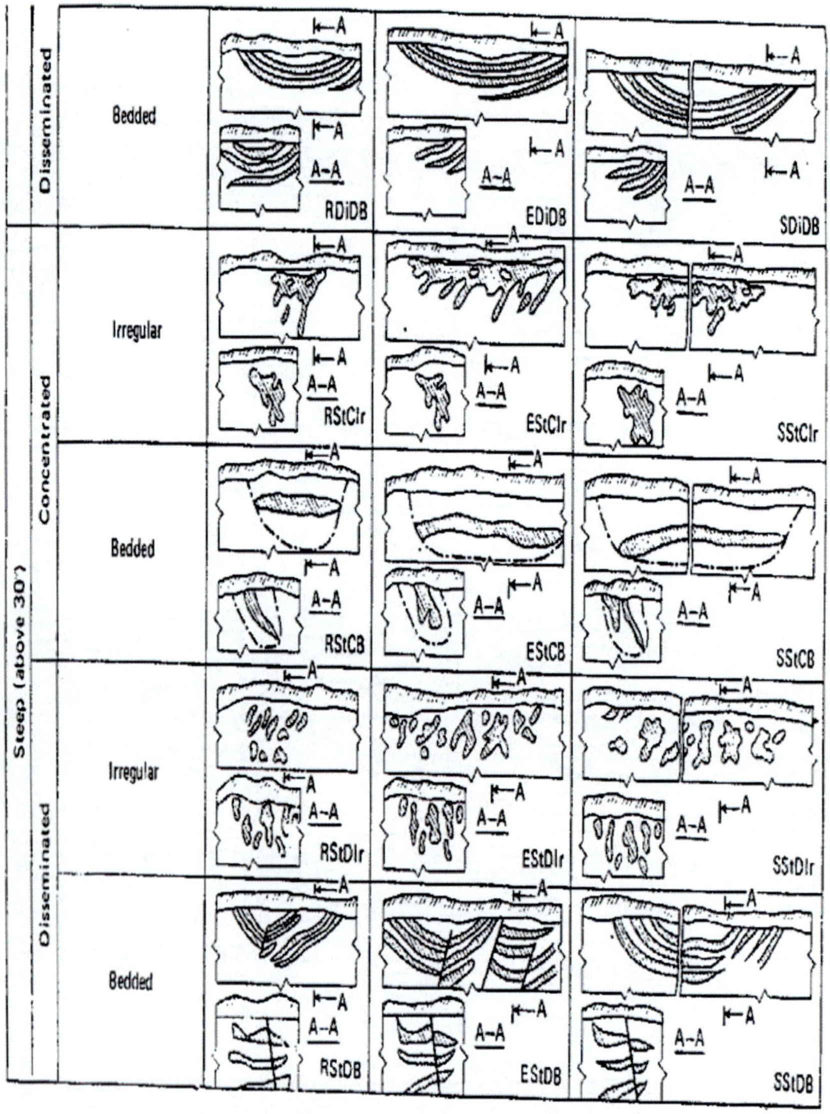

Fig. 2.6 Shapes of horizontal and gently dipping deposits (part II)

- Increase the reserve of mineral resources within the limits of an open-pit field and thus increase the capacity of a quarry.
- Reduce the average and current stripping coefficients due to a smaller effect of the volume of quarry flank spacing; with a flat-land relief, the volumes of spacing of longitudinal flanks, V_I (m^3), and end flanks, V_e (m^3) of a quarry can be found using Eqs. 2.2 and Eq. 2.3, Fig. 2.7.

Table 2.1 Shape of ore deposits terminology

Symbol	Terminology
R	Rounded
E	Elongated
S	Stretched
H	Horizontal
C	Concentrated
Ir	Irregular
B	Bedded
D	Disseminated
EHCIr	Elongated horizontal concentrated irregular (iron-ore deposit; apatite's)
EHCB	Elongated horizontal concentrated bedded (coal fields; limestone deposit; sand-gravel quarry)
EHDIr	Elongated horizontal disseminated irregular (gypsum deposit; limestone deposit)
RHDB	Rounded horizontal disseminated bedded (coal fields; oil shale)

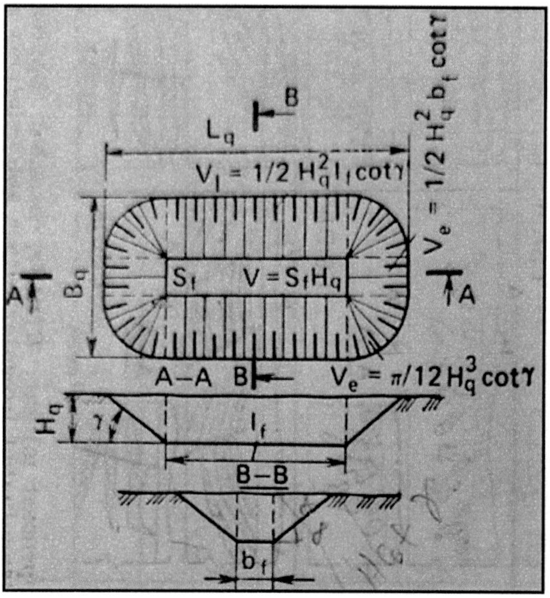

Fig. 2.7 Dimensions of an open-pit field

$$V_I = H_q^2 L_f \cdot \cot \gamma_{avg.} \tag{2.2}$$

$$V_e = H_q^2 b_f \cot \gamma_{avg} + \left(\frac{\pi}{3(H_q^2 \cot \gamma_{avg})} \right) \tag{2.3}$$

where:

$\gamma_{avg.}$ is the averaged angle of slope of flanks, degrees.

- Reduce the total quarry expenditures per ton of mineral yes and the cost of extraction of the mineral.

A large length of an open-pit field, especially with the use of railway transport, incurs the following drawbacks:

- Increases the total run of transport vehicles on working berms, the time of a single run, and the haulage expenditures.
- Worsens the conditions of train interchange on benches, which causes the necessity to increase the effective mass of trains and employ additional railway switches.
- Increases the scope of development work for ensuring the growth of quarry capacity by increasing the length of working trenches, the number of inclined trenches.

The length of open-pit fields varies from a few hundred meters to 6 km and their width may be up to 5 km depending on the type of deposit and the kind of surface mining. At large quarries, $L_q = 2–2.5$ km. The total volume of the mined rock within the limits of a quarry determines the productive capacity of the plant, its age, etc. With flat land topography, the total volume of mined rock can be determined quite accurately from the Eq. 2.4:

$$V_q = S_f H_q + 0.5 P_f H_q^2 \cot \gamma_{avg} \left(\frac{\pi}{3(H_q^2 \cot \gamma_{avg})} \right) \tag{2.4}$$

where:

S_f : The surface area of the quarry floor, m^2, and
P_f: The floor perimeter, in. the area of an open-pit field on the surface (m^2) can he found approximately as given in Eq. 2.5:

$$S_q = K L_q B_q \tag{2.5}$$

where:

K: A coefficient which considers the shape of an open-pit field (usually k = 0.8–0.9).

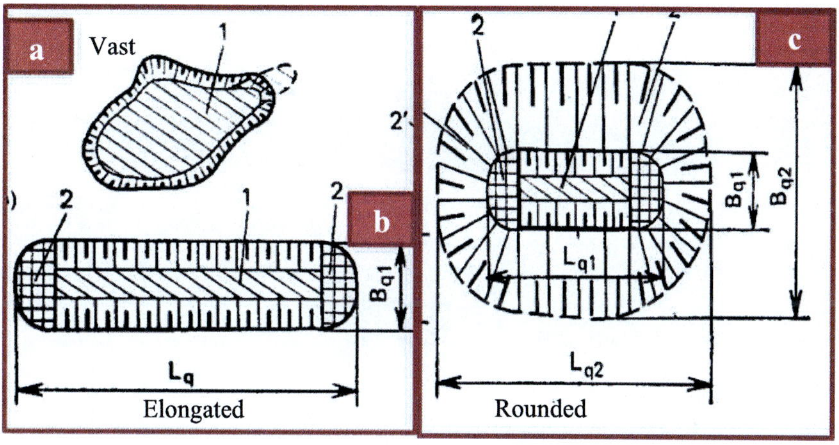

Fig. 2.8 Shapes and dimensions of open-pit fields

By their shape and size, open-pit fields can be divided into vast (expanded), elongated and rounded, Fig. 2.8.

*1: quarry floor.

*2 and 2', end portions of elongated and rounded quarry.

a. Vast open-pit fields, employed mainly in surface-type mining, are characterized by a relatively small depth (Hq up to 100 m), a large surface area of a quarry in plan (up to 10–40 km^2), and relatively little differing dimensions L q and B.
b. Elongated open-pit fields have large dimensions along the strike (L_q up to 3–5 km), exceeding several times the dimensions across the strike B_q. They are typical for deep-kind open-cast mining with Hq up to 150–200 m and for surface-kind mining of stretched narrow deposits.
c. Rounded open-pit fields are typical for mining of columnar deposits of any depth and of deep deposits (200–800 m) of any shape in plan; the bench spacing over the perimeter of a quarry of such an appreciable depth predetermines the round or oval shape of the open-pit field in plan irrespective of the shape of quarry floor which usually is also of an oval shape.

The division of open-pit fields by their shape in plan into elongated and round is done depending on the ratio of the total volume of a quarry, V_q, to the volume of its end portions, V_e: open-pit fields are related to the elongated type if V_e is not more than (0.15–0.20) V_q. For rough calculations of the total volumes of mining work, an open-pit field can be regarded as elongated if L_q: B_q > 4:1. During mining of surface-type quarries, their plan dimensions vary gradually; with deep-type quarries, their depth and plan dimensions are changed simultaneously. The width of a quarry usually increases more quickly than the length, the L_q: B_q ratio decreases and the open-pit field become more and more rounded, even on sheet deposits. Depending on the size of open-pit fields and deposits and the sequence in which a deposit is involved into mining, the following types of open-pit fields may be distinguished.

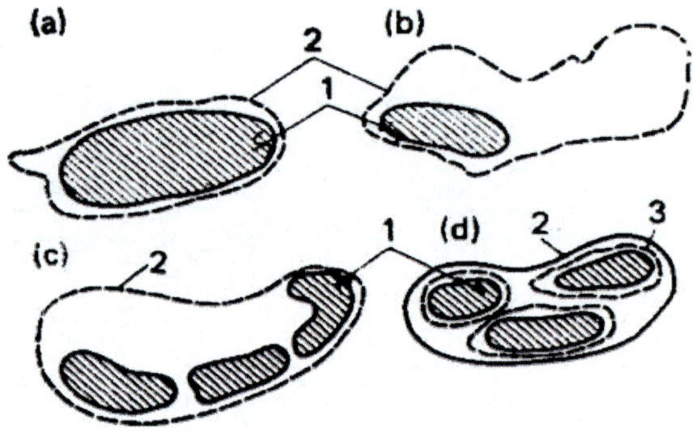

Fig. 2.9 Schemes of open-pit fields

An open-pit field covers the entire deposit, Fig. 2.9a, the plan dimensions and shape of the field are determined by those of the deposit and by the conditions of flank spacing. The numbers in Fig. 2.9 represent the following: (1) Contour of deposit or individual open-pit field. (2) Contour of open–pit field (deposit). (3) Contour of off-balance reserves; see Vagenas et al. [16–18] and Yazici et al. [19, 20].

An open-pit field covers only a portion of the deposit which is allotted initially for mining, Fig. 2.9b, the dimensions of the field are determined by the production utilization (capacity) and age of the quarry. An open-pit field belongs to a system (group) of simultaneously mined fields into which the given deposit is divided, Fig. 2.9c. Such open-pit fields are possible on large deposits where a single quarry of a high capacity would be unfeasible or economically un-favorable; a system of open-pit fields usually has a common network of haulage lines on the surface and in waste dumps, see Bley et al. [3]; Caccetta and Hill [5]; Hustrulid and Kuchta [8] and Osanloo et al. [12].

An open-pit field consists of independently mined individual portions, Fig. 2.9d, such fields are common in mining of deposits broken down into a number of independent districts, the individual portions of an open-pit field can be regarded as small independent fields. Classification of open-pit fields by their plan dimensions and the ultimate depth or height of the working zone (on slopes) is given in Table 2.2. When a quarry is being designed, it is important to find the quarry dimensions at which the total expenses per ton of output will be at a minimum (considering the time of investment). These expenses per ton of output include running (operating) expenses and capital costs (spent at once initially or portion wise at different time), calculated for a working year.

2.5 Variations of Open Pit Mining

Are depending on natural, geologic, and technological factors, several variations of open pit mining are possible, Fig. 2.10. The differences are mainly in pit design and equipment; the basic sequence of development and cycle of operations is very similar. Figure 2.10a, is flat-lying seam or bed, flat terrain such as Iron, Taconite.

Table 2.2 Kinds of open-pit fields

Field size	Kind of Surface mining	Quarry plan area (km^2)	Quarry depth (m)	Total volume of rock × 10^6 (m^3)	Quarry age (years)
Very small	Surface	up to 0.4	up to 20	up to 10	up to 10
	On-slope	up to 0.3	up to 40		
Small	Surface	0.4–2.0	up to 40	10–400	10–25
	On-slope and deep	0.3–1.5	40–100		
Medium	Surface	2.5–6.0	up to 60	100–500	25–30
	On-slope and deep	1.5–5.0	100–200		
Large	Surface	4–20	up to 80	500–2000	80–60
	On-slope and deep	4–12	100–250		
Very large	Surface	10–40	up to 120	2000–10,000	60–100
	Deep	10–30	200–800		

Figure 2.10b, is a massive deposit, flat terrain like Iron. Figure 2.10c, is a pitching seam or bed, flat terrain such as anthracite. Figure 2.10d, is a massive deposit, high relief such as copper. Figure 2.10e, is a thick bedded deposits, little overburden such as nonmetallic, western U.S. Coal.

2.6 Surface Mining Economics

The cost per ton for surface mining can be considerably lower than for underground mining if the conditions are right. In practice there is seldom scope for competition between the two systems. Where a deposit is accessible to surface mining it is often a case of determining the "cut-off limit" when open pit operations are uneconomic and underground methods (if feasible) must be adopted, i.e. steeply dipping deposits, advancing into thickening overburden. In surface mining barren material must be stripped—"waste', overburden, sterile inclusions, "dead work" at pit ends, spoil re-handling. The stripping ratio, (S) is defined as per Eq. 2.6; see Putra [15]:

$$S = \frac{\text{Volume of overburden removed}}{\text{Volume of mineral recovered}} \tag{2.6}$$

The basic production, Eq. 2.7, cost is:

$$\mathrm{C_{v0}} = \mathrm{m} + \mathrm{S\ O} \tag{2.7}$$

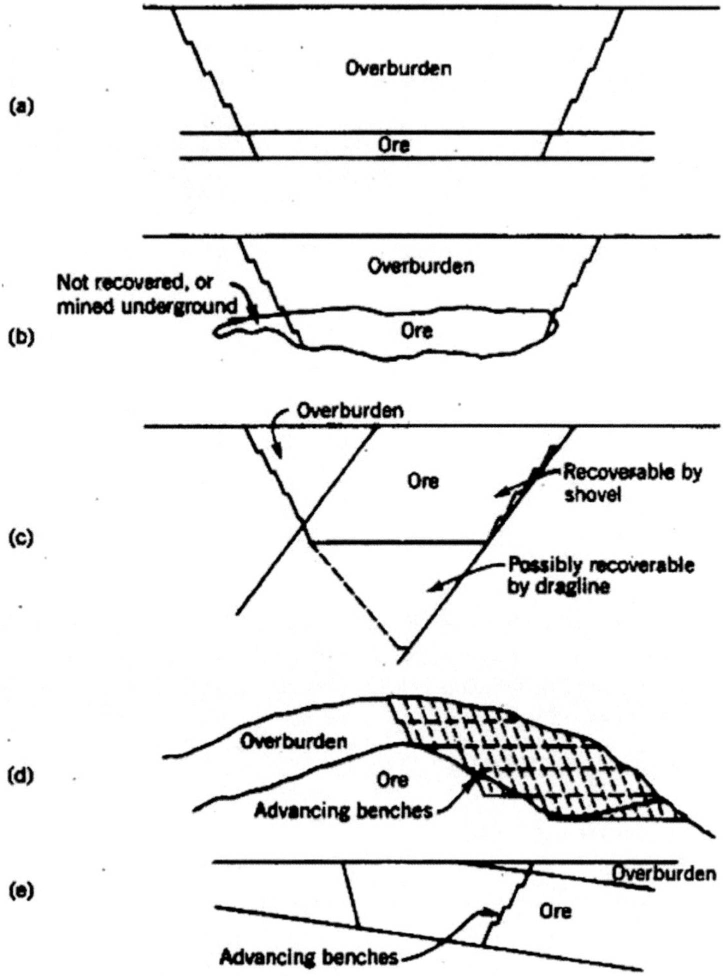

Fig. 2.10 Variations of open pit mining

where:

C_{v0} Cost/unit volume of mineral at pit limits.
m Cost of excavating unit volume of mineral and transporting it to pit limits.
O Cost of excavating unit volume of overburden, waste, transporting it to the spoil heap and disposing of it.

(All volumes are expressed as solid measure, i.e. .n "bank").

The costs are inclusive of all charges, i.e. capital and operating (labor, fuel and lubricants, power, maintenance, depreciation, interest, etc.). The production cost at a pit limits is then, Eq. 2.8:

$$C_{vp} = \mathrm{m} + \mathrm{S\,O} + \mathrm{A} \quad (2.8)$$

where:

A: Total fixed charges due to development and administration expressed per unit volume of mineral.

If:

U: Production cost per unit volume of mineral mined by underground methods then, Eq. 2.9:

$$\mathrm{m} + \mathrm{S\,O} + \mathrm{A} \leq \mathrm{U} \quad (2.9)$$

For economic surface mining occurs when:

$$\mathrm{m} + \mathrm{S\,O} + \mathrm{A} = \mathrm{U}$$

Hence the maximum stripping ratio when, the economic cut-off when, Eq. 2.10:

$$\mathrm{S_e} = \frac{(\mathrm{U} - \mathrm{m} - \mathrm{A})}{\mathrm{O}} \quad (2.10)$$

where:

S_e maximum economic stripping ratio.

The mineral often requires preparation before loading out as a saleable product and additional cost such as:

- Transport from pit Limits to preparation plant.
- Preparation costs.
- Stockpile and loading costs.
- Cost of preparation plant waste disposal.

All must be added to C_{vp} to arrive at the cost of mineral loaded into the transport system. During these processes, Losses of saleable product occur, due to the recovery factor of the preparation plant etc., and the loaded cost (which is invariably expressed on a mass basis) of saleable product is given in Eq. 2.11:

$$C_{vp} = \left[\left(\left(\frac{1}{R}\right) \times \left(\frac{m + S\,O + A}{D}\right)\right) + B\right] \quad (2.11)$$

where:

C_{vp} cost per unit mass (usually tonnes) of saleable product loaded.
R recovery of mineral per unit in the preparation plant.
D density of prepared mineral per unit mass.
B ost of transport from pit limits to preparation plant + cost of preparation plant input + cost of Loading out.

If the loading out station is the point of sale and if:

p minimum acceptable profit per unit mass of saleable product then the minimum selling price P_s is, Eq. 2.12:

$$K_{ind} = \frac{V_{O.B.}}{V_O},\ m^3/m^3 \tag{2.12}$$

2.6.1 The Concept of "Cut-Off"

Various authorities use the term "cut-off" for stripping ratio or for mineral grade (hence "cut-off grade"). In general the following are used: Cut off -the limit of economic working involving stripping ratio, grade (G), variable costs. Then; the Sc. ratio "break-even stripping ratio" and "economic stripping" is given from:

$$S_c = \left[\frac{(D\,(P_s \times R\ \times G - B - P) - (m + A))}{O}\right] \tag{2.13}$$

The cut-off grade, Eq. 2.14, is the break-even cost between sending material to the preparation plant or the spoil heap; see Baidowi et al. [2]; Henning [7]; Johnson et al. [9]; Lane [10]; Thompson and Barr [14] and Putra [15].

$$G_c = \left[\frac{\left(\frac{\frac{m+S\times O+A}{D}}{B+P}\right)}{P_s \times R}\right] \tag{2.14}$$

where:

G_c: is the cut-off grade is the term applied in deciding which material must be moved as mineral or as waste.

It can be seen from Eqs. 2.13 and 2.14 that, S_c, is some function of G and G_c is some function of S. Thus both grade and stripping ration are of importance in determining the point of "cut-off" in open pit mines. The term cut-off is somewhat loosely used, depending on the scale of the operation, i.e. it may be a considerable time before the break-even stripping ratio is reached. In this case grade is most important.

2.6.2 *Profit Margin*

Generally it is possible to project a surface mining scheme within narrower confidence limits than is the case for an underground mine (although there are exceptions, e.g. when complex grade variations occur). As a consequence of this, the projected minimum return on capital may be lowered, in other words the margin, which necessarily must be allowed for error is smaller. The profit per tonne of mineral mined may be very small and the success of the operation depends upon the optimization of the small differences between two very large numbers. It is for this reason that sophisticated methods of numerical analysis and control are so essential, see Dagdelen and Johnson [6].

2.7 Maximum Versus Overall Stripping Ratio

It is on the basis of calculating stripping ratios that we are able to locate pit limits and to express volumes of overburden to be moved per unit weight of ore, coal, or stone uncovered. We must distinguish between two stripping ratios (units: yd^3/ton or m^3/tonne).

- Maximum allowable stripping ratio, Eq. 2.15:

$$SR_{\max.} = \frac{Volume\ of\ overburden}{Weight\ \ of\ ore\ at\ economic\ pit\ limit} = V_{lw} \tag{2.15}$$

Overall stripping ratio, Eq. 2.16:

$$\mathrm{SR_0} = \frac{\text{Volume of overburden}}{\text{Weight of ore for entire ore body or cross-section}} = \frac{\mathrm{V}}{\mathrm{W}} \tag{2.16}$$

In only one instance does SRmax not establish the pit limit or exceed SR0 in magnitude, and that occurs when (1) the surface is flat, and (2) the deposit is flat, tabular, and of constant thickness. In that singular case, SRmax lacks significance, and the pit limits are located at the property lines. Another distinction is that SR_0 is an actual numerical ratio of yd^3/ton (m^3/tonne), whereas SR_{max} is expressed in units of equivalent yards (see the next section).

Under long-range planning, we discovered that the maximum stripping ratio, while a physical quantity, is determined solely by economics. The overall stripping ratio, on the other hand, has mainly physical significance. It is because of its economic basis that we can employ SR_{max} to locate the pit limits of a deposit in the general case, that is, an ore body of varying thickness, dip, or grade occurring beneath an inclined or horizontal surface.

2.8 Different Stripping Ratios

Excavation (stripping) ratio is defined as: the relationship between the amounts of overburden (O.B.) to be removed and the ore to be extracted. Usually is taken in volume of O.B. in cubic meters, related to the unit of weight of ore extracted (m^3/ton), sometimes it is taken in m^3/m^3or even in ton/ton. This ratio is one of the most important factors to determine the effectiveness of surface mining. There exists more than a single excavation ratio for the same surface mine. The following section gives a brief account about the main excavation ratios.

2.8.1 Industrial Excavation Ratio

The industrial excavation ratio is the relationship between the volume of O.B. included in the project-borders of the surface mine and the amount of ore (industrial reserves) included in the same borders as depicted in Fig. 2.11 and given from Eq. 2.17.

$$K_{ind} = \frac{V_{O.B.}}{V_O},\ m^3/m^3 \tag{2.17}$$

It must be noted that ore reserves are divided into:

Geological
Balanced
Unbalanced
Industrial
Un-industrial.

A special case of the industrial excavation ratio is the "geological excavation ratio", which is based on geological data.

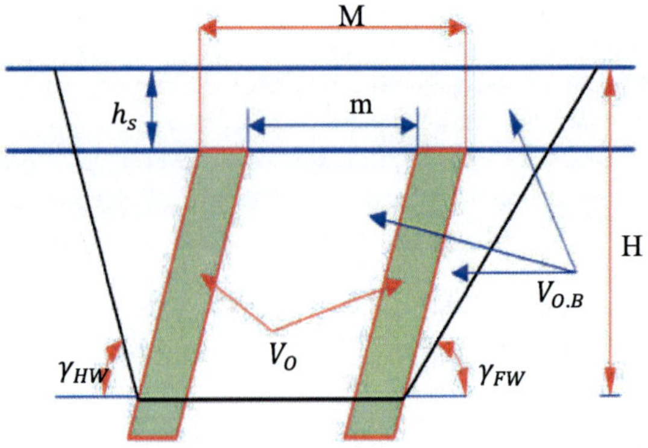

Fig. 2.11 Industrial excavation ratio

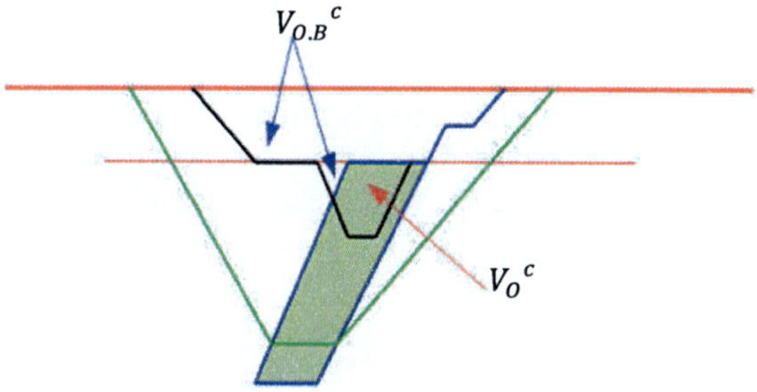

Fig. 2.12 Exploitation stripping ratio

2.8.2 Exploitation Excavation Ratio

The exploitation excavation ratio is the ratio between the total volume of rocks ($V_{O.B.}$), with the exclusion of that volume which has been removed during the construction period of the surface mine, and the extracted (industrial) reserves of the ore, with the exclusion of the amount of ore which has been extracted during the same construction period. This may be calculated to the surface mine as a whole, or to a part of it, Fig. 2.12 and Eq. 2.18.

$$K_{\exp l.} = \frac{V_{O.B.} - V_{O.B.}^{C}}{V_O - V_O^{C}},\ \mathrm{m^3/m^3} \quad (2.18)$$

2.8.3 Current Excavation Ratio

It is similar to the above (e.g. exploitation excavation ratio) but during a definite period of time as given from Eq. 2.19.

$$K_{current} = \frac{V_{O.B.}^{T}}{V_O^{T}},\ \mathrm{m^3/m^3} \quad (2.19)$$

The current excavation ratio depends on the following:

Climatic conditions
Organization factors.

Deviations in the current excavation ratio are due to organization and project values, as well as the designed parameters of the given surface mine.

2.8.4 Expansion Excavation Ratio

Expansion excavation ratio is the relationship between the amount of overburden (O.B.) to the amount of extracted ore resulting due to the expansion of the borders of the surface mine, both in plan and in depth. It is a summation of values, Eq. 2.20 and Fig. 2.13.

$$K_{\text{exp}} = \frac{\sum \Delta V_{O.B.}}{\sum \Delta V_O}, \ \text{m}^3/\text{m}^3 \tag{2.20}$$

The expansion excavation ratio is especially used in problems related to inclined as well as steeply inclined deposits. It increases with the increase of the depth of the surface mine. It usually equals the final (border) ratio, at the last period of exploitation of the surface mine. Expansion ratio for complicated forms of deposits and relief is given from Eq. 2.21 and as depicted in Fig. 2.14.

Fig. 2.13 Expansion excavation ratio

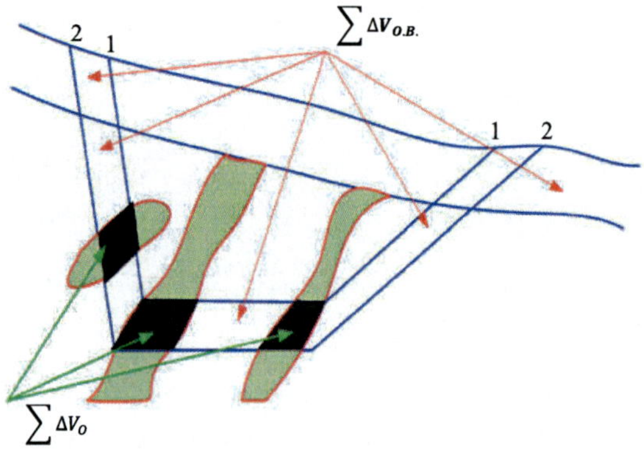

Fig. 2.14 Expansion excavation ratio for complicated forms of ore deposits

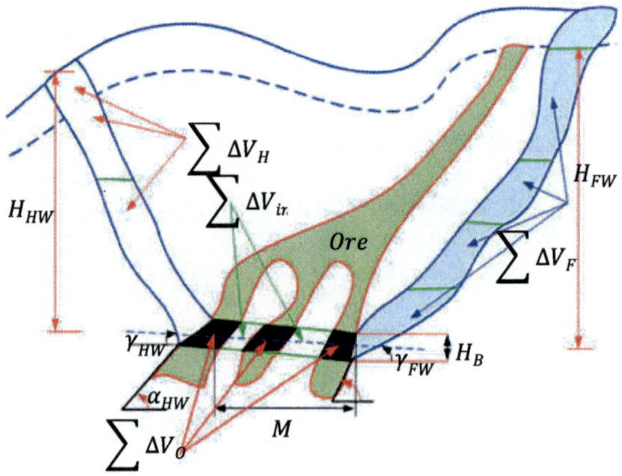

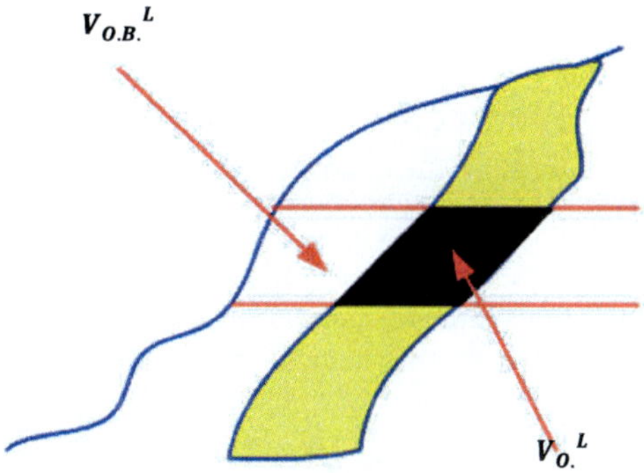

Fig. 2.15 Layer excavation ratio

$$K_{\exp} = \frac{\sum \Delta V_{FW}^{O.B.} + \sum \Delta V_{HW}^{O.B.} + \sum \Delta V_{inclusions}^{O.B.}}{\sum \Delta V_O}, \ \text{m}^3/\text{m}^3 \quad (2.21)$$

2.8.4.1 Special Case

Expansion excavation ratio due to an increase of 1 m in the depth of the surface mine:

> These values of the expansion excavation ratio are valid for the given geologic cross-sections are considered as a "block of reserves".
> The value, thus calculated, does not take into consideration the side amounts, so it may be applied for these reserves having considerable extensions.

2.8.5 Layer Ratio

It is defined as the relation between the volumes of removed overburden to the volume of extracted ore within a certain layer as shown in Fig. 2.15. The layer excavation ratio is usually taken as the bench, and hence the term "bench ratio" may be applied.

2.8.6 Border Excavation Ratio (Critical Ratio)

It is defined as: the maximum admitted ratio, which is determined according to mine economic factors, as well as to the gains obtained from surface mining as given in Eq. 2.22.

$$K_b = \frac{C_u - C_s}{C_o} \quad (2.22)$$

where:

C_u Costs of underground extraction of 1 m^3 of ore (extraction).
C_s Costs of surface extraction of 1 m^3 of ore (extraction).
C_o Costs of the removal of 1 m^3 of the present overburden (removal).

The borders of surface mine fields are mainly determined according to the value of K_b. The value of this ratio has a wide range (from 60 to 10) in m^3/m^3, and more, for "direct removal" methods of nearly horizontal beds of wide extension.

2.9 Difficult Parts

They happened when the stripping ratio is high or when the extractive is very expensive because of hard overburden. It may be extracted by surface or underground method.

2.9.1 Difficult Part Near to One of the Borders (Case I)

The amount of reserves, near to one of the borders of the surface mine field, having a volume Q, where $K > K_b$ as depicted in Fig. 2.16.

In this case the difficult part can be included within the mine fields if U.G. ≥ S.M. Alternatively:

$$Q\,C_u + V_T C_{ot} \geq Q\,C_s + K\,Q\,C_0 \tag{2.23}$$

where:

Q Volume of ore reserves near to one of the borders, (having $K > K_b$, m^3).
C_u Costs of underground mining of 1 $m^3$of the ore, \$/ m^3.
C_s Costs of surface mining of I $m^3$of ore \$/ m^3.
C_o Costs of removal of 1 m3of overburden.
C_{ot} Costs per 1 m^3, trench construction \$/ m^3.

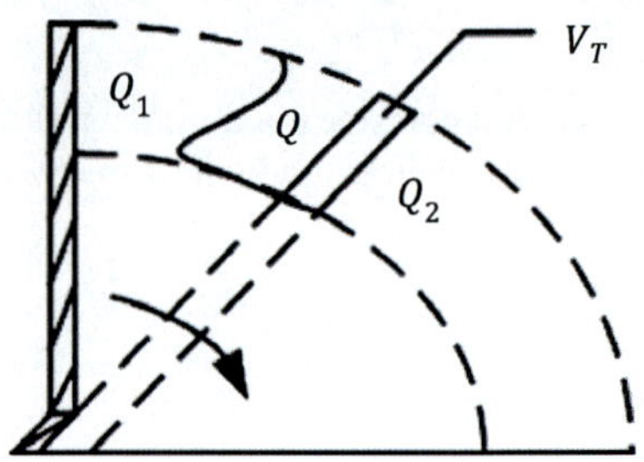

Fig. 2.16 Large difficult part near to one of the borders of the surface mine field (case I)

V_T Volume of additional trench, m^3.
K The excavation (stripping) ratio of Q (K > Kb).

Now, dividing by Q C_o and rearranging, we will have, Eq. 2.24:

$$K \leq \frac{C_u - C_s}{C_o} + \frac{V_{1T} C_{oT}}{Q\ C_o} \tag{2.24}$$

where:

$\frac{C_u - C_s}{C_o}$ Border (excavation) stripping ratio or the critical ratio (K_b) if $K > K_b$ is higher than $K_b$by a maximum value of:

$$\frac{V_T C_{oT}}{Q\ C_o}.$$

2.9.2 Difficult Part near to Two Borders (Case II)

The difficult part, Q_1, in this case having $K < K_b$ with a limited (small) amount of reserves which may not cover the expenses (costs) of trench construction (VT). Taking into consideration that the part of reserve, Q, ($K > K_b$) is to be mined out by underground mining methods, or is added to Q1 and extracted by surface mining method, Fig. 2.17.

Here we make the comparison between the expected rise and economy. If both amounts of reserves $Q + Q_1$ are to be worked out by a surface mine, therefore:

Rise (Surface) ≤ Economy (Surface), Eq. 2.25:

$$V\ C_{oT}\ \leq\ Q_1\ C_u - (Q_1 C_s + K\ Q_1 C_o) \tag{2.25}$$

Dividing by $Q_1 C_o$ and rearranging, Eq. 2.26:

$$K \leq \frac{C_u - C_s}{C_o} - \frac{V_T\ C_{oT}}{Q_1\ C_o} \tag{2.26}$$

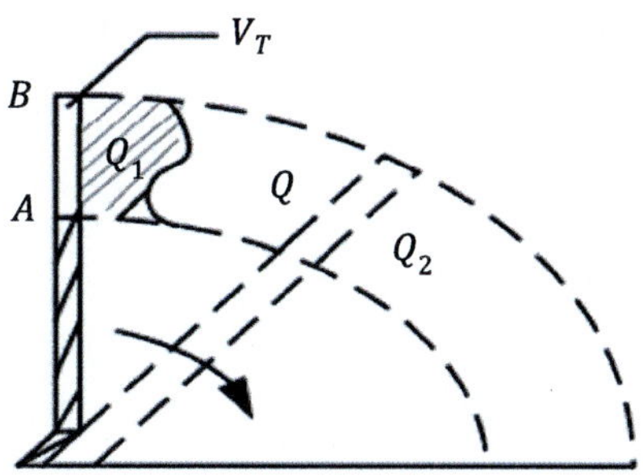

Fig. 2.17 Difficult part of a small amount of reserve (case II)

Which is the limiting value of the excavation ratio (E.R.), i.e. the part of the reserve, Q_1, is to be worked out by surface mining if its excavation ratio (E.R.) is less than the border of excavation ratio (K_b) by a value having a minimum limit of:

$$\frac{V_T\,C_{oT}}{Q_1\,C_o}.$$

2.9.3 *Turning Point (Case III)*

This case discusses the effect of the construction of a turning point (dressing plant) in addition to the necessary capital trench, Fig. 2.18. The case where the additional turning plant is effective and the difficult part $Q_d(K < K_b)$ can be worked by surface mining methods can be expressed as follows:

S.M. $\leq$ U.G. (Eq. 2.27)

$$V_T C_{oT} \pm Q\Delta q + Q_d C_s + K\, Q_d\, C_o \leq Q_d C_u \tag{2.27}$$

where:

- V_T The volume of a new capital trench, m^3.
- Q_d The difficult part having $K < K_b$ with changing distance to the dressing plant (D. P.), m^3.
- Δq The difference in the transportation cost per ton or cubic meter in normal case and new condition through the original and the new capital trenches due to the construction of the turning point, $\$/m^3$.

Dividing by $Q_d\,C_o$ and rearranging, Eq. 2.28:

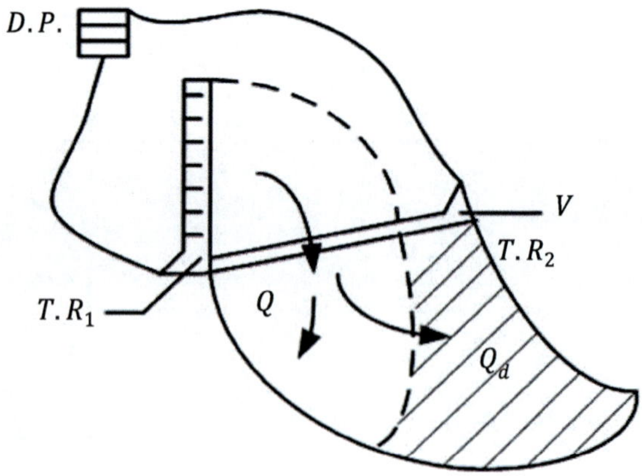

Fig. 2.18 Difficult part shows the effect of construction of dressing plant in addition to the necessary capital trench (case III)

$$K \leq \frac{C_u - C_s}{C_o} - \frac{V_T C_{oT} \pm Q\,\Delta q}{Q_d C_o} \quad (2.28)$$

As:

$$\frac{V_T C_{o.\,T} \pm Q\,\Delta q}{Q_d C_o} = \Delta K \quad (2.29)$$

Then:

$$K \leq \frac{C_u - C_s}{C_o} - \Delta K \quad (2.30)$$

Therefore; the difficult part may be extracted within the surface mine field if it is less than the border ratio by a minimum value ΔK.

2.9.4 Intersection (Case IV)

The intersection of the surface mine field is defined by a difficult part ($K > K_b$), Fig. 2.19. Here, there are two alternatives:

1. Working the deposit by a single (big) surface mine, including the difficult part within its borders.

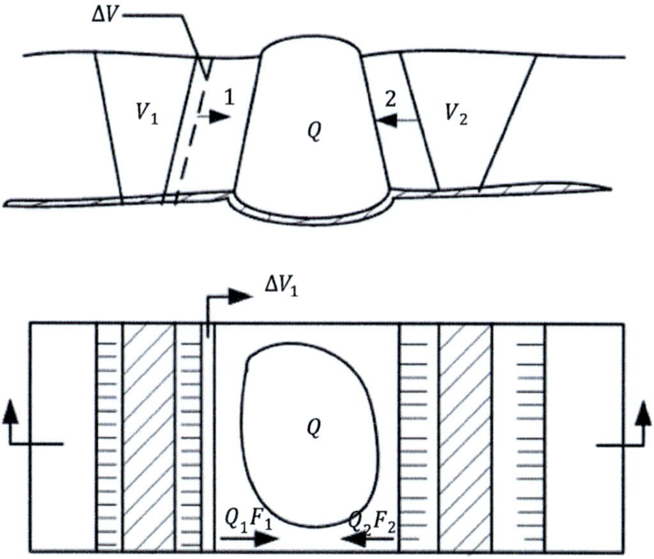

Fig. 2.19 Difficult parts located in the middle or centre of the mine (case IV)

2. Working the deposit by two small surface mines, having reserves Q_1 and Q_2 where K < K_b plus an underground mine to work out the difficult part having reserves Q (K > K_b).

To start with, the costs of surface mining in a small surface mine are higher than those of a big surface mine. This fact must be taken into consideration. The first alternative (a single big surface mine) is to be applied if economy due to underground working out of the reserves difficult part Q (K > K_b) is less than the rise in the costs which result due to the division to two small surface mines. This can be expressed as:

Eu: economy due to underground mining of the difficult part.

$$Eu = Q\ C_s + Q\ K\ C_o - Q\ C_u \tag{2.31}$$

Rs (u): the rise in the costs due to the division of surface mine field into two small fields resulting from working Q by underground mine, Eq. 2.32:

$$\mathrm{Rs(u)} = [\mathrm{F_1C_1} - (\mathrm{F_1} - \Delta \mathrm{V_1})\ \mathrm{C}_o] - \Delta \mathrm{V_1}\ \mathrm{C_{o.T}} + \mathrm{F_2(C_2} - \mathrm{C_o}) + V_2\ (\mathrm{C_{o.T}} - \mathrm{C_o}) \tag{2.32}$$

Or simply as given in Eq. 2.33:

$$\mathrm{Rs(u)} = \mathrm{F_1(C_1} - \mathrm{C_0}) + \mathrm{F_2(C_2} - \mathrm{C}_o) + \mathrm{V_2}\ (\mathrm{C_{o.T}} - \mathrm{C_0}) - \Delta \mathrm{V_1}\ (\mathrm{C_{oT}} - \mathrm{C_0}) \tag{2.33}$$

where:

F_1, F_2 The volumes of removed overburden in the two small open-pits,m^3.
V_1, V_2 The volumes of trenches of small surface mines, m^3.
ΔV_1 The additional volume of the capital trench of the first small open-pit in case of mining the ore deposit as one large open-pit.
C_1, C_2 The costs of removal of 1 m^3 of overburden in case of extracting the deposit with two small open-pits and the difficult part mined by underground method ($C_1,\ C_2 > C_0$).
C_0 The cost of removal of 1 m^3 of overburden in case of extracting ore as one large open-pit.
C_0T Costs of overburden removal of trench construction, $/$m^3$.

Now, let us take the rise in costs per 1 m^3 removal of overburden (in comparison with that of the single mine) for the two small mines to be $C_1 - C_o$ and $C_2 - C_o$, or ΔC_1 and ΔC_2 respectively. Then we have:

$$\mathrm{Rise} = \mathrm{F_1 \Delta C_1 + F_2 \Delta C_2 + (V_2 - \Delta V_1)(C_{oT} - C_0)}.$$

The final conclusion will be as follows:

Working the deposit by alternative (1), i.e., big a single surface mine will be have advantage if: $R_{s(u)} \le E_{(u)}$.

Or:

$$Q\ (C_s + K\ C_o) - Q\ C_u\ \le R_{s(u)} \tag{2.34}$$

Dividing by QC_o and rearranging:

$$K \le \frac{C_u - C_s}{C_o} + \frac{R_{s(u)}}{Q\ C_o} \tag{2.35}$$

Or, in other words, the difficult part Q ($K > K_b$) may be economically included in the borders of surface mine field (the single big one) if its excavation (stripping) ratio (K), is higher than the border (critical) E. R. = K_b by a maximum value, ΔK, where:

$$\Delta K = \frac{F_1 \Delta C_1 + F_2\ \Delta C_2 + (V_2 - \Delta V_1)(C_{o.T} - C_o)}{Q\ C_o} \tag{2.36}$$

2.10 Important Coefficients in Surface Mining

Table 2.3 gives the most widely important coefficients used in surface mining.

Table 2.3 The most important coefficient in surface mining

No	Coefficient	Index	Explanation	Equation	Characteristics
1	Coefficient of swelling	K_S	$\frac{\text{Vo.l of rock in exc. bucket}}{\text{Vol. of rock in situ}}$	$K_s = \frac{V_b}{V_s}$	$K_S \ge 1$
2	Coefficient of filling	η_f	Vol. of rock in exc. bucket	$\eta_f = \frac{V_b}{E}$	$\eta_f < 1$ $\eta_f > 1$
3	Coefficient of excavation	η_e	$\frac{\text{Coeff. of filling}}{\text{Coeff. of swelling}}$	$\eta_e = \frac{\eta_f}{K_S}$	$\eta_e < 1$ $\eta_e > 1$
4	Coefficient of job condition	β	$\frac{\text{Theor. no. of buckets/cycle}}{\text{actual no. of buckets/cycle}}$		$\beta \ge 1$
5	Coefficient of operation	K_o	$\frac{\text{Coeff. of excavation}}{\text{Coeff. of job condition}}$	$K_o = \frac{\eta_e}{\beta}$	$K_o \le 1$
6	Coefficient of time	η_t	$\frac{\text{Useful time}}{\text{Total time}}$	$\eta_t = \frac{T_u}{T_e}$	$\eta_t \le 1$
7	Coefficient of efficiency	K_{eff}	Coefficient of operation × coefficient of time	$K_{eff} = K_o \times \eta_t$	$K_{eff} \le 1$

References

1. Abdel Sabour SA, Dimitrakopoulos RG, Kumral M (2008) Mine design selection under uncertainty. Min Technol 117(2):53–64
2. Baidowi N, Rosyidi CN, Aisyati A (2021) A cut-off grade optimization model in open pit mining considering reclamation cost and revenue. IOP Conf Ser: Mater Sci Eng 1096 012021
3. Bley A, Boland N, Fricke C, Froyland G (2010) A strengthened formulation and cutting planes for the open pit mine production scheduling problem. Comput Oper Res 37:1641–1647
4. Bullivant DA (1987) Current surface mining techniques. J Transp Mater bulk: Bulk Solids Handling 7:Ch.2
5. Caccetta L, Hill SP (2003) An application of branch and cut to open pit mine scheduling. J Global Optim 27:349–365
6. Dagdelen K, Johnson TB (1986) Optimum open pit mine production scheduling by Lagrangian parameterization, application of computers and operations research in the mineral industry: 19th AIME, Society of Mining Engineers, vol 19, pp 127–142
7. Henning U (1963) Calculation of cut-off grade. Can Min J 84(3):54–57
8. Hustrulid W, Kuchta K (2006) Open pit mine planning and design—Volume 1: fundamentals, 2nd edn. Taylor and Francis
9. Johnson PV, Evatt GW, Duck PW, Howell SD (2011) The determination of a dynamic cut-off grade for the mining industry. In: Electrical engineering and applied computing, pp 392–403
10. Lane KF (1964) Choosing the optimum cut-off grade. Colo Sch Mines Q 5(5):350–353
11. Mario AM (2001) Underground Hardrock mine design and planning—a system's perspective. Ph. D. thesis, Queen's University, Kingston, Ontario, Canada
12. Osanloo M, Gholamnejad J, Karimi B (2008) Long-term open pit mine production planning: a review of models and algorithms. Int J Min Reclam Environ 22(1):3–35
13. Rzhevsky VV (1987) Opencast (Technology and integrated mechanization). Mir Publishers, Moscow
14. Thompson M, Barr D (2014) Cut-off grade: a real options analysis. Resour Policy 42:83–92
15. Teuku Andika R Putra (2014). Stripping ratios, pit limits & cutoff grade optimization. Chapter 8. Department of Mechanical Engineering in Materials & Minerals, Syiah Kuala University, Malaysia
16. Vagenas N et al (1995) Simulation for design, planning and control in the automated mine. In: Singhal et al (eds) Proceedings of of the 1995 Mine Plannina and Eauioment Selection Co., Balkema, Rotterdam, pp 271–276
17. Vagenas N et al (1996) Simulation of underground Hardrock mining using AutoMod. In: Proceedings of the 1st international spp. of mine simulation via the internet, Dec 2–13, Bakema, Rotterdam
18. Vagenas N et al (1998) Simulation of teleremote mining systems. In: Proceedings of the CIM'98 Cod. Montreal, Canada
19. Yazici HJ et al (1999a) Simulation-based planning of tele-operated mining processes. In: Proceedings of the 5th symposium on mine mechanization and automation—Proceedings of Telemin 1, June 14–1 7, Sudbury, Canada
20. Yazici HJ et al (1999b) Simulation of mining method alternatives in underground Hardrock mining. In: Proceedings of the CIMY99 Cod Calgary

Chapter 3
Slope Stability

3.1 Introduction

Slopes in soils and rocks—in nature or man-made structures—are: Highways, dams, levees, Canals, and stockpiles. They are constructed by sloping the lateral faces of the soil. Slopes are generally less expensive than Constructing walls. Natural forces (wind. water. snow. etc.) change the topography on Earth and other planets, often creating unstable slopes, see Abramson et al. [1] and Alfreds [2].

Slope failure (landslides) has resulted in much death and destruction. Failures are: sudden (catastrophic) or insidious. Failures are: widespread or localized. Earth slides, ruptures of cut/fill slopes, supported banks, endanger the safety of buildings-bridges-highways-railroads-canals in anywhere regions of the world. It can be said that almost no year has passed without some great failure in foundation and earthwork engineering.

Figure 3.1a–e illustrates the nature of failures of some engineering structures. Figure 3.1a shows a possible danger of lateral expulsion of soil from underneath the foundation of a two-hinged frame structure. Figure 3.1b, illustrates the failure of a 1,000,000-bushel grain elevator, weighing 20,000 tons, at North Transcona, Manitoba, Canada, in 1914. Figure 3.1c shows extensive ruptures of slopes in the Panama Canal cuts are also often cited as examples of extensive soil failure. Figure 3.1 d gives rupture of slope of the embankment of U.S. Route 101 south of Scotia, California. Figure 3.1e shows a rupture of slope of the Delaware-Chesapeake Canal bank at the north pier of the Summit Bridge, Delaware.

Geotechnical engineers have to pay particular attention to: Geology, Surface drainage, groundwater, and the shear strength of soils in assessing slope stability. However, we are handicapped by:

- The geological variability of soils.
- Methods of obtaining reliable values of shear strength.

The analyses of slope stability are based on simplifying assumptions and the design of a stable slope relies heavily on experience and careful site investigation. In

M. M. Ali Elbeblawi et al., *Surface Mining Technology*, Topics in Mining, Metallurgy and Materials Engineering, https://doi.org/10.1007/978-981-16-3568-7_3

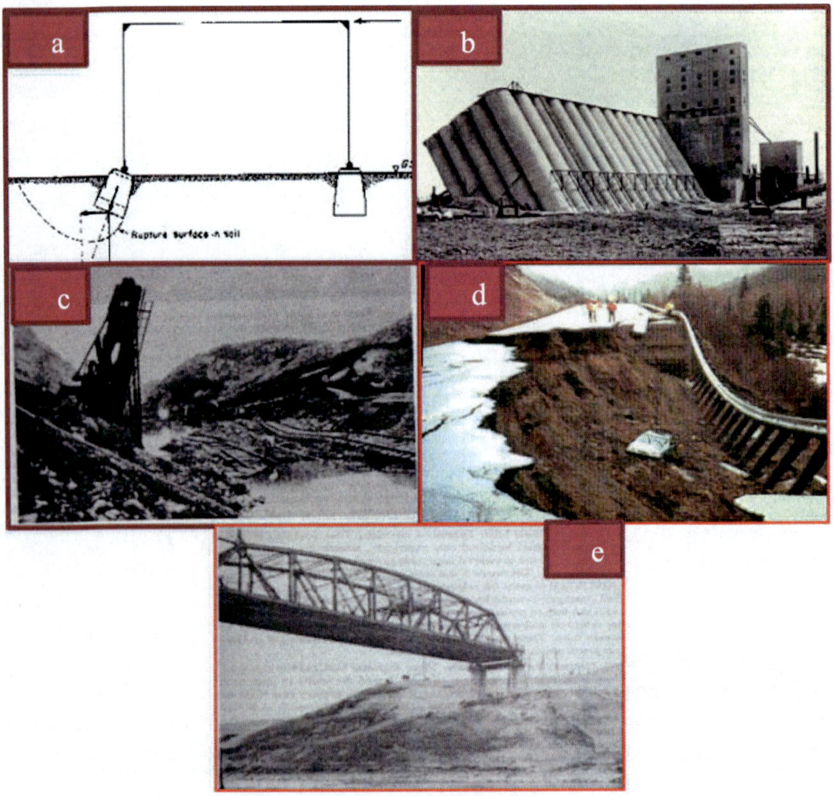

Fig. 3.1 a–e Nature of failures of some engineering structures

this chapter, we will discuss few simple methods of analysis from which you should be able to:

- Estimate the stability of slopes with simple geometry and geological features.
- Understand the forces and activities that provoke the slope failures.
- Understand the effects of geology, seepage, and pore water pressures on the stability of slopes.

We would make use of the following: Physical properties of the soil slope material, Shear strength of the soil slope material, Seepage and Seismic effects.

3.2 Soil Slope Physical Properties

3.2.1 Formation of the Soil

As already indicated, soil, which is a complex mixture of inorganic matter and sometimes organic residues which blankets the earth's crust. Soil is formed by weathering (disintegration and decomposition of rocks and minerals at or near the earth's surface through): Natural physical, Mechanical, and Chemical agents into smaller and smaller particles. The two latter kinds of weathering processes are concomitant, and occur simultaneously. The factors of weathering: Atmospheric, Chemical, Man's activities, and Time, see Bishop and Morgenstern [5] and Taylor [19].

3.2.2 Soil Types

One distinguishes between: Glacial-, Residual-, Alluvial-, Windborne-soils. Because there are no definitions giving accurate characterizations of a soil, in practice there still are some descriptive terms of soils based on soil profile-appearance-properties short description of some soil types, which are used in practice: Bedrock, Boulders, Calcareous Soil, Caliche, Cobbles, Gumbo, Hardpan, Humus, Lime soils, Loam, Loess, Marl, Muck, Mud, Peat, Pebbles and Till.

One of the distinctive properties of the various kinds of soils is bond or cohesion between the individual particles. The degree of cohesion-or sticking together is compared by subjecting the soil sample to drying, and later attempting to rupture the sample. Observations show that in such a case sand ruptures immediately. Clayey sands can be crushed easily by means of the fingertips, but fat clays become hard. Because of their cohesive properties, soils may be classified into two main groups, namely: (a) non-cohesive soils, and (b) cohesive soils.

3.3 Physical Properties of Soil

The fundamental soil properties characterize, to a certain extent, the quality of the soil as a construction material or the degree of stability of soil under Foundations of engineering structures. Some of the fundamental physical properties of soil are:

- Color.
- Mechanical composition.
- Structure.
- Particle size and shape.
- Texture, Specific gravity.
- Unit weight, and

- Porosity.

Most of these physical properties have direct or indirect effect on the slope stability, of which or in which is made.

3.3.1 Soil Moisture Content

Soil moisture content is defined as that amount of water, which is contained in the voids of the soil. The amount of water contained in the voids of a soil in its natural state is termed the natural moisture content of the soil. The natural moisture content characterizes in general the strength properties of that soil, as well as its performance under the action of load and temperature. The knowledge of moisture content in soil is necessary in soil compaction control, in determining consistency limits of soil, and for the calculation of the stabi1ity of all kinds of earthworks and foundations. The soil moisture content is determined by drying the soil sample in a drying oven at 105°C to 110 °C until a constant weight of the soil sample is attained. The moisture content, w, is expressed in percent of the dry weight of the soil, and is called the absolute moisture content of the soil, Eq. 3.1.

$$W = \frac{\text{Weight of water}}{\text{Weight of solids}} \times 100\% = \frac{W_w}{W_d} \tag{3.1}$$

where:

W_w: Weight of water in the voids of the soil, and
W_d: Weight of solids (weight of oven dry soil).

$$W_d = \frac{W}{(1 + w)} \tag{3.2}$$

The wet weight, dry weight, and moisture content relationship, Eq. 3.2, is applied in soil compaction operations for highway and runway fills and for the construction of earth dams.

3.3.2 Permeability

Any material having continuous voids would allow the passage of fluids through it. This property of material, which permits fluids to percolate through, is called its "permeability". The permeability of soils greatly varies; some of the clays are less

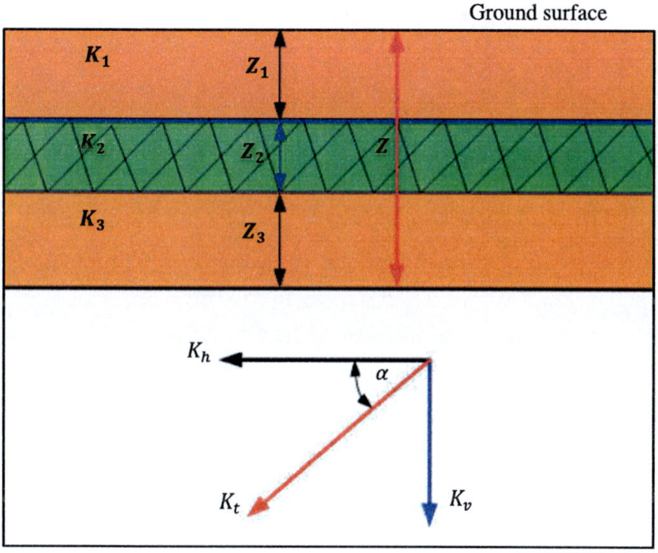

Fig. 3.2 Natural deposit

pervious whereas coarse sand and gravel are more pervious. The degree of permeability of a soil is expressed as the coefficient of permeability 'k'. This coefficient can be determined in field or lab for one bed or the whole natural deposit.

In case of natural deposits which are stratified, Fig. 3.2, from the known thickness and permeability of the individual layers, the average permeability parallel and normal to bedding planes can be estimated as:

3.3.2.1 Flow Parallel to Bedding Planes

$$K_h = \frac{1}{Z}[K_1 \times Z_1 + K_2 \times Z_2 + \cdots + K_n \times Z_n] \tag{3.3}$$

3.3.2.2 Flow Normal to Bedding Planes

$$K_v = \frac{Z}{\left[\frac{Z_1}{K_1} + \frac{Z_2}{K_2} + \cdots + \frac{Z_n}{K_n}\right]} \tag{3.4}$$

The coefficient of permeability for the whole formation, K_t.
Where:

$$K_t = \sqrt{(K_h)^2 + (K_v)^2} \tag{3.5}$$

The direction with respect to horizontal is given from:

$$\tan(\alpha) = \frac{K_v}{K_h} \tag{3.6}$$

The value of K_t and its direction are very important to decide drainage system in this formation or not.

3.3.3 *Capillarity*

The ground water table or the "phreatic surface" is the level to which water will rise in an observation well in a soil deposit. The water at the phreatic surface is at atmospheric pressure. It is a well-known fact that soils are fully saturated for some height above the water table and for some further height they are partially saturated. This rise of water above the water table is due to surface tension which combined with attraction between water and solid substances form a force opposed to gravity.

This force is termed- as 'capillarity'. Capillarity may be demonstrated by immersing the lower end of a small diameter tube into water. The rise of water in the tube known as height of capillary rise, 'be' depends primarily upon the radius of the tube, 'r' and cleanliness of its inner surface i.e., a the contact angle between the surface of water and wall of tube, Fig. 3.3.

Height of capillary rise:

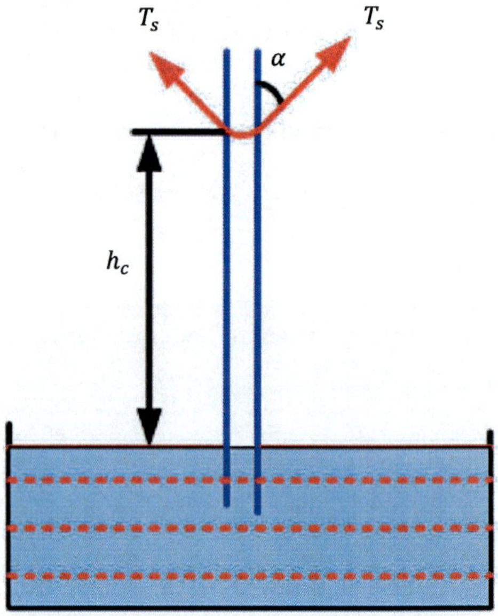

Fig. 3.3 Rise of water in capillarity tube

$$h_c = \frac{2T_s \cos\alpha}{(r \times \gamma_w)} \tag{3.7}$$

where:

T_s: is the surface tension in g per centimeter. The value of T_s at room temperature is 0.075 g/cm.

Hence, the Eq. 3.7 can be re-written as:

$$\mathrm{h_c} = \frac{\cos\alpha}{\mathrm{r}} \tag{3.8}$$

As the pores of soil provide interconnected irregular passages of varying sizes, the condition for capillary rise in a soil mass are not strictly analogous to those in glass tubes. An approximate relation for capillary rise h_c in soils is:

$$\mathrm{h_c} = \frac{\mathrm{C}}{(\mathrm{e} \times \mathrm{D_{10}})} \tag{3.9}$$

where:

D_{10} : Hazen's effective size, cm.
C: empirical constant (0.1–0.5 cm^2).

Capillarity height is important in determining the digging level in surface mine free from its effect.

3.3.4 Soil Shear Strength

The shear strength of soils can be determined by field or laboratory tests. No matter what tests are used, it is necessary to conduct an overall geologic appraisal of the site, followed by a planned subsurface investigation. The purpose of the subsurface investigation is to determine the nature and extent of each type of material that may have an effect on the stability of the slope. A detailed knowledge of the slope from toe to crest is essential. Fills situated over a deep layer of clays and silts may merit expensive drilling. Auger holes, pits, or trenches will suffice for smaller fills or those with bedrock only a short distance below the surface.

3.3.4.1 Undrained and Drained Shear Strength

Drained condition occurs when the excess pore water pressure developed during loading of a soil dissipates, i.e., $\Delta u = 0$. Undrained condition occurs when the excess pore water pressure cannot drain, at least quickly, from the soil; that is,

$\Delta u \neq 0$. The existence of either condition-drained or Undrained depends on: The soil type, Geological formation (fissures, sand layers in clays, etc.), and the rate of loading. Undrained condition does not apply to clean coarse-grained soils under static loading but only to fine-grained soils and to mixtures of coarse and fine-grained soils. Dynamic loading, such as during an earthquake, is imposed. So, quickly those even coarse-grained soils do not have sufficient time to dissipate the excess pore water pressure and undrained condition applies.

3.3.4.2 Field Tests

There are a variety of field tests for determining the shear strength of soils. However, only the standard penetration test and the Dutch cone test, which are used in conjunction with subsurface explorations, and the vane shear test will be discussed here. These tests are applicable to soils free from substantial gravel or cobble-sized particles.

3.3.4.3 Laboratory Tests

Laboratory tests complement field tests to give a more complete picture of the materials within the slope and their engineering properties. Furthermore, it is possible in the laboratory to establish the changes in soil behavior due to the changes in environment. For example, the construction of an embankment will certainly affect the shear strength in the foundation soils. Field tests before construction cannot establish these changes, while laboratory tests can simulate these changes as they occur in the field. The major laboratory tests for determining the shear strength of soils include the direct shear test, the triaxial compression test, the unconfined compression test and the laboratory vane shear test. The direct shear test is very' easy to operate because of its simplicity Due to the thin specimen used. Drained conditions exist for most materials except for the highly plastic clays. Therefore, the direct shear strength is usually in terms of the effective stress. The triaxial compression test can he used for determining either the total strength or the effective strength. The unconfined compression and the vane shear test can only be used for determining the undrained shear strength.

Direct Shear Test

Figure 3.4 shows a schematic diagram of the direct shear box. The sample is placed between two porous stones to facilitate drainage. The normal load is applied to the sample by placing weights in a hanger system. The shear force is supplied by the piston driven by an electric motor. A horizontal dial and the shear force measure the horizontal displacement by a proving ring and load dial, which are not shown in the figure. Form granular materials and silty clays, such as coal refuse and mine spoils,

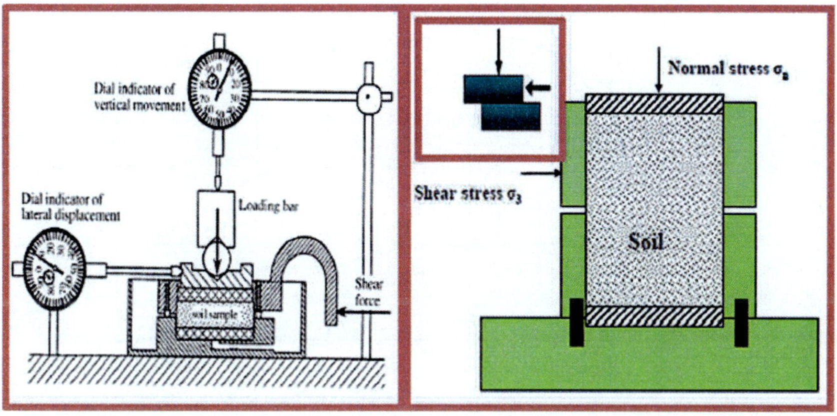

Fig. 3.4 Direct shear test

Huang (1978b) found that their effective strength could be easily determined by the direct shear tests, which check closely with the results of triaxial compression tests with pore pressure measurements. His suggested procedure is as follows:

The soil is air-dried and sieved through a no. 4 sieve (4.75 mm (. The material retained on no. 4 is discarded because the specimen is only 2.5 in 63.5 mm) in diameter and is not adequate for large particles. The material passing the no. 4 sieve is mixed with a large amount of wafer to make it very plastic and then is, placed in the direct shear box. To prevent the sample from squeezing out, a Teflon ring is used to separate the two halves of the shear box. After a given normal stress is applied for about 10 mm. the shear stress is applied at a rate of 0.02 in. (0.5 mm) per mm until the specimen fails, as indicated by a decrease in the reading of the proving ring dial. If the specimen does not fail, the test is stopped at 25 min, or a horizontal deformation of 0.5 in. At least three tests involving three different normal stresses must be performed.

Figure 3.5 shows the stress-displacement curves of a fine refuse fewer than three different normal stresses: 0.52 tsf (49.8 kPa), 1.55 tsf (148.5 kPa) and 2.58 tsf (247.2 kPa). In all three curves the shear stress increases with the horizontal displacement up to a peak and then decreases until a nearly constant value is obtained. (Fig. 3.6)

The shear stress at the peak is called the peak shear strength, while that at the constant value is called the residual shear strength. Because of progressive failure, the average shear strength actually developed along a failure surface is somewhere between the peak and the residual strength. If the two strengths are not significantly different, as is the case shown in Fig. 3.5, the peak strength can be used.

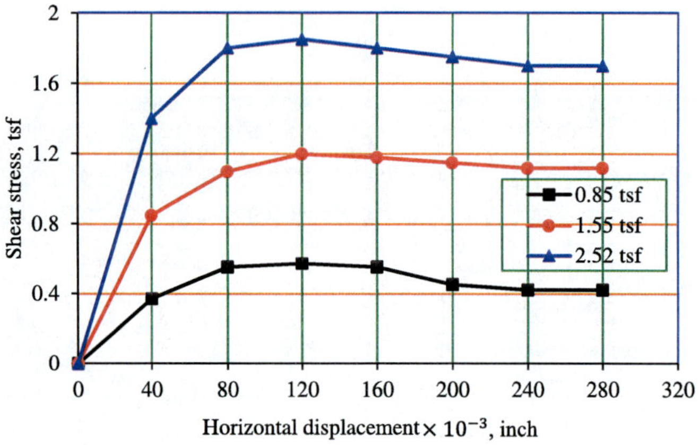

Fig. 3.5 Force displacement curve

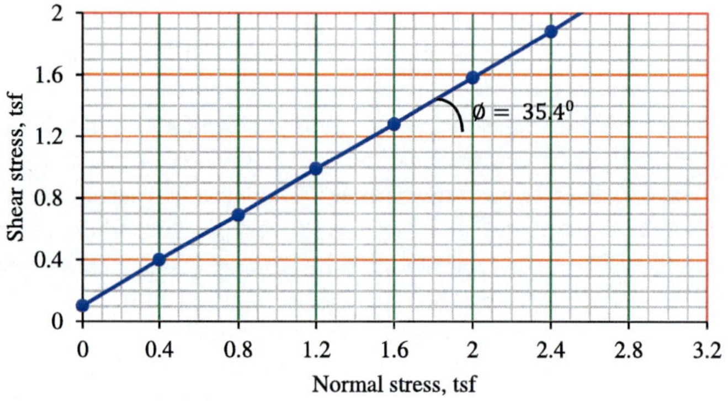

Fig. 3.6 Shear strength parameters

Triaxial Compression Test

The triaxial compression test can be used to determine either the total peak strength parameters or the effective peak strength parameters. Figure 3.7 shows a schematic diagram of a triaxial chamber. The specimen is covered with a rubber membrane and placed in the triaxial chamber. Water is introduced into the chamber and a given confining pressure is applied. A vertical axial stress is then applied, and the deformation and loading dials are read until the specimen fails, as indicated by a decrease in reading of the loading dial. If the specimen does not fail, the test continues until a strain of 15% is obtained. One simple way to determine the total strength parameters of unsaturated soils is to prepare two identical specimens, and then subject one to the unconfined compression test and the other to the triaxial compression test. The

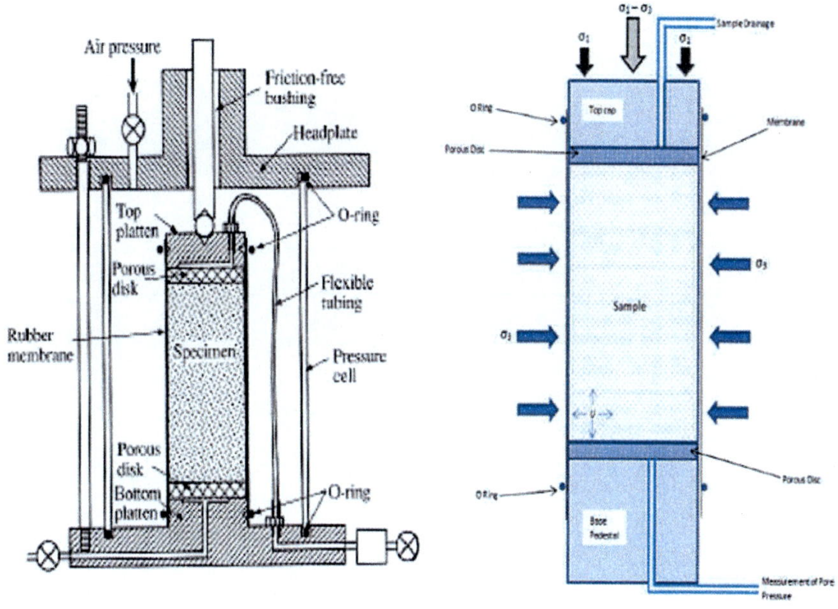

Fig. 3.7 Triaxial Cell

confining pressure used for the triaxial test should nearly equal the maximum overburden pressure expected in the field, see Bishop and Bjerrum [4] and Chowdhury [7].

The procedure for the unconfined compression test is similar to the triaxial test, except that the specimen is not enclosed in the rubber membrane and that no confining pressure is applied. To prevent any drainage in the triaxial test, the drainage valves must be closed. After both tests are completed, two Mohr's circles are drawn and a straight line tangent to these two circles is the Mohr's envelope. The vertical intercept of the envelope at zero normal stress is the cohesion, and the angle of the envelope with the horizontal is the friction angle as shown in Fig. 3.8. The total strength parameters generally exhibit a large cohesion and a small friction angle. If the specimen is completely saturated, the envelope will be horizontal with an angle of internal friction equal to zero.

The effective strength parameters can be determined by a consolidated drained test or consolidated undrained test with pore pressure measurements. Instead of using the total normal stress as shown in Fig. 3.8, the shear stress is plotted versus the effective normal stress, and a Mohr's envelope in terms of effective stress is obtained. The vertical intercept of the envelope at zero effective normal stress is the effective cohesion, and the angle of the envelope with the horizontal is the effective friction angle. The effective strength parameters always exhibit a small effective cohesion and a large effective friction angle. Another procedure to obtain the effective strength parameters is by the use of stress path method (Lambe and Whitman 1969).

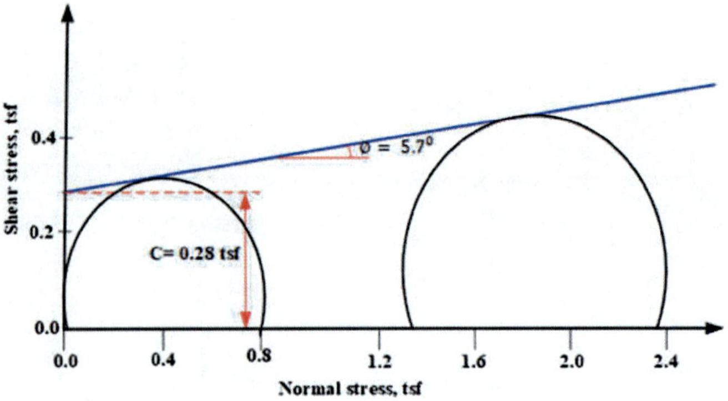

Fig. 3.8 Shear strength determination

$$P' = \left(\frac{\sigma_1' + \sigma_3'}{2}\right) \tag{3.10}$$

$$q = \left(\frac{\sigma_1 - \sigma_3}{2}\right) = \left(\frac{\sigma_1' - \sigma_3'}{2}\right) \tag{3.11}$$

Note that: in which σ1 is the major principal stress and σ3 is the minor principal stress. The corresponding effective stresses, σ1' and σ3', can be determined by subtracting the measured pore pressure from the total stresses, σ1 and σ3. The triaxial compression tests with pore pressure measurements were made on two specimens, one subject to an effective confining pressure of 0.6 tsf. or 58 kPa and the other to 1.4 tsf. or 134 kPa. To saturate the specimens, filter strips were placed around the sample and a back pressure was applied. A straight line tangent to the failure part of the stress path is called the K-line. The angle of the K-line with the horizontal is called α', which is related to $\text{Ø}'$ by Eq. 3.12:

$$\text{SinØ}' = \tan\alpha' \tag{3.12}$$

The intercept of K-line with the q-axis is called a', which is related to c' by Eq. 3.13:

$$a' = \frac{C'}{\cos \text{Ø}'} \tag{3.13}$$

Unconfined Compression (UC) Test

The purpose of this test is to determine the undrained shear strength of saturated clays quickly. In the UC test, no radial stress is applied to the sample ($\sigma_3 = 0$). The plunger load, P_z is increased rapidly until the soil sample fails, that is, cannot support any

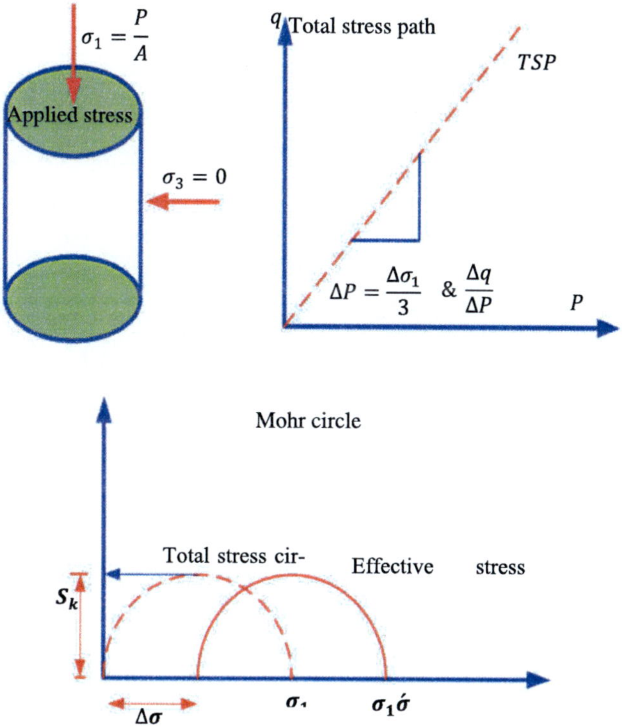

Fig. 3.9 Stresses, stress paths and Mohr's circle for the uniaxial test

additional load. The loading is applied quickly so that the pore water cannot drain from the soil; the sample is sheared at constant volume. The stresses applied on the soil sample and the total stress path followed are shown in Fig. 3.9a, b.

The effective stress path is unknown since pore water pressure changes are not normally measured. Mohr's circle using total stresses is depicted in Fig. 3.9c. If the excess pore water pressures were to be measured, they would be negative. The theoretical reason for negative excess pore water pressures is as follows. Since $\sigma_3 = 0$, then from the principle of effective stresses, $\sigma_3' = \sigma_3 - \Delta u = 0 - \Delta u = -\Delta u$. The effective radial stress, σ_3'cannot be negative because soils cannot sustain tension. Therefore, the excess pore water pressure must be negative so that u is positive. Mohr's circle of effective stresses would be to the $r_i \times 2h \times t$ of the total stress circle as shown in Fig. 3.9c. The undrained shear strength is:

$$S_u = \frac{P_Z}{2A} = \frac{\sigma_1}{2} \tag{3.14}$$

The undrained elastic modulus E_u is determined from a plot of Є1, versus σ1. The results from UCS tests are used to:

- Estimate the short-term bearing capacity of fine-grained soils for foundations.
- Estimate the short-term stability of slopes.
- Compare the shear strengths of soils from a site to establish soil strength variability quickly and cost-effectively (the UC test is cheaper to perform than other triaxial tests).
- Determine the stress–strain characteristics under fast (undrained) loading conditions.

3.4 Slope Stability Analysis

A slope is an inclined boundary surface between air and the body of an earthwork such as a dam, or highway cut, and/or fills, for example. If the shear stress exceeds the shear strength of the soil, or, in other words when the driving forces are larger than those resisting the motion of the soil mass, then rupture (failure in shear) of the soil material located beneath the slope of the earthwork takes place. Such a failure in shear is termed a slide. After rupture the overlying mass of soil can move by gravity; until rupture occurs motion is opposed by the shear resistance of the soil. When this is overcome, resistance is diminished and the soil mass slides down; see Bishop [3], Hoek and Bray [10] and Ireland [13].

A slide of an earthwork slope is the perceptible translational or rotational movement of the ruptured soil wedge downward and outward from the embankment, depending upon the set of combinations of the various factors partaking in the shear deformation of the soil. The rupture of a slope may set in almost every conceivable manner: slowly or suddenly, and the perceptible movement of the ruptured soil mass may be characterized in most instances as rapid. The failure of a slope-forming soil constitutes the loss of its stability.

The stability of any slope made of or in a soil material depends upon the shear strength of the soil. The shear strength, in its turn, is a function of friction and cohesion of the soil. Here friction is proportional to the effective normal stress on the rupture surface at the instant of failure. In soil mechanics, the topic "stability of slopes" is dealt with from two engineering viewpoints, namely:

1. The design of slopes of cuts and fills in advance of new earthwork construction in accordance with prescribed safety requirements, and
2. The study of stability of existing slopes of earthworks slopes which are potentially unstable, or which have failed, or which have to be redesigned.

In soil engineering, stability analysis of a slope of an earthwork is concerned with the computing of the degree of stability relative to a sudden rupture of slope characterized by the so-called factor of safety, η, of the soil slope-load system relative to the most dangerous, or disadvantageous stability condition of the slope-a sudden, deep-seated gravity slide. Stable slopes; whose factor of safety, η, with respect to failure in shear has been calculated to be greater than a unity, i.e., $\eta > 1.0$. For slopes of new earthworks it is advisable to maintain the factor of safety at $\eta = 1.5$ at least.

3.4.1 Factors Contributing to Slope Failures

The stability of slopes depends upon the following factors:

- The type of soil of which or in which the slope is made.
- The geometry of the cross section of the slope (height, slope angle).
- Weight and loads, and weight and load distribution (gravity is the principal cause of all slides).
- Increase in moisture content of the soil material.
- Decrease in shear strength of soil for reasons other than water.
- Vibrations and earthquakes.

Thus one possibility of reducing the degree of stability in slopes comes more to the fore: stability varies principally with the variation in the water regime in the soil. Therefore, stability analysis of slopes of earthworks should be performed for the most dangerous conditions of the slope.

3.4.2 Classification of Slides

Slides may be classified from various view-points, see Morgenstern and Price [17], Whitman and Moore [20]:

1. Analysis of stability of slopes, slides may be subjected to the following rough classification:
   - Slides over plane rupture surfaces.
   - Slides over curved rupture surfaces.
2. The type of the sliding soil material, such as:
   - Slides in pure cohesive soils (c-soils),
   - Slides in pure non-cohesive or frictional soils (φ-soils), and
   - Slides in cohesive-frictional soils (φ-c soils).
3. The condition of the sliding soil material:
   - Homogeneous soils, and
   - Non-homogeneous soils (irregular texture or density, layered soil systems, partly submerged slopes, seepage through soil and out of it through the slopes, and other possible conditions).

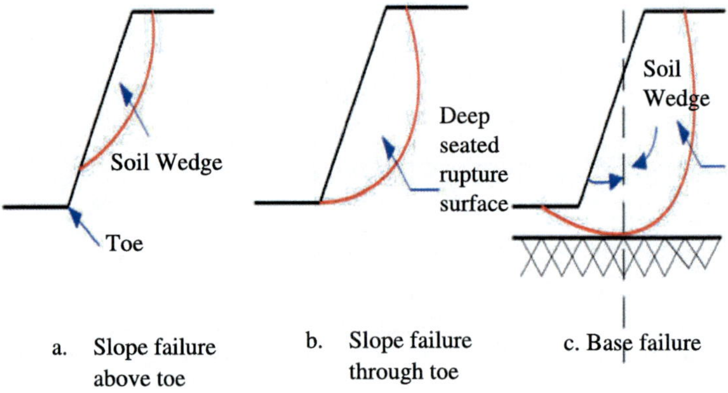

Fig. 3.10 Modes of slope failures

3.4.3 Mode of Rupture

The slope may fail basically in the following two ways:

- The rupture surface sets in above the toe of the slope, Fig. 3.10, and the rupture surface passes through the toe of the slope. Such failures are known as slope failures.
- The rupture surface is deep-seated and passes through the embankment to support the soil below the toe of the slope. This mode of slope failure is known as the base failure; see Schuster and Krizek [18].

3.4.4 Plane Rupture Surfaces

In order that an analysis of a slope can be made one must know the shape of the rupture surface (plane, or curvilinear), or it should be assumed-the assumption being then based on a thorough analysis of all natural and artificial conditions at the site of the slope. In stability analyses of slopes of earthworks over plane rupture surfaces two principal cases are to be distinguished, namely:

- Slides of unconsolidated material over a pre-existing inclined sliding plane of firm surface, and
- Slides taking place over a plane rupture surface sheared by the driving forces within the mass of an unconsolidated soil material.

3.4.4.1 Stability of Mass of Soil on an Inclined Plane

To start out with the stability analysis of the triangular, homogeneous soil mass, ABC, Fig. 3.11, assume that it rests upon a firm, inclined surface, AC, inclined at an

Fig. 3.11 Soil mass on an incline

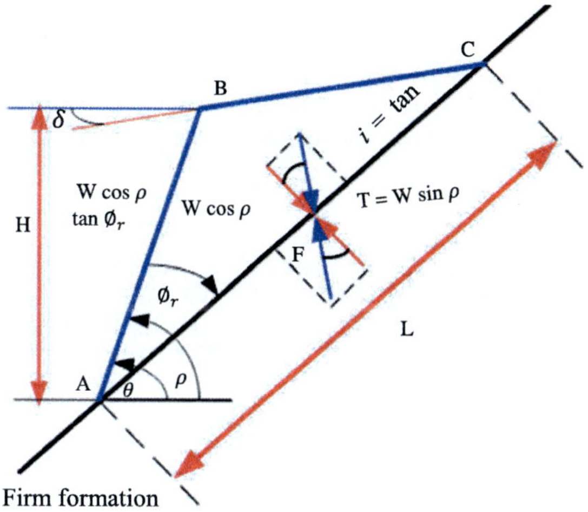

angle ρ, with the horizontal. In this problem two kinds of stabilities must be checked, namely:

- The overall stability of the entire soil mass against sliding down the incline, and
- The stability of the slope, AB, proper against rupture in shear if the overall stability of the earth mass against sliding down is satisfactory.

The principle underlying stability calculation of the triangular soil mass is the failure in shear along the inclined plane, AC, when the driving forces exceed the resisting forces. In such a calculation, it is assumed that the shear stress, s, is distributed uniformly over the plane sliding surface, AC; see Hoek, [9]; Hungr et al. [12] and Yang [21].

The driving force (tangential force), T, Eq. 3.15:

$$T = W \sin \rho \tag{3.15}$$

where:

W = weight of soil wedge = γ A (ABC).

The resisting force, R, Eqs. 3.16 and 3.16a:

$$R = F \sin \emptyset_r + C \times L = \left(\frac{N}{\cos \emptyset_r}\right) \sin \emptyset_r + C \times \mathrm{L}$$
$$= \mathrm{N} \tan \emptyset_r + \mathrm{C} \times \mathrm{L} \tag{3.16}$$

$$= \mathrm{W\ COS}\rho \tan \emptyset_r + \mathrm{C} \times \mathrm{L} \tag{3.16a}$$

where:

$Ø_r$: Angle of external frictional resistance in plane, AC, between the soil material and the firm surface of the inclined sliding plane.
tan $Ø_r$: coefficient of external friction.
C: external cohesion in plane, AC, and
L: length of triangular soil wedge resting on incline = AC.

The $Ø_r$ and C values must be ascertained by test in the laboratory. In a stable system driving forces, Σ FD, should constitute only a fraction of the resisting forces, Σ FR. The factor of safety, η, of the system, characterizing the degree of stability of the soil mass on the incline is expressed as, Eq. 3.17:

$$\eta = \frac{\sum \mathrm{FR}}{\sum \mathrm{FD}} = \frac{\tan Ø_r \sum \mathrm{N} + \mathrm{CL}}{\sum \mathrm{T}} \tag{3.17}$$

where:

Σ N =W cos p = the sum of all effective normal forces to the sliding plane.
Σ T = the sum of all shear forces.
If: $\eta > 1$, the soil mass on the incline may be considered as stable.
If: $\eta = 1$, the system is at equilibrium.
If: $\eta < 1$, the system should be regarded as instable as concerns pure static calculations.

The factor of safety, η, is usually required to be ≥ 1.5. If in Eq. 3.17, C = 0 as for a non-cohesive soil, then, Eq. 3.18:

$$\eta = \frac{\tan Ø_r \sum N}{\sum T} = \frac{\tan Ø_r W \cos \rho}{W \sin \rho} = \frac{\tan Ø_r}{\tan \rho} \tag{3.18}$$

where:
tan ρ = i = slope of the inclined, firm, plane sliding surface. If a factor of safety, η, and φr are known, then the angle of incline should be, Eq. 3.19:

$$\rho = \text{arc } \tan\left(\frac{\tan Ø_r}{\eta}\right) \tag{3.19}$$

If ρ cannot be changed, then the coefficient of friction for a given η should be at least, Eq. 3.20:

$$\tan Ø_r = \mathrm{i} \tan \rho \tag{3.20}$$

Up to the slope, i, of the sliding plane of 1: 5 and ε < 10o, the fill is usually stable. When $\tan \rho > 1/5$, then it is the practice to make so-called saw teeth in the firm, inclined surface, or steps to provide for support against the sliding down of the unconsolidated mass of soil, Fig. 3.12. The steps are bulldozed out 3 ft. and more in

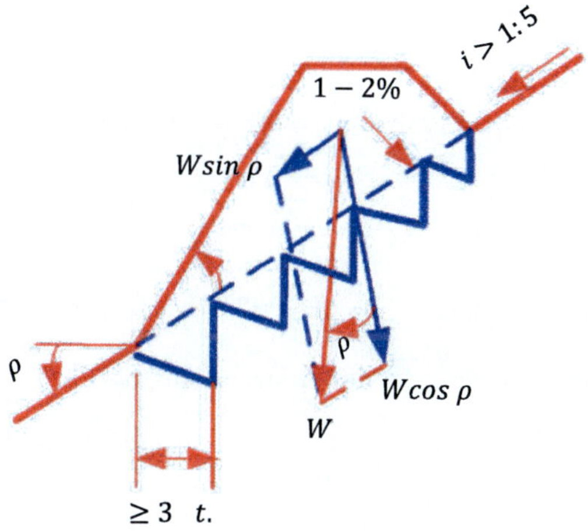

Fig. 3.12 Steps for stability against sliding down of the unconsolidated mass of soil

length, and with a back-slope of 1–2% to cope with the water problem here, drainage should be arranged.

3.4.4.2 Stability of Slopes with a Plane Rupture Surface

In a homogeneous soil failure in shear along a plane rupture surface may be assumed to take place in a surface such that the acting shear stresses are greater than the shear strength of the soil. The stability calculation is thus based on a shear plane, AC, through the toe-line (A), and finding the so-called critical angle of the slope, ω, which corresponds to the most dangerous rupture surface, Fig. 3.13, at its maximum stressed surface. The weight of the ruptured soil wedge, ΔABC, is, Eq. 3.21; see Janbu [14]; May and Brahtz [15]; Moore [16]:

$$W = 0.5\gamma(AC)h = 0.5\gamma H^2 \frac{\sin(\theta - \delta)\sin(\theta - \omega)}{\sin^2\theta \sin(\omega - \delta)} \quad (3.21)$$

The force equilibrium on plane AC, Eq. 3.22:

$$c\ (AC) + N \tan \emptyset - T = 0 \quad (3.22)$$

Here: c (AC) is the so-called acting cohesive force, and N tan Ø is the acting frictional force, not to be confused with the soil resistance forces. But, Eq. 3.23–3.25:

$$Ac = \frac{H \sin(\theta - \delta)}{\sin\theta \sin(\omega - \delta)} \quad (3.23)$$

$$N = W \cos\omega = 0.5\gamma H^2 \frac{\sin(\theta - \delta)\sin(\theta - \omega)\cos\omega}{\sin^2\theta \sin(\omega - \delta)} \quad (3.24)$$

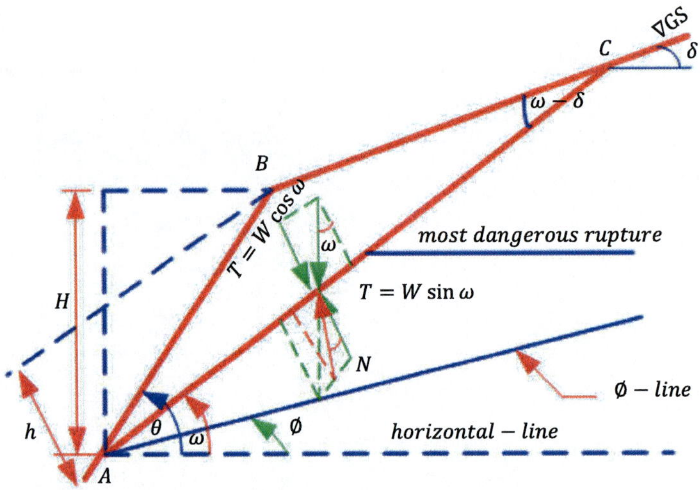

Fig. 3.13 Slope system with plane rupture surface

$$T = W\sin\omega = 0.5\gamma H^2 \frac{\sin(\theta-\delta)\sin(\theta-\omega)\sin\omega}{\sin^2\theta\sin(\omega-\delta)} \tag{3.25}$$

With the aforementioned expressions, the acting cohesion from Eq. 3.22 becomes as given per Eq. 3.26:

$$c = \frac{T - N\tan\emptyset}{AC} = \frac{(W\sin\omega - W\cos\omega\tan\emptyset)\sin\theta\sin(\theta-\delta)}{H\sin(\theta-\delta)}$$
$$= \left(\frac{0.5\gamma H\sin(\theta-\delta)(\sin\omega - \cos\omega\tan\emptyset}{\sin\theta}\right) \tag{3.26}$$

In the most dangerous rupture surface the induced, acting cohesion attains a maximum value, i.e., the trigonometric nominator should be, Eq. 3.27–3.28:

$$\sin(\theta-\delta)(\sin\omega - \cos\omega\tan\emptyset) = \text{Maximum}$$
$$\frac{d}{d_\omega}[\sin(\theta-\delta)(\sin\omega - \cos\omega\tan\emptyset)] = 0$$
$$tan(\theta-\omega) = \frac{sin\omega - cos\omega tan\emptyset}{cos\omega + sin\omega tan\emptyset} = tan(\omega-\emptyset) \tag{3.27}$$

$$(\theta-\omega) = (\omega-\emptyset) \tag{3.28}$$

The critical angle ω, Eq. 3.29, is:

$$\omega = \frac{(\theta + \emptyset)}{2} \tag{3.29}$$

This angle determines the position of the most dangerous rupture surface, A C. The substitution of $\omega = \frac{(\theta+\emptyset)}{2}$ into the c-equation, Eq. 3.26, yields the mathematical expression for the maximum value of developed cohesion, $c_{max.}$ in the most dangerous rupture plane, AC, upon rupture, Eq. 3.30:

$$c_{max.} = 0.5\gamma H \frac{\frac{\sin^2(\theta-\omega)}{2}}{\sin\theta\cos\emptyset} \tag{3.30}$$

In other words, $c_{max.}$ is the necessary value a soil material with an angle of internal friction, φ should possess in order that the slope, AB, with a slope angle of θ still stays at a height, H, Eq. 3.31:

$$H = \frac{2c\sin\theta\cos\emptyset}{\frac{\gamma\sin^2(\theta-\emptyset)}{2}} \tag{3.31}$$

Note that c as well as H is independent of the slope, δ, of the ground surface. When $\varphi = 0$, then Eqs. 3.30 and 3.31 transform into, Eqs. 3.32 and 3.33:

$$c_{max.} = 0.25\gamma\ H\tan\left(\frac{\theta}{2}\right) \tag{3.32}$$

and

$$H = 4\left(\frac{c}{\gamma}\right)\cot\left(\frac{\theta}{2}\right) \tag{3.33}$$

This equation shows that the greater the slope angle θ, the smaller the height of the slope. The overall factor of safety, η, for the critical plane is, Eq. 3.34:

$$\eta = \frac{\sum \text{Resisting forces}}{\sum \text{Driving forces}} = \frac{c(AC) + N\tan\emptyset}{T} \tag{3.34}$$

If $c = 0$, then R.F. = N tan φ, and tan ω = tan φ or ω = φ. For a factor of safety η; see (Eqs. 3.35 and 3.36).

$$\eta = \frac{\tan\emptyset}{\tan\omega} \tag{3.35}$$

or

$$\tan\emptyset = \eta\tan\omega \tag{3.36}$$

3.4.5 *Circular Sliding Surface*

There are several theories available for analyzing the stability of earthwork slopes: circular rupture surfaces; circular rupture surfaces with the so-called friction, or φ-circle; logarithmically spiraled rupture surfaces; cycloids, logoids and possibly other kinds of rupture surfaces. Nowadays one of the most commonly used types of curved rupture surfaces in stability analyses of slopes is the circular, cylindrical rupture surface. The circular, cylindrical rupture surface is, of course, merely a conventional one in order to simplify mathematical computations involved in the stability analysis. The use of the circular rupture surface may probably be justified for the three following reasons, see Bishop [3]:

- It approximately coincides with the real shape of the rupture surfaces observed in nature.
- It is necessary to make a number of other assumptions for the mathematical analyses anyway.
- The circular rupture surface is easy to draw by using a compass.

Because of the early extensive studies of ruptures of slopes of earthworks made by Swedish engineers, the circular rupture surface is often referred to as the Swedish circle method. The stability analysis of slopes on circular rupture surfaces is actually a grapho-analytical method. One distinguishes between the grapho-analytical methods with circular sliding surfaces: one, as applied to homogeneous, pure cohesive soils; and the other as applied for frictional-cohesive soils (φ-circle method). The method applied to pure cohesive soils, in its turn, can be divided into two methods: stability analysis of slopes en masse and stability analysis of slopes by the method of slices. Besides, the analyses may be performed on circular sliding surfaces passing through the toe of the slope, and on circular sliding surfaces passing through the base of the dam.

3.4.5.1 Pure Cohesive Soils-Stability En Masse-Slope Failure

The principle of stability analysis of a slope en masse, as explained by Fellenius, is now described as follows, Fig. 3.14. In his works Fellenius expressly states that stability calculations of earth masses have for their object the determination of the properties of soil (cohesion and friction) necessary for equilibrium in different, assumed, circular sliding surfaces see Fellenius [8].

The method of analysis is based on the consideration that that sliding surface which requires the greatest amount of cohesion, c, and the greatest angle of internal friction, φ, for equilibrium of the earth slope is the most dangerous one of all, and that the degree of stability of the slope is expressed as the ratio of the shear strength of the soil, τ, to needed strength of the soil.

In pure, homogeneous, cohesive soils the stability analysis of slopes is relatively simple. The required amount of cohesion is calculated from the equilibrium condition

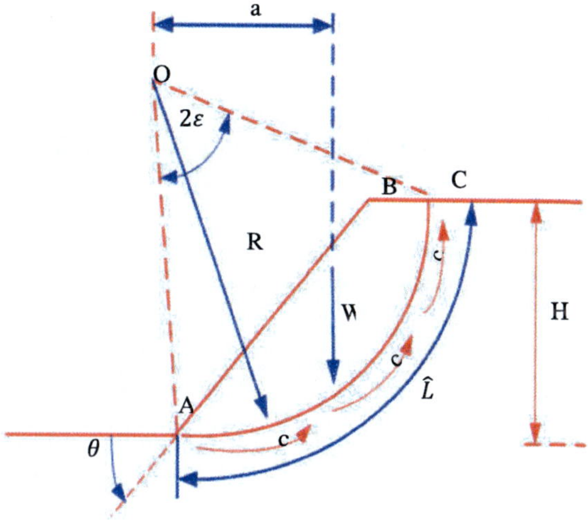

Fig. 3.14 Fellenius' system for cohesive soils

of the driving moment, MD = W a; and the resisting moment, MR = c L′R about the axis of rotation, 0, which is the axis of the circular cylindrical rupture surface, AC, Eqs. 3.37 and 3.38:

$$M_D = M_R \tag{3.37}$$

or

$$Wa = cL'R \tag{3.38}$$

where:

W: weight of sliding soil wedge, ABCA.
a: moment arm of W about 0.
c: necessary amount of cohesion per unit area in the rupture (sliding) surface.
L′ = R (2ε) = length AC of circular arc of the rupture surface.

The coordinates of the point of application, c.g., of the weight, W, are determined by the rules of statics for finding the position of the centroid of irregular geometric figures analytically, graphically, or experimentally. The necessary amount of cohesion for equilibrium, from Eq. 3.38, is:

$$c_1 = \frac{Wa}{L'R} \tag{3.39}$$

Factor of safety, η, Eq. 3.40:

$$\eta = \frac{\tau}{C_1} \tag{3.40}$$

where:

τ: shear strength of soil.

When MD > MR, the slope-forming soil mass ruptures and slides down. In these stability calculations of slopes it is tacitly assumed that the ruptured soil wedge, ABC, does not deform upon its sliding down the circular rupture surface, and that the cohesion is uniformly distributed throughout the entire area of the rupture surface. If the circular arc, A C, represents the most dangerous rupture surface, then the factor of safety, η, can also be formulated in Eq. 3.41:

$$\eta = \frac{M_R}{M_D} = \frac{\tau L'R}{Wa} = \frac{\tau(2\varepsilon)R^2}{Wa} \tag{3.41}$$

When η = 1, rupture is just impending.
When η > 1, the required ≥ 1.5. Thus, the slope is usually considered as stable.
When η < 1. Hence, the slope is considered as instable.

The position of the most dangerous rupture surface is found by the method of trial and error. This method consists of drawing several arcs through the toe of the slope by assuming different positions of the centers of the rupture circles. Each of the rupture circles thus assumed should be analyzed about each center chosen for the factor of safety. That circle which gives the least factor of safety among the circles analyzed is the critical rupture surface. The slope is considered to be stable if the least factor of safety, η, for the most dangerous rupture surface is η > 1.5.

Note that the "factor of safety" is an arbitrary quantity because the center of rotation, 0, and likewise the most dangerous rupture surface forming the circular arc, AC, are chosen by trial and error. This random procedure is the reason why several rupture surfaces and not merely one surface should be analyzed in order to obtain the position of the most probably dangerous rupture surface resulting in the least acceptable factor of safety. Obviously, the trial-and-error method requires a certain amount of time. Fellenius' graphs, however, decrease this deficiency of the grapho-analytical method considerably.

Using the system as shown in Fig. 3.15, Fellenius calculated the necessary cohesion for equilibrium as a function of the slope angle, O, the central half-angle, ε, and the angle, ω, of the chord subtending the most dangerous rupture surface, AC, Eqs. 3.42–3.44:

$$c_0 = 0.25\,\gamma H f(\theta, \varepsilon, \omega) \tag{3.42}$$

where:

$$\mathrm{f}(\theta, \varepsilon, \omega) = \left[\frac{\left(2\sin^2\ \varepsilon \sin^2 \omega\right)}{\varepsilon}\right] \times \left[\cot\varepsilon\cot\omega - \cot\varepsilon\cot\theta + \cot\theta\cot\omega - 2/3\cot^2\theta + 1/3\right] \tag{3.43}$$

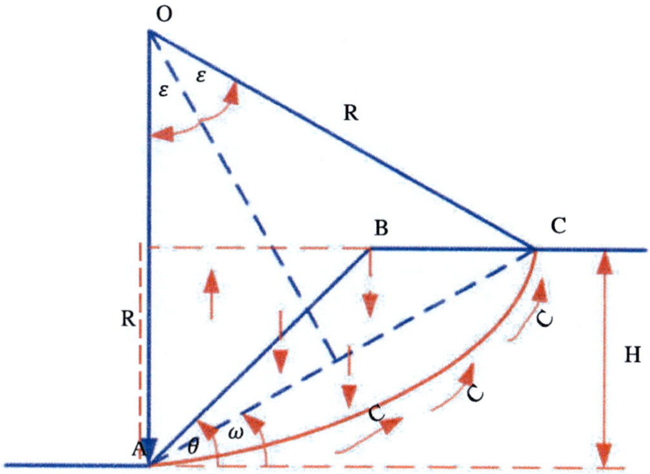

Fig. 3.15 System for calculating cohesion

The converse: if c is known, then at equilibrium with a slope angle of 0 would support itself to a height of H:

$$H = \frac{4C}{[\gamma f(\theta, \varepsilon, \omega)]} \tag{3.44}$$

Table 3.1 contains for various slope angles θ, auxiliary directional angles, BA and BB for finding the position of the center of the most dangerous rupture surface through the toe of the slope made in a homogeneous, pure, cohesive soil. The method of finding the center, 0, is illustrated in Fig. 3.16, by plotting the angles, BA and BB, as shown. Should θ be different from those given in Table 3.1, then BA and BB may be interpolated for the angle, θ, in question from Table 3.1. To facilitate interpolation, BA and BB graphs based on values as shown in Table 3.1, can be plotted as functions of θ.

Table 3.1 Directional angles for critical rupture surface through the toe of a slope in pure c-soil

Slope 1:n	Slope angle θ	Directional angles	
		BA	BB
$\sqrt{3}$:1	60	~ 29	~ 40
1:1	45	~ 28	~ 38
1:1.5	33 41′	~ 26	~ 35
1: 2	25 34′	~ 25	~ 35
1:3	18 26′	~ 25	~ 35
1:5	11 19′	~ 25	~ 37

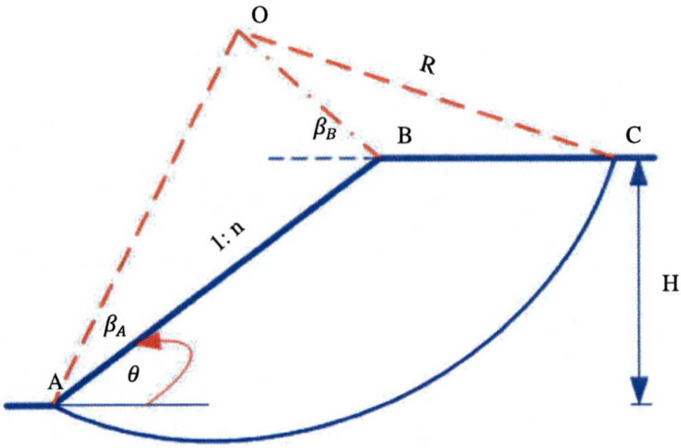

Fig. 3.16 Locating the center of the most dangerous rupture surface in pure cohesive soils for slope failure

3.4.5.2 Tension Cracks

If there are tension cracks present in the cohesive soil, then below the depth, z, of the crack, rupture would occur along a curved rupture surface after the shear strength of the cohesive soil is exhausted. The finding of the most dangerous rupture surface is accomplished by the method of trial and error. The most dangerous condition in the case of tension cracks would occur during rainy seasons when the tension cracks crack fill up with water and exert a hydro-static pressure horizontally of the magnitude of, Eq. 3.45:

$$P_H = 0.5\gamma Z^2 = \frac{2C}{\gamma} \tag{3.45}$$

This pressure contributes to the total driving moment, see Fig. 3.17. In a (φ-c) soil, Eq. 3.46:

$$Z = 2c\tan\left(\frac{\pi}{4} + \frac{\emptyset}{2}\right) \tag{3.46}$$

Frost, evaporation, and shrinkage affect the development of tension cracks and their depth.

3.4.5.3 Pure Cohesive Soils-Stability En Masse-Base Failure

If the slope supporting soil below the toe line of the slope is the same material and has the same properties as the slope-forming material, then there may occur more

Fig. 3.17 Tension crack

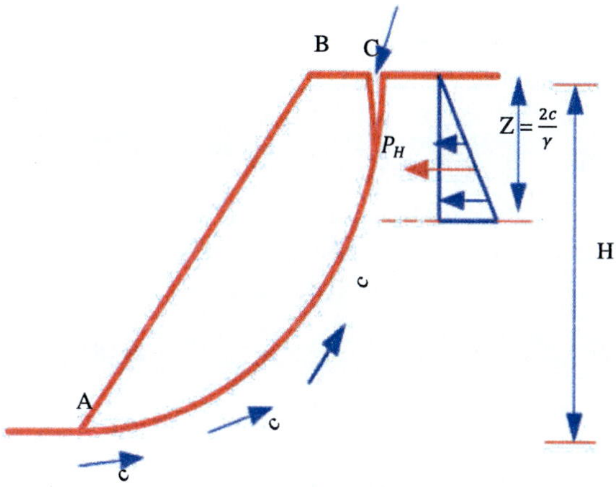

dangerous rupture surfaces than that passing through the toe of the slope. The "more dangerous" rupture surfaces are deep-seated and are termed base failures, Fig. 3.18. When AB is a slope of a straight line, then, according to Fellenius, the center, O, of the rupture surface, AC, is situated on a vertical line drawn through the midpoint, M, of the slope, AB, so that a = a, where 2 a = horizontal projection of the slope.

When AB is a vertical wall, then the center, O, of rupture surface is located on a vertical line drawn through the vertical wall. The stability analysis is performed by trial and error. Fellenius did find that for pure cohesive soils the position of the most dangerous rupture surface associated with deep-seated base failure is located very deep (theoretically infinitely deep), and has the value of the central angle of the circular rupture surface of $2\varepsilon = 133° 34'$. However, the cohesion, cnec., needed in

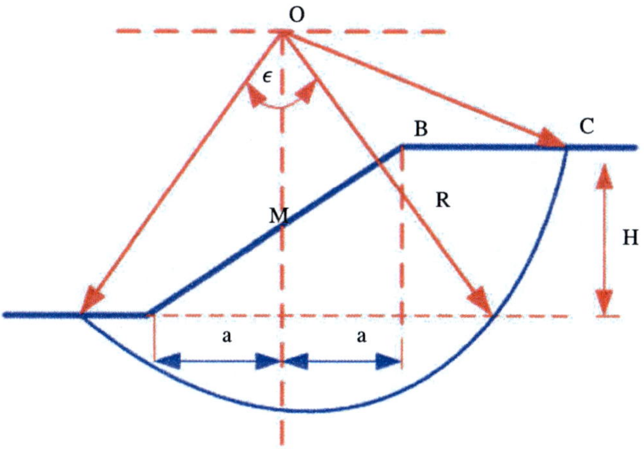

Fig. 3.18 Base failure

the most dangerous rupture surface through a point a little under the toe of the slope, is only slightly less than that needed in the infinitely deep-seated rupture surface, Eq. 3.47:

$$C_{nec.} = 0.25\gamma H \sin 133°34' = 0.181\gamma H \tag{3.47}$$

If the value of the cohesion, c, of a cohesive soil is known, then the critical height, of the slope can be calculated by Eq. 3.48 as:

$$H_{crit} = \frac{c}{0.181\ \gamma} \tag{3.48}$$

3.4.5.4 Stability Number

To facilitate calculations, Ehrenberg, uses for the cohesion a factor of safety, η, and calculates an auxiliary value, Eq. 3.49:

$$\frac{c}{\eta \times \gamma \times H} \tag{3.49}$$

And, using Krey's c/γ H graphs, determines the height of the fill, or the angle of the slope, θ, whichever the case is. To obviate the cumbersome method of trial and error of the grapho-analytical procedure for evaluating the stability of slopes, Taylor, used a simplified method of analysis similar to that given by Ehrenberg, namely, Eq. 3.50:

$$S = \frac{c}{\eta \times \gamma \times H} \tag{3.50}$$

where:

S: is termed by Taylor the "stability number".

By means of graphs the various problems connected with stability determinations of regular, uniform slopes in homogeneous soil materials can be solved. It appears, however, that the method of trial and error for fixing the center point of the most dangerous rupture surface is more profitable than the stability number method because most of the soils encountered in earthworks are not homogeneous, Fig. 3.19.

3.4.5.5 Frictional-Cohesive Soils, (φ-c)-Soils-Stability En Masse-Slope Failure

If the slope-forming soil is a (φ-c)-soil, one proceeds, in principle, similarly as described for pure c-soils, Fig. 3.20. Upon rupture, frictional forces, N tan Ø, and a

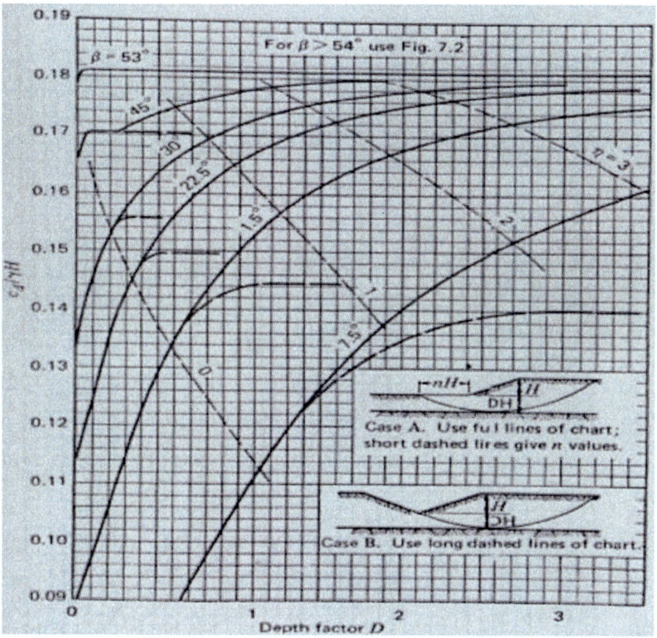

Fig. 3.19 Taylor's graphs

Fig. 3.20 Stability system for a (φ-c) soil

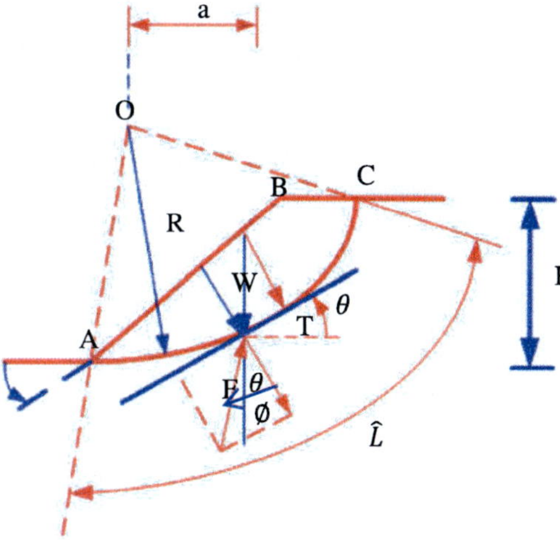

cohesive force, C L′, are acting in the circular rupture surface. The driving force is T = W sin θ, and the resisting force is P = N tan φ + C L. Again, the approximate degree of stability of a (φ-c)-slope can be judged from the factor of safety, η, which is calculated based on the position of the most dangerous rupture surface, Eqs. 3.51a-d to Eq. 3.52.

$$\eta = \frac{M_R}{M_D} = \frac{PR}{TR} = \frac{(N\tan \emptyset + cL')R}{TR} \tag{3.51}$$

$$= \frac{N\tan \emptyset + cL'}{T} \tag{3.51a}$$

$$= \frac{W\cos\theta \tan \emptyset + cL'}{W\sin\theta} \tag{3.51b}$$

$$= \cot\theta \tan \emptyset + \frac{cL'}{W\sin\theta} \tag{3.51c}$$

$$= \frac{\tan \emptyset}{\tan\theta} + \frac{cL'}{W\sin\theta} \tag{3.51d}$$

Or else:

$$\eta = \frac{M_R}{M_D} = \frac{\tau cL'}{Wa} \tag{3.52}$$

where:

$T = \sigma_e \tan\varphi + c$ is the general equation for the shear strength of the soil.
σ_e = Effective normal stress on the rupture surface.

Simple as the foregoing equation, Eq. 3.52 appears to be, in reality the stability calculations of slopes with circular rupture surfaces are very complex indeed. For this reason, and in order to obtain a clear insight into the partaking force system in the analysis, the calculations are best to be carried out graphically. The equilibrium equation of moments, W a ≈ [(tan φ)N + cL'] R, when η = 1 in the case of φ, and c being simultaneously present, holds only approximately, particularly for small angles of slopes, O, and, hence, may be used only for rough, preliminary calculations, Fig. 3.20.

3.4.5.6 Pure Cohesive Soils-(φ-c)-Soils-Stability Calculations by Method of Slices

If the cross section of a slope-forming body of soil is composed of pure cohesive soil layers, each layer of which has different shear strength properties; if a homogeneous slope is partially submerged; if through a homogeneous dam seepage takes place, or if the earthwork-forming slope is a broken one with steps and berms, then stability calculations of slopes over circular rupture surfaces can be more conveniently performed by the method of slices as originally shown by Petterson, Fig. 3.21.

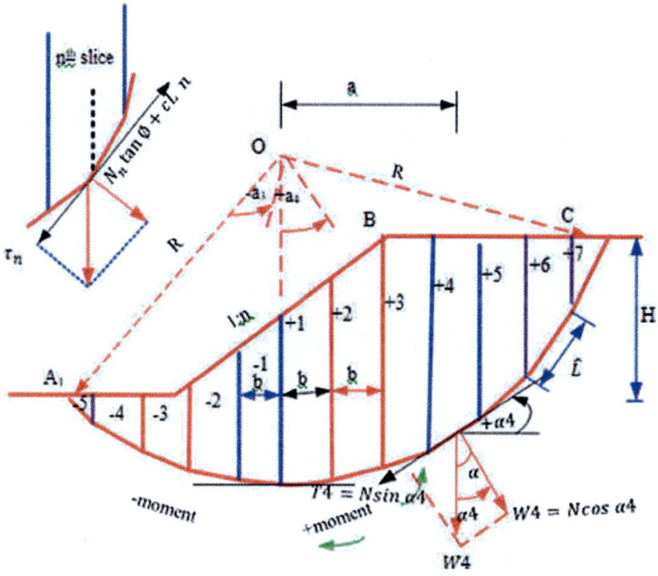

Fig. 3.21 Method of slices

The method of slices is as follows. The cross section of the slope-forming mass of soil encompassed by the circular rupture surface is subdivided into a number of vertical, parallel elements or slices of equal width, such as shown in Fig. 3.22. The width of the slices is usually taken as b ≈ (0.1) R. The weight of each slice, W, for

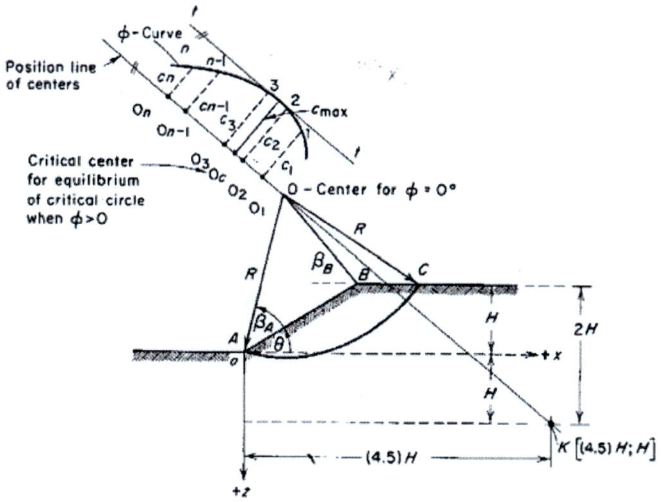

Fig. 3.22 Locating center, Oc, for critical circle

example, slice No. + 4, is applied at the center of gravity of the slice, and is projected on and allowed to act on the circular arc, GE' at point, M, for example. The weight, W, is then resolved into normal and tangential components, N and T, respectively, where N = W cos α and T = W sin α. Note that N passes through point M, as well as through O, the center of rotation. The weight of the slices is calculated by simplified geometric areas (rectangles, trapezoids, and triangles) as given per Eq. 3.53.

$$W_n = A_n\,(1)\gamma \tag{3.53}$$

where:

A (1) = V = volume of slice, and γ = unit weight of soil.

The algebraic sum of the tangential forces of all slices, Σ T, tends to shear the soil wedges along the impending circular rupture surface, and is resisted by the total resistance of the soil, tan φ ΣN = cL′, whereby the normal force, ΣN, must be the effective normal force. The advantage of the method of slices lies in the convenience of evaluation of the magnitude of the normal weight component, N, of each slice which permits summing them up as ΣN to be used in stability calculations. Note that the weights of the slices to the right of the vertical through the center of rotation, point 0, contribute to driving moments, whereas the slices to the left of the vertical contribute to resisting moments. The degree of the stability of slope is evaluated by comparing driving moments with resisting moments about the center, O, of rotation, Eq. 3.54.

$$\eta = \frac{M_R}{M_D} = \frac{\left(\tan \emptyset \sum N + \sum cL'\right)R}{R\sum T} \tag{3.54}$$

The factor of safety, η, should satisfy in each case the prescribed requirements for η, a magnitude of which should be at least η > 1.5. When φ, = 0, then η becomes, as given per Eq. 3.55.

$$\eta = \frac{\left(\sum cL'\right)R}{\sum T} \tag{3.55}$$

Several circles must be analyzed, and for each circle the factor of safety, η, computed. The least factor of safety among them indicates the most dangerous rupture surface. The approximate coordinates of the center of rotation, Oc, of the most dangerous circular sliding surfaces through the toe of the slope for (φ-c)-soils, for which the factor of safety is a minimum, may be found by trial and error, starting out with Fellenius' directional angles, βA and βB, for pure cohesive soils, Fig. 3.23, for which φ = 0. The center O_o of such a circular sliding surface, AC, having its center at point O_o, is also shown in Fig. 3.23.

In order to find the critical center, Oo for a (φ,-c)-soil, one may proceed as follows. From Fellenius' graphs it can be noted that the line on which all of the centers O_o,

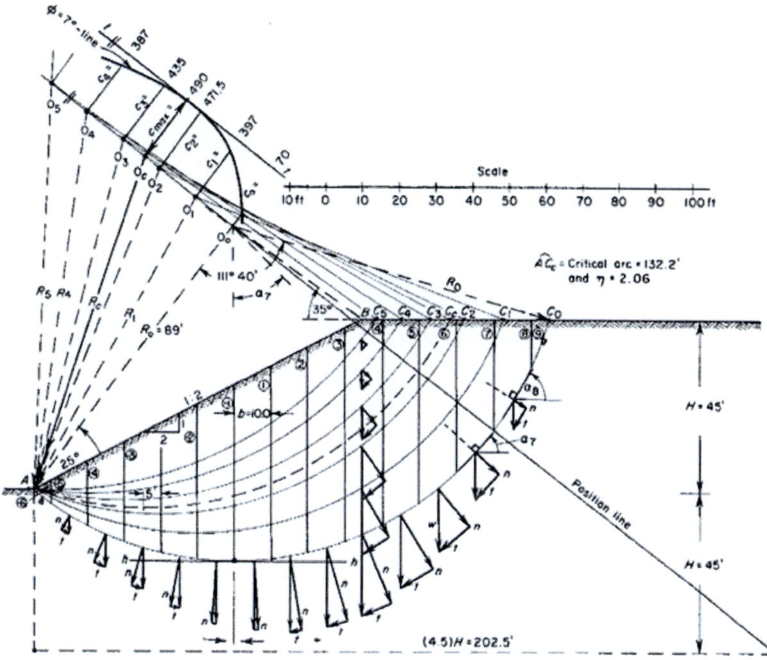

Fig. 3.23 Slicing method example

O_1, O_2, O_3, …O_n, line up passes through point O_o and point O_K, Fig. 3.24. Point K has the approximate coordinates of x = (4.5) H and z = H. Hence, in order to establish the position line 00-K, find point 00 by means of directional angles, βA and βB. Then plot point K with the coordinates x = (4.5) H, and z = H as indicated on Fig. 3.24. Then draw through points O_o and K the position line on which the centers,

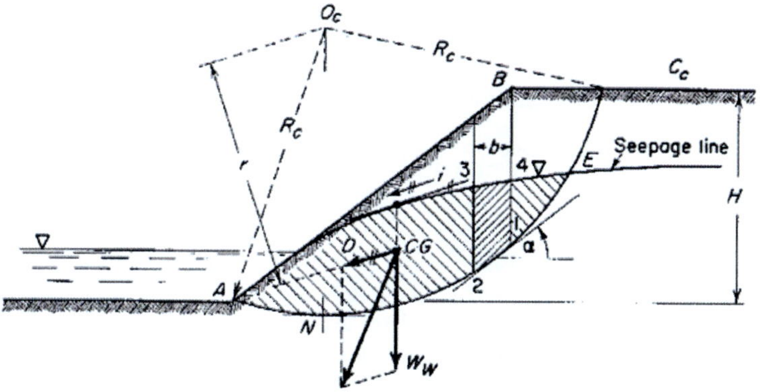

Fig. 3.24 Seepage force, D

$O_o, O_1, O_2, O_3, \ldots O_n$ of the trial circles lie. Fellenius' graphs show that as the value of φ, increases, the center O_n, of the circular rupture surface moves up from point O_o ($\varphi = 0$) along the position line $O_{o\text{-}k}$.

The stability computation may now be performed as follows. Select on the position line (above O_o) several, say equally spaced, centers, $O_1, O_2, O_3, \ldots$, and draw n circles through the toe-point, A. Calculate for each circle the amount of cohesion required to maintain equilibrium for the given value of φ, of the soil, Eqs. 3.56 and 3.57.

$$\sum T = \tan \emptyset \sum N + cL' \tag{3.56}$$

Or:

$$C = \frac{\sum T - \tan \emptyset \sum N}{L'} \tag{3.57}$$

Then plot these calculated c-values as ordinates to a certain scale on the position line at points of centers for which these c-values were found. For example, plot c_1 at point O_1, c_2 at point O_2, … and c_n, at O_n. Connect the c-ordinates with a curvilinear line 1-2-3-(n-1)-n, the so-called φ,-curves, and draw a tangent, t-t, to the curve and parallel to the position line 00-K. The tangent to the 4,-curve then scales off the maximum ordinate = c_{max}. This is the maximum value of the cohesion for maintaining equilibrium of the slope for a given φ. This maximum c-ordinate gives on the position line (O_0-K) a point O_c which is the critical center of the critical rupture surface for φ and for which the factor of safety, η, is a minimum. With a radius of R = (Oc − A) draw the critical rupture surface, AC′, along which the rupture would most probably take place, divide the slope forming mass of soil on this circle into slices, determine Σ T tan φ, Σ N and cL' (use φ, and c obtained from tests of this particular soil), and calculate the factor of safety, η, Eq. 3.58.

$$\eta = \frac{\tan \emptyset \sum N + cL'}{\sum T} \tag{3.58}$$

If this factor of safety is $\eta \geq 1.5$, the slope may be considered as stable, if $\eta <$ 1.5, the slope, or the height of the slope has to be redesigned. The calculations can best be performed in a tabular form.

3.4.5.7 Example

A slope 1: 2, the height of which is H = 45 ft, is to be made in a (φ-c) soil the unit weight of which is γ = 110 lb. /ft^3. the angle of internal friction, φ = 7o and the cohesive strength is found to be c = 1200 lb/ft2. Compute the factor of safety, η, against rupture of slope. The minimum factor of safety should be η = 2.00.

Solution

Referring to Fig. 3.23, the critical circle for a pure cohesive soil ($\varphi = 0$) is drawn from the corresponding critical center, O_o, ($\eta = 1$). Center Oo is found by means of directional angles $\beta A = 25°$ and $\beta B = 35°$ Draw circle C_0, with point O° as the center. Establish point K, and draw position line $O_{o\text{–}k}$, extending it beyond point O°. The critical center, C_0, for the (4,-c) soil, when $\varphi = 7°$ must be found by the method of trial and error. Assume, therefore, on the position line five points, O_1, O_2, O_3, O_4, and O_5, arbitrarily spaced at equal distances from each other. Draw curves C_1, C_2, C_3, C_4, and C_5. Slice up the soil mass above the curves with slices of equal width, b = 10 ft to scale, for example. For each curve, determine the weight, W, of each slice, and resolve it into normal and tangential components, n and t, respectively. Because weight is proportional to area, i.e., $w = a\,\gamma$, where w = weight of slice, and a = areas of slice, calculate the area of each slice, Eqs. 3.59 and 3.60, and find:

$$n = a\cos\alpha \tag{3.59}$$

And:

$$t = a\sin\alpha \tag{3.60}$$

Find the values of the a-components graphically. The sum of the normal and tangential components of the areas of all slices on one circle is then Σ N and Σ T, Table 3.2. The normal and vertical forces, Σ N and Σ T, respectively, resulting from the weight of the rupturing soil wedge, Eqs. 3.61 and 3.62, ABC, is:

$$N = \gamma \sum n \tag{3.61}$$

And:

$$T = \gamma \sum t \tag{3.62}$$

For circles C_1, C_2, C_3, C_4, and C_5, calculate the maximum necessary cohesion C_1, C_2, C_3, C_4, and C_5, plot them at points O_1, O_2, O_3, O_4, and O_5 as ordinates and draw the $\varphi = 7°$. For example, by Eq. 3.42, the necessary maximum cohesions for η = 1.0 was found to be:

$$C_1 = 397\ \text{lb/ft}^2 C_2 = 471.5\ \text{lb/ft}^2 C_3 = 435.0\ \text{lb/ft}^2 C_4 = 387.0\ \text{lb/ft}^2$$

Because from O_2 the cohesion drops from $C_2 = 471.5$ lb/ft^2 to $C_3 = 435$ lb/ft^2 at point O_3, the analysis of the fifth planned circle was omitted, since the maximum co-ordinate must be between points O_1 and O_3. The necessary maximum cohesion, C_{max} for a soil with $\varphi = 7°$ is found to be located at point, O_c, as shown in Fig. 3.23. The value, 490 lb/ft^2, was scaled off the drawing. Through point A is now drawn the critical circle, C_c, with point O_c as the center. This critical circle, with C = 1200 lb/ft^2.

Table 3.2 Normal and Tangential forces

Center of Curve		O_0		O_1		O_2		O_3		O_4		O_c
Length of Arc		171		153.5		139		126		116		
Forces		n	t	n	t	n	t	n	t	n	t	n
Slice number												
1		2		3		4		5		6		7
−6	12.9	−10.5	6.9	−4.0	7.0	−3.1	7.8	−2.1	5	−1	9.2	−3.2
−5	107.2	−63.0	88.0	−39.1	69.5	−20.6	59.6	−9.5	54.8	−2.8	71.5	−16.2
−4	206.0	−88.5	168.0	−48.8	145.0	−24.6	124	−6.6	114.9	5	140	−15.6
-3		−84.6	3.60	−43.1	224.0	−12.4	174	9.2	149	22.5	(109.0	−16
-2	364.0	63.4	322.0	−18.1	276.0	15.1	224	35.6	175	45	242	27
-1	423.0	−24.2	368.0	20.7	307.0	50.8	244	65.7	183	68.3	271	61.5
1	468.0	27.8	404.0	68.9	324.0	91.8	248	97.5	184.8	92.1	287	97.9
2	510.0	88.7	423.0	123.0	324.0	134.4	3.0.0	120.6	169.5	110	284	134.8
3	522.0	155.0	396.4	176.0	309.0	173.1	223	150.5	127	113.8	3.9.0	165.5
4	508.5	218.5	394.0	228.0	268.0	202	169	157	92.5	102	212	179.6
5	447.0	263.0	298.0	234.0	177.0	178	75	94.6	5.9	8.2	187	151.2
6	334.1	273.0	195.0	210.0	71.5	103	3.8	5.6		–	107.7	49.6
7	236.4	3.60	86.30	50.3	8.3	16.5	–	–	–	–	–	–
8	111.5	186.0	4.10	13.8	–	–	–	–	–	–	–	–
9	8.5	23.2	–	–	–	–	–	–	–	–	–	–
$\sum n, \sum t$	4544.1	1187	3409	9971.6	310.3	904	1802	2718.1	1261.4	563.1	22 74.2	832.5
η		1.63		2.15		2.11		2.22		3.46		2.06
2 ε	111°40′		99°		86°		76°		66°		82°	

(continued)

Table 3.2 (continued)

Center of Curve		O_0		O_1		O_2		O_3		O_4		O_c
Length of Arc		171		153.5		139		126		116		
Forces		n	t	n	t	n	t	n	t	n	t	n
Slice number												
1		2		3		4		5		6		7
−6	12.9	−10.5	6.9	−4.0	7.0	−3.1	7.8	−2.1	5	−1	9.2	−3.2
c_{nec}	69.8		397.0		471.5		435.0		387.0		C_{max} 490	

(*$C = 1200$ lb. /ft^2; $\varphi = 7°$; $\gamma = 110$ lb./ft^3)

(Given) is analyzed, and its factor of safety, as well as the factors of safety for the other circles is computed by Eq. 3.58. The critical rupture surface (for $\varphi = 7°$ and c = 1200 lb/ft^2) is revealed to have the least factor of safety, namely, $\eta = 2.06 \approx 2.00$. All other factors are > 2.00.

This indicates that the slope satisfies the prescribed value of factor of safety of 2.0. Therefore, with respect to $\eta = 2.0$, the slope for the given conditions may be considered safe. The factor of safety for the pure cohesive soil, shown in Column 2, Table 3.2 is $\eta = 1.63$, but such a soil was not given for the analysis. The $\eta = 1.63$ is here shown only for comparison: note that, theoretically, friction adds to the factor of safety. The slide-rule calculations in this problem are now given to elucidate some of the details involved in the foregoing analysis, and to illustrate the technique of the routine.

Necessary maximum cohesion for equilibrium when $\eta = 1$. Call Eqs. 3.42 and 3.43:

$$c_0 = 0.25\gamma H f(\theta, \varepsilon, \omega) \tag{3.63}$$

$\Theta = 22°$; $\varepsilon = 55°\ 50'$, $\omega = 18°\ 30'$, $\gamma = 110$ lb/ft^3, H = 45 ft.

$$f\ (\theta, \varepsilon, \omega) = \left[\frac{\left(2\sin^2\ \varepsilon \sin^2 \omega\right)}{\varepsilon}\right] \times [\cot\varepsilon \cot\omega - \cot\varepsilon\cot\theta$$

$$+ \cot\theta\cot\omega - 2/3\cot^2\theta + 1/3] = 0.564 \tag{3.64}$$

Thus:

$$c_0 = 0.25\gamma H f(\theta, \varepsilon, \omega) = 0.25(110)(45)(0.564) = 69.70\ \text{lb/ft}^2$$

The various angles may be calculated analytically from the geometry of the problem, or, if the drawing is made to a large scale, the angles may be scaled off by means of the protractor. Likewise, the radii, R, of the circles may be established, the lengths of the arcs, L′, are obtained by calculation. Call Eq. 3.57 and with $\tan\varphi = \tan 7° = 0.1228$ and L′ as given in Table 3.2 for the corresponding circles:

$$C = \frac{\sum T - \tan \text{Ø} \sum N}{L'} \tag{3.57}$$

$C_1 \approx 397$ lb/ft^2, $C_2 \approx 471$ lb/ft^2, $C_3 \approx 435$ lb/ft^2, and $C_4 \approx 387$ lb/ft^2.

The maximum necessary cohesion, C_c, for the critical circle, C_c, is scaled off the drawing as $C_{max} = 490$ (lb/ft^2). The factors of safety, η, for circles C_1 through C_4, as well as for C_c, are calculated using Eq. 3.43 as per Eq. 3.65:

$$\eta = \frac{\tan \text{Ø} \sum N + cL'}{\sum T} \tag{3.65}$$

$\eta_1 = 2.15$, $\eta_2 = 2.11$, $\eta_3 = 2.22$, $\eta_4 = 2.46$, and $\eta_c = 2.06$.
Where:
c = 1200 lb. /ft^2. is the actual, tested shear strength of the soil.

3.4.6 Seepage Force

In analyzing stability of slopes and the seepage which takes place through the soil, one should not forget to consider in the stability equations the seepage force, D, Fig. 3.24 and Eq. 3.66.

$$D = \gamma_w i A(1) \tag{3.66}$$

where:

γ_w : unit weight of water,
i: hydraulic gradient,
A: area AMDNA = cross-sectional area of the dam below the seepage line, AMD.

This hydrodynamic force, D, is applied through the centroid of the soil mass below the seepage line, and is directed parallel to the tangent to the seepage line at a point on the seepage line above the centroid. The factor of safety in the case of the seepage force present is calculated for the critical circle as per Eq. 3.67:

$$\eta = \frac{\tan \emptyset \sum N + cL'}{\sum T + D\left(\frac{r}{R}\right)} \tag{3.67}$$

where:

r: moment arm of seepage force, D, with respect to point of rotation.

Because part of the slices are below water and part above water, Fig. 3.25, the weight of slices, W, is calculated as per Eq. 3.68:

$$W = \sum W = B \sum (\gamma_d h_1 + \gamma_{subm.} h_2) \tag{3.68}$$

where:

γ_d: unit weight of soil above the uppermost flow line.
$\gamma_{subm.} = (1 - n)(G - 1)\gamma_w$ = unit weight of the submerged soil.
n = porosity of soil.
G = specific gravity of soil particles.
γ_w = unit weight of water.

Therefore the terms Σ N and Σ T in the η-equation should be formed according to these partially submerged conditions of soil. The resisting moment is then generally

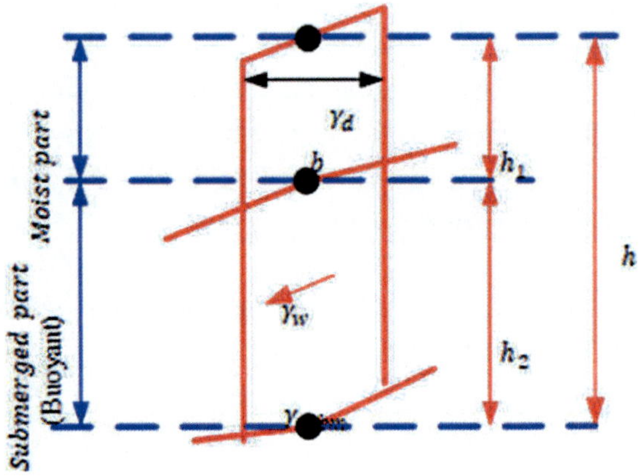

Fig. 3.25 Partly submerged slice

indicated as given in Eq. 3.69:

$$MR = [\Sigma s(c_1\,L_1) + \sum d(c_2\,L_2) + \tan\varphi 1 \sum s(w_1 + w_2)\cos\alpha s + \tan\varphi_2 \Sigma d(w_2 \cos\alpha d]R \tag{3.69}$$

The driving moment, Eq. 3.70:

$$MD = \left[\sum s(w_1 + w_2)\sin\alpha + \sum dw_2 \sin\alpha\right]R + \left(\sum sD\right)r \tag{3.70}$$

where:

s: number of submerged slices;
d: number of dry slices, unsubmerged;
$w_1 + w_2 =$ submerged plus dry weight of one slice, respectively.

Factor of safety: $\eta = \frac{MR}{MD}$.

The stability of slopes of an earth dam must be evaluated for both the upstream and the downstream sides of the dam. Fields and Wells describe a modified circular arc method applied in the analysis of the Pendleton Levee failure, where the ruptured soil mass was of an irregular shape differing from those shapes discussed in this book.

3.4.7 *Seismic Forces*

In regions of seismic activity the stability calculations of slopes of a dam should also include the seismic forces, because they reduce the margin of safety, or may even bring about the collapse of a structure. Seismic forces are applied at the center of gravity of the body, or at the center of gravity of the section of the body above any horizontal plane passed through the structure. Both horizontal and vertical seismic forces must be considered. The magnitude of a horizontal seismic force, F, is calculated as, see Calder and Blackwell [6], Hoek and Brown [11]:

$$F = \frac{Wa}{g} = ma \tag{3.71}$$

where:

m: mass of structural body above the horizontal plane being studied.
g: acceleration of gravity.
a: seismic acceleration, such as the acceleration of the earth quake wave.

In the United States, the value of the seismic factor for a, is taken as 0.75, or a = (0.75) g for rock foundations and a = (0.l0) g on sand and other soil foundations, where g is the acceleration due to gravity. If the dam is constructed near a live, or active fault, or, if it is founded on loose soil material, higher values of horizontal acceleration are in order.

3.4.8 Friction-Circle Method

The friction-circle method of stability analysis of slopes is particularly applicable to (φ-c) soils. However, the method applies to slopes which are just at equilibrium (η = 1.0). The stability analysis by the φ-circle method can be performed in two ways, namely:

- Use cohesion of soil, c, and find the necessary angle of friction, φnec, for maintaining equilibrium.
- Use friction angle, φ, to determine the necessary cohesion, cnec, for maintaining equilibrium.

The factor of safety, η relative to friction, in the first case is then expressed as the ratio of the actual angle of friction, φ, to the necessary angle of friction, φnec, Eq. 3.72:

$$\eta = \frac{\emptyset}{\emptyset_{nec.}} \tag{3.72}$$

In the second case the factor of safety, η_c, is expressed relative to cohesion as the ratio of the actually available cohesion of the soil to the amount of cohesion, cnec, necessary for maintaining equilibrium, Eq. 3.73:

$$\eta_c = \frac{c}{c_{nec.}} \tag{3.73}$$

The working of the φ-circle method, again; is by trial and error, whereby one tries to find for the same slope the most dangerous rupture surface. The principal reasoning in this method is that for a definite φ-value (viz., c-value) that rupture surface is the most dangerous one which requires for equilibrium the greatest c-value (viz., φ-value). With reference to Fig. 3.26, the forces of reaction, F, at point N, for example, are directed against the direction of motion of the sliding soil wedge, ABCNA. In such a case, it is assumed, based on statics principles, that at the instant of the impending rupture all reactive forces, F, along the circular rupture surface, AC, form with the normal, n–n (point N), to that rupture surface an angle, φ, which is the angle of internal friction of the slope-forming soil. Therefore, in the case of a circular-cylindrical rupture surface, all reactions would touch (tangent) a circle, the so-called friction or φ-circle, the radius of which is $R\varphi = R \sin \varphi$.

The φ-circle is concentric with the rupture circle of radius R. When $\varphi = 0$, there is no φ-circle. In the φ-circle system, with a known φ, for example, the following quantities are known:

- Magnitude of weight of sliding wedge and its direction of action.

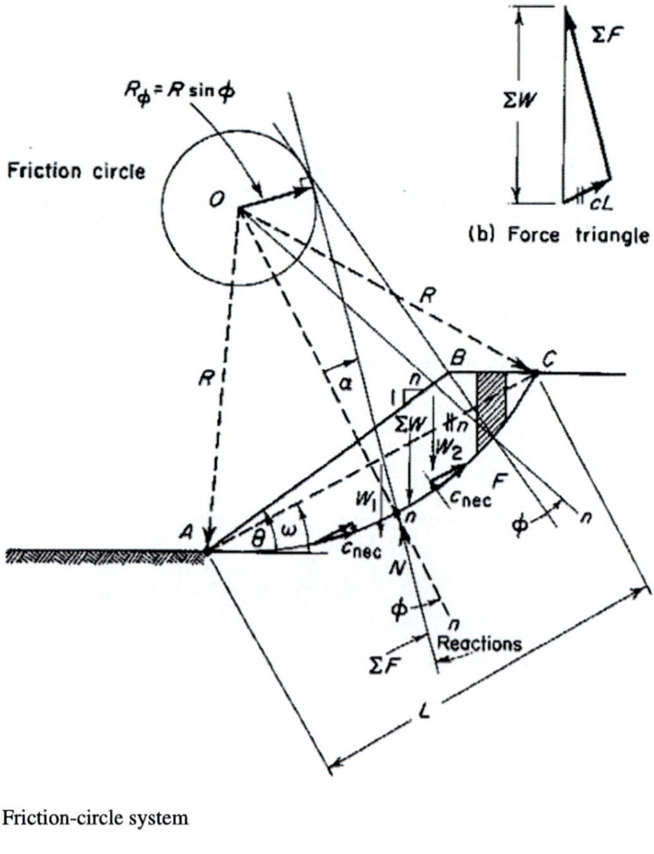

Fig. 3.26 Friction-circle system

- Direction of reaction, F (under the known angle φ with the normal of the rupture surface), though the magnitude of F, is not yet known.

The direction of the total cohesion, cL′ (along the rupture surface, viz., parallel to the chord, L′ = AC′; its magnitude is not yet known. To find the above-mentioned unknown quantities at equilibrium, the force triangle, viz., force polygon, is used, from which the magnitudes of reactions, F, and the necessary cohesion, cnec, for equilibrium can be determined. In the case of equilibrium, the force triangle must close. From the magnitude of cnec L′, one determines the necessary cohesion for equilibrium, and compares it with the available cohesive strength of the soil in question, Eq. 3.74:

$$\eta_C = \frac{c_{nec}}{c_{avail.}} \tag{3.74}$$

The direction and point of application of the total resultant, necessary cohesive force, cnec L′, acting along the circular sliding surface, AC′, is found as follows: assuming a uniform distribution of cohesion over the entire sliding surface, the sum of the components of cnec, of elementary cohesive forces parallel to AC′ = L′, form a total moment, Mc, about point O, Fig. 3.27 and Eq. 3.75. The cnec-components perpendicularly to AC = L are oppositely directed, Fig. 3.28, and do not contribute any moment.

$$M_c = C_{nec.} L r_c \tag{3.75}$$

where:

r_c : Arm of the cohesive moment.

This moment, M_c, is, therefore, also equal to the general expression of, Eqs. 3.76–3.78:

$$M = \text{Cnec}\ L'R \tag{3.76}$$

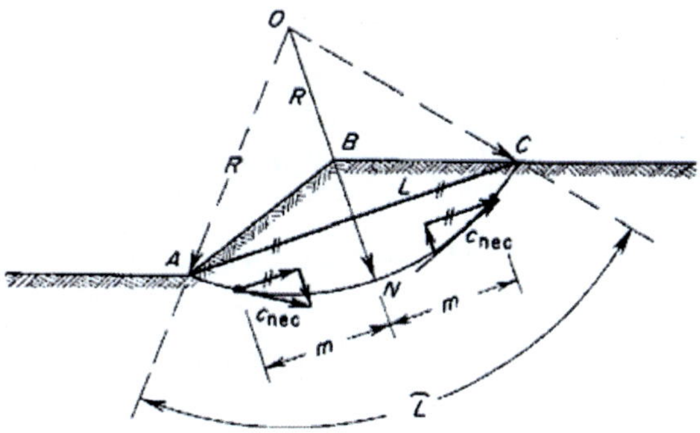

Fig. 3.27 Resolution of elementary force of cohesion

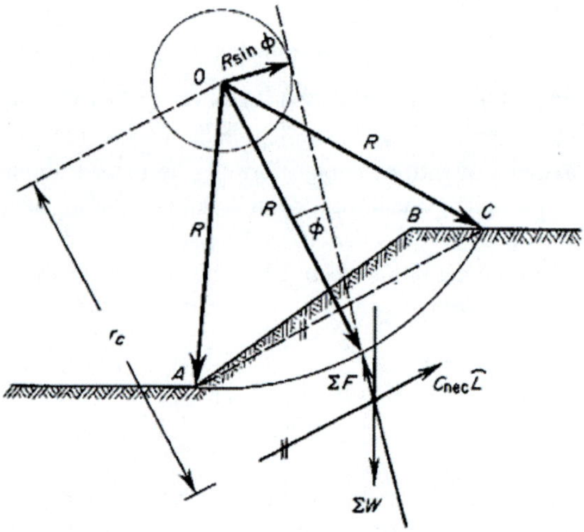

Fig. 3.28 Position of total cohesive force

Or:

$$\text{cnec L rc} = \text{cnec L}'\text{R} \quad (3.77)$$

Or:

The moment arm, rc, of the cohesive force is:

$$r_c = \frac{L'}{L}R \quad (3.78)$$

Because $L' > L$ by geometry, then $r_c > R$. Hence, the direction of the total cohesive force, cL′, to use in the force triangle or polygon, whichever the case is, is parallel to the direction of the chord, AC. The position of the total cohesive force, cL′, in the slope-friction angle system can now be determined by means of rc, and the direction of cL′. The point of application, M, of cL′, is the point of intersection of cL′ with Σ W, through which point, M, the third force, ΣF, the reaction, passes so that ΣF is tangent to the friction-circle.

3.4.9 *Remedial Work Against Failures of Slopes*

The scope of remedial work against the failures of earthwork slopes comprehends a variety of practices, Fig. 3.29. Some such remedial means are:

- Reducing weight by forming a berm.
- Flattening the slope.
- Adjustment of steepness of slopes according to strength of soil layers.

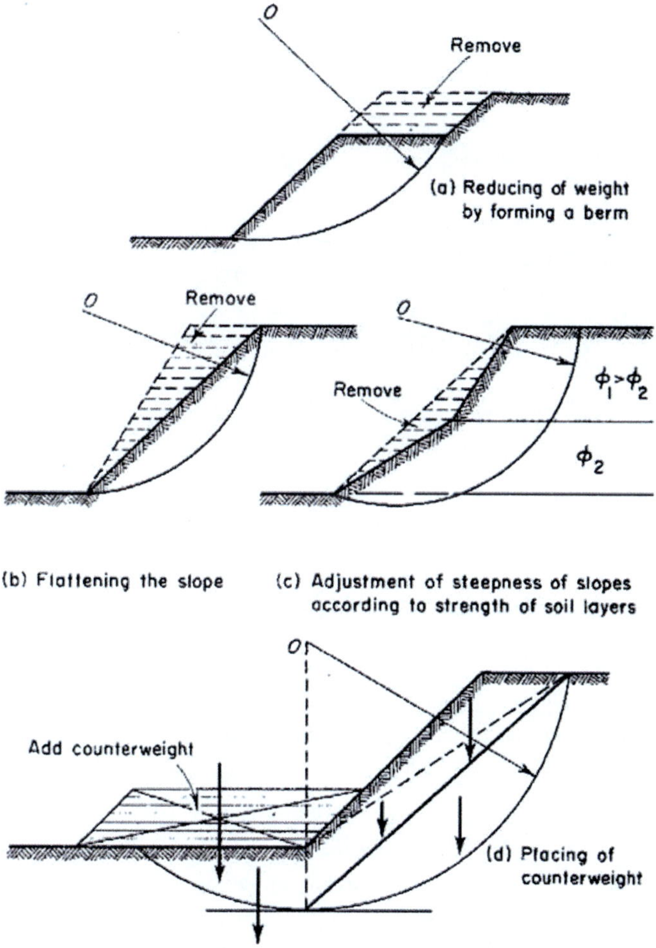

Fig. 3.29 Some means or remedy against rupture of slopes

- Placing of counter weight.
- Removing some weight of the slope-forming soil material which tends to cause failure, thus reducing the slope angle and weight of the slope forming earthwork body as well, Fig. 3.29. This will reduce the driving moment, thus increasing the calculated factor of safety.
- Providing some external support to hold back the toe of the slope by sheet piling, retaining wall, or counterweight at the toe, Fig. 3.29.
- Protection against undercutting of the slope.

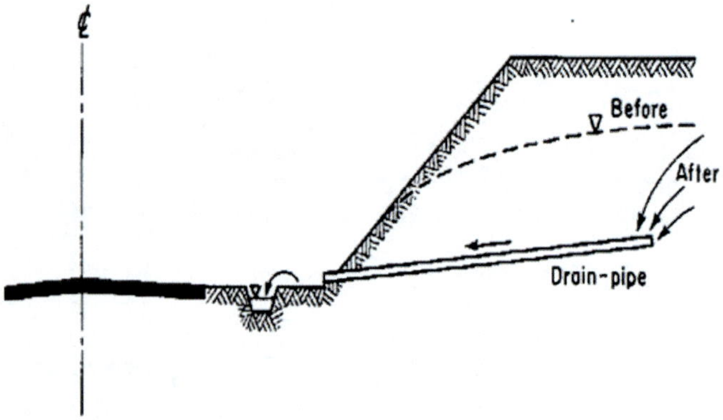

Fig. 3.30 Deep drainage of slope

- Providing a good and appropriate drainage system, thus preventing the water from entering the earthworks. Deep drainage of slope, Fig. 3.30, under some conditions, may be very effective.

- Practicing good maintenance concerning slopes, Fig. 3.31, repair activities, water and snow.
- Consolidation.
- Increasing the shear strength of the soil within which the rupture surface may develop (drainage, electrosmosis, soil stabilization by various mechanical and chemical means).
- Surfacing.

Since water seems to be the worst factor affecting the stability of slopes, drainage seems to be the most effective and practical method of controlling slides. Drainage facilities should be installed for the interception of the run-off waters and for diversion of water. Deep vertical and lateral drainage facilities for keeping the slopes dry can also be installed within the slope-forming body of earth-works.

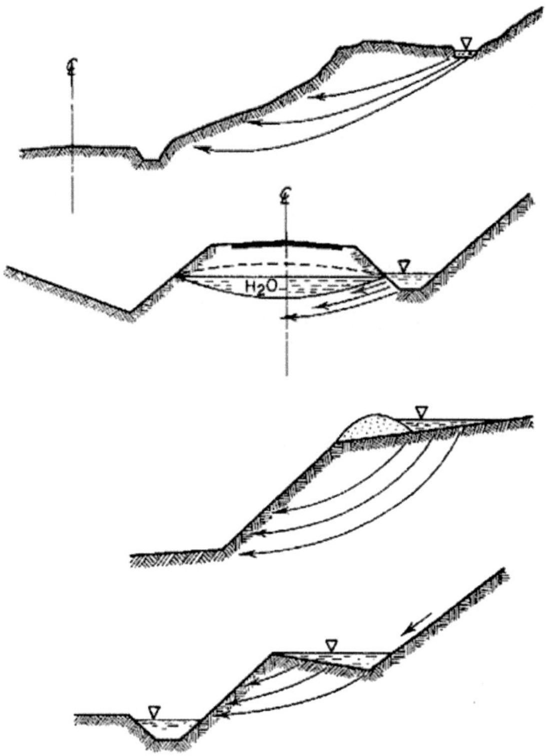

Fig. 3.31 Some maintenance problems of slopes relative to drainage

References

1. Abramson LW, Lee TS, Sharma S, Boyce GM (1996) Slope stability and stabilization methods. Wiley, New York
2. Alfreds RJ (1962) Soil mechanics. D. van Nostrand Company, Inc.
3. Bishop AW (1955) The use of the slip circle in the stability analysis of slopes. Geotechnique 5(1):7–17
4. Bishop AW, Bjerrum L (1960) The relevance of the triaxial test to the solution of stability problems. In: ASCE conference on strength of cohesive soils, pp 437–501
5. Bishop AW, Morgenstern N (1960) Stability coefficients for earth slopes. Geotechnique 19(4):129–150
6. Calder PN, Blackwell GH (1980) Investigation of a complex rock slope displacement at brenda mines. Can Inst Min Bull 73(820):73–82
7. Chowdhury RN (1978) Slope analysis. Elsevier Scientific Pub. Co., Amsterdam, p 423
8. Fellenius W (1936) Calculation of the stability of earth dams. In: Transactions. 2nd Congress on Large Dams, vol 4, p 445
9. Hoek E (1986) General two-dimensional slope stability analysis. In: Brown ET (ed) Analytical and numerical methods in rock engineering. George Allen and Uñwin, London, pp 95–128
10. Hoek E, Bray JW (1981) Rock slope engineering, 3rd edn. Institution of Mining and Metallurgy, London, p 402
11. Hoek E, Brown ET (1980) Underground excavations in rock. Institution of Mining and Metallurgy, London, p 527

12. Hungr O, Salgado FM, Byrne PM (1989) Evaluation of a three-dimensional method of slope stability analysis. Can Geotech J 26(4):679–686
13. Ireland HO (1954) Stability analysis of the congress street open cut. Geotechnique 4:163
14. Janbu N (1973) Slope stability computations. In: Hirshfeld, Poulos (eds) Embankment dam engineering. Wiley, New York
15. May DR, Brahtz HA (1936) Proposed methods of calculating the stability of earth dams. In: Transactions. 2nd congress on large dams, vol 4, p 539
16. Moore PJ (1970) The factor of safety against undrained failure of a slope. Soils Found Jpn Soc Soil Mech Found Eng 10(3):81–91
17. Morgenstern N, Price VE (1965) The analysis of the stability of general slip surfaces. Geotechnique 15(1):79–93
18. Schuster RL, Krizek RJ (eds) (1978) Landslides—analysis and control Transportation Research Board, Special Report 176, Washington D.C.
19. Taylor DW (1948) Fundamentals of soil mechanics. Wiley, 700 p
20. Whitman RV, Moore PJ (1963) Thoughts concerning the mechanics of slope stability analysis. In: Proceedings of the Second Pan-American Conference on Soil Mechanics and Foundation Engineering, Brazil, vol 1, pp 391–411
21. Yang HH (1983) Stability analysis of earth slopes. van Nostrand Reinhold company, Inc.

Chapter 4
Prevention of Slides in Surface Mines

4.1 Characteristics of Slides and Falls in Opencast Mines

One of the most serious safety problems in opencast mining is how to maintain stability of the walls and faces of the pits and waste banks. Correct solution of this problem will ensure complete safety in the working conditions of the pit, and economical efficiency of the mining operations. The seriousness of this problem can be proved by several exam-pies. For instance, in November 1930 in the copper mining pit in. Bingham Canyon, Utah, an enormous mass of barren overburden slid down into the pit from the upper levels, filling it to almost half its depth and along a considerable length. The volume of the overburden in this slide amounted to some 10 million cubic meters. From the data available in the literature, the initial slope angle of the pit walls ranged from 56 to 35°; but as the depth of the pit increased, it became necessary to reduce the angle to as little as 26 to 22°. Formerly the height of the faces, their slope angle, and the width of the working benches were selected solely on economic grounds so as to attain efficient production levels at each given elevation. The mistake in determining the slope angles of the faces and the general slope of the pit walls did not then influence the conditions in the pits; this occurred later when the depth of the pit exceeded a certain limit. The above example fully illustrates the significance of the correct selection of the general wall slope angle. An excessive angle gives rise to the danger of slides; while too small an angle results in extra cost of removing the overburden. An error of one degree in determination of the wall slope angle will change the volume of overburden removal by about four per cent.

In USSR mining practice, a typical example was the large slide of a lying wall in the pit of the Bakal mining administration. Schematically, this case is illustrated in Fig. 4.1. The slide completely covered the excavator at the face and a train of cars awaiting loading. Fortunately, the movement of the slide was detected early enough to remove all persons working in the area of danger. A slide that occurred in 1931 in the United Verde mine (USA) had a volume of 380,000 cubic meters, and to prevent ally further slides 3,800,000 cubic meters of rock had to be removed to

M. M. Ali Elbeblawi et al., *Surface Mining Technology*, Topics in Mining, Metallurgy and Materials Engineering, https://doi.org/10.1007/978-981-16-3568-7_4

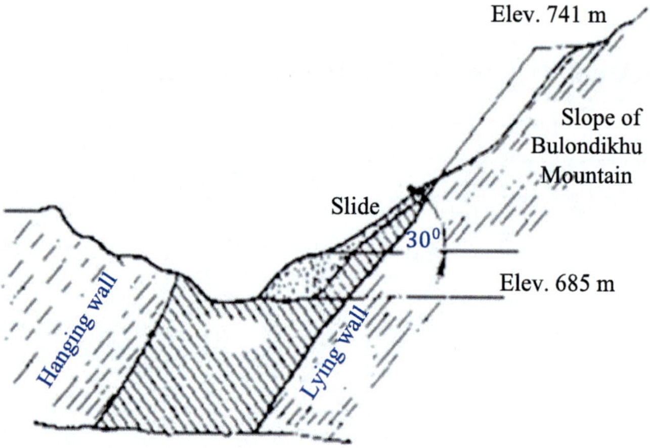

Fig. 4.1 Slide occurred on the slope of the Bad/cal. mine

give the walls a safe slope of 45°, according to conditions existing in the pit strata. In the Tyulenevsk nickel milling pit in the Urals a slide occurred in 1941 on the west side along the slip surface. Large slides have also occurred in the Magnitogorsk iron ore mine. As a result of a slide that swept across eight benches and had a width of 200 m, almost 2 million cubic meters of rock had to be removed in 1946. For further reading: Álvarez-Fernández [1], Call and Savely [2], Gao-Lie et al. [3], Wyllie and Mah [4].

The operation of coal deposits by opencast mining methods also involves mass-movement hazard. For example, the large slide that occurred in 1946 in the Baturin pit of the Korkinugol Trust had a width of 500 m across the lying wall and a volume of over one million cubic meters. The slide that occurred in pit No. 4 of the Volchanskugol Trust in 1950 was a mass-movement of about 5 million cubic meters, and on further deepening of the pit doubled in volume. In the Bogoslovsk pits, slides occurred on both the hanging and lying walls, involving a mass-movement of 5–10 million cubic meters and more. Large slides have also occurred in the Baidakovsk and Yermolayevsk pits.

Notwithstanding the large number of slides and falls in the pits, it would be incorrect to assume that slides are unavoidable in open-cast mining and in opencast coal mines in particular. Many investigations in this field have definitely indicated that the majority of mass-movements of rock and earth could have been averted if timely measures had been taken with due allowance for the basic factors that govern such slope and face movements. In maintaining the stability of bench slopes and faces, pit walls and waste banks, it is essential to base one's considerations upon the strength and stability of the strata being worked and the deformations they suffer during the working. For further reading see Karam [5].

G. L. Fisenko distinguishes four types of deformation that pit walls and faces may undergo: debris slide, rock fall, rockslide and mudflow. Debris or scree is formed by the effects of weathering on a slope or face, during which some of the larger segments of material that become detached roll and slide down to the foot of the slope. It is well known that such a process can be observed in mountainous areas where scree gathers in large amounts at the foot of mountain slopes. As meant here, rock falls comprise a considerable mass of strata in which the separation of the mass from the parent rock takes place at an angle exceeding the angle of internal friction; as a result the detached mass begins to move and acquires a relatively high velocity under the forces of gravity. This type of slide is therefore characterized by rapid mass-movement of material which, in its fall, separates into segments, lumps and smaller fragments and, as a result of its travel and impact, produces a dynamic shock whose magnitude is determined by the weight of the material and the height of fall. Cases are on record where these falls produced serious damage to mining equipment and transport facilities, and resulted in injury to the personnel. Rock slides differ from rock falls in that they are slow mass-movements that occur along a curvilinear surface (slip or failure surface) lying at an angle less than the internal angle of friction. The stresses along the slip surface, being greater than those corresponding to the limits of elastic deformation, lead to plastic deformation of the rock and soil resulting in a slide.

During the initial stage, the mass of the rock slide retains the structure existing prior to its movement, but at a further stage breaks appear in the solid mass and form cracks that lead to its separation into several parts. Under certain conditions a rock slide may attain considerable velocity and acquire the properties of a fall. According to the classification of Academician A. P. Pavlov, rock slides can be divided into slipping (delapsive) and pushing (detrusive) slides. The first kind of slide begins at the bottom section of the mass, gradually extending to a higher point as the part just below it begins to slump. The second, detrusive type of slide, on the contrary, begins at the top and sets up a pressure that acts to push the mass downward. Rock slides may last from several tens of minutes to as long as several months; sometimes, in conjunction with new masses ~ the movement may continue as long as several years. In those cases when clays and sandy clays (loam) become saturated with water and acquire a fluid condition, mass-movement may occur on surfaces having an incline of 4–6°. When a large amount of such fluidized material begins to gather on a bench, it may flow over the edge and slide down onto the lower face, and when it becomes very fluid with water it may turn into a mudstream. Malyushitsky proposed a more detailed list of the forms of disturbance in pit wall and bench face stability and the conditions causing them, Table 4.1, and also the corresponding schematic pictures of these forms of slope failure, Fig. 4.2. For further reading: Zharikov [6]

The deformations that may occur in opencast mine pit walls and faces owe their origin and nature to factors that may be both natural and artificial. Among the basic natural factors are: the physical and mechanical properties of the soil or rock being worked, the presence of weak surfaces in the strata and the nature of such areas (cracks, inter-layers, etc.), the hydrogeological features of the given mineral deposit and the climatic conditions (annual precipitation, alteration of thaws and frosts, etc.). The artificial factors, which depend on the activity of man, may include: the dimensions and form of the opencast pit, the slope angle of the pit walls and faces,

Table 4.1 Failure forms of pit walls and bench slopes

Stability disturbance form	Deformation characteristics	Deformation velocity	Characteristic conditions
Rock falls, general and local	Detachment, unsteady segments of large size	Sudden (catastrophic)	Excessive wall slope in block-structure rock safety berms too narrow
Scree accumulation	Sliding of small fragments, within limits of a separate bench	From very low to catastrophic	Slope angle exceeds repose angle in weathered rock, slates, sands and conglomerates
Rock falls with shear and turning	Movement of mass along a curvilinear slip surface	Reaches very high value	Excessive wall or separate face slope, mainly I loose homogeneous strata
Contact slide-movements	Predominantly horizontal mass movement, partly or wholly taking place in sharply weakened zones	Very low rate	Mainly in soft strata suffering sharp loss in shear strength on wetted clayey contact surface
Flat slip of rock with break-off of strata	Flat slip of sheared masses lacking restraint on face side	Reaches relatively high value	Break-off of slate and stratified rocks, or freed blocks. Mass deprived of restraint on face side
Flat slip of rock without break-off of strata	Flat slip along strata with buckling and bulging of layers	Reaches relatively high value high value	Excessive wall height on lying wall side without sufficiently large unloading berms in weak, stratified material (clay, talc, chlorite,)
Wash (diluvia) slide	Slide of overburden lying on bed rock	Relatively low value	Overburden lies on bed rock steeply pitching in direction of pit
Slide due to fall into u.g. workings and caves	Sudden sinking with sidewise movement	catastrophic	Presence of u.g. workings or caves under near pit wall
Shearing due to subsidence	Gradual sinking with sidewise movement	Reaches relatively high value	Presence in underlying strata, water- laden sands

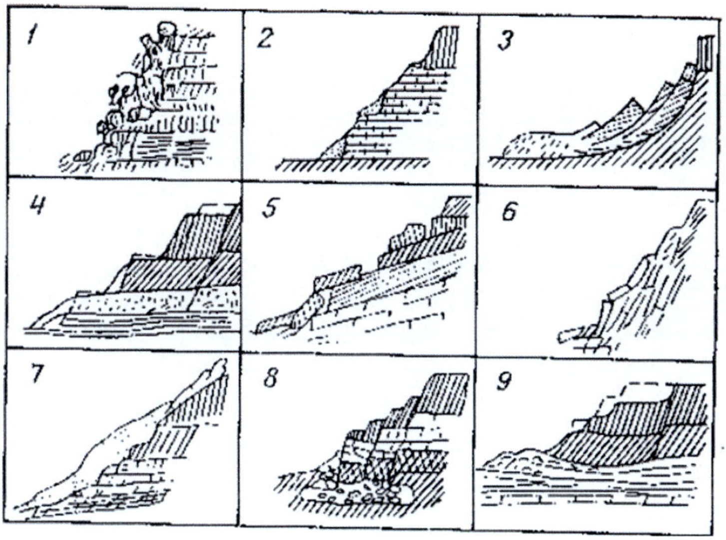

Fig. 4.2 Failure forms of pit walls and bench slopes

the size of the benches, the height of the faces and the depth of the workings, the period the pit is kept in operation and the rate of production, the kind of drilling and blasting, the method of hydromechanization, the presence and magnitude of any external loads on the walls of the pit. For further reading: Xia [7].

Comprehensive characteristic of conditions and factors, which give rise to wall and face disturbances in opencast mining pits, after Malyushitsky is given in Table 4.2. The most important of the natural factors mentioned above are the physical and mechanical properties of the overburden and deposits being worked, while the important artificial factors are the slope angles of the bench faces and the pit walls, the height of the faces and the depth of the pit. Also very significant is the method used in breaking down the rock, in particular, the use of explosives and the method of control of the blasthole firing. The main physical and mechanical characteristics that determine the stability of rocks are: volume weight, cohesion arid angle of internal friction. Besides these, other properties of importance ale, for example, moisture content, porosity, density, tensile strength, hardness, compressive strength, elasticity, and plasticity, all of which must be taken into account in considering problems involving the stability of pit faces and walls. For further reading: He et al. [8], Hoek et al. [9], Marschalko [10].

The prevention of slides and falls in pits is a complex problem. For convenience of discussion it is best to single out the problems of face stability, pit wall stability, and waste bank stability.

Table 4.2 Factors cause stability deterioration of pit walls and faces

Conditions in pit	Ground water conditions	1st Category Increase in disturbing forces		2nd Category Lowering of resistive forces		
		Class I. Increase in steepness of stope	Class II. Increase in weight of burden	Class III. Additional forces of disturbance	Class IV. Lowering volume and weight of counterbalancing mass	Class V. Lowering of resistive forces of strata
Opening up and mining of deposit	Level: lowered in drainage of pit	Bench: Stopes and walls allowed to be steep	Overburden cast onto top of pit walls (within zone of active earth pressure) or near them	Seismic action of blasting	Undercutting and removal of material at foot of wall or bench slope (in zone of restraint)	Strata fissuring from nearby blasting
	Hydrostatic gradient raised by lowering datum of drainage	Steep Slopes formed by falls and slides resulting from rapid drainage	Removal of up-lifting forces of hydrostatic pressure during drainage	Dynamic loading due to travel of haulage units and operation of heavy equipment	Undercutting and Shearing of down-dipping strata	Cracks from intensive drying
					Piping (suffusion) and floating of sands during fast rates of drainage Squeeze of weak material in slope and from under floor	Hydration and dehydration Change in properties of rock due to action of water
					Underground workings below floor at pit walls	Weathering (disintegration)

(continued)

Table 4.2 (continued)

Conditions in pit	Ground water conditions	1st Category Increase in disturbing forces		2nd Category Lowering of resistive forces		
		Class I. Increase in steepness of stope	Class II. Increase in weight of burden	Class III. Additional forces of disturbance	Class IV. Lowering volume and weight of counterbalancing mass	Class V. Lowering of resistive forces of strata
Unsuitable conditions of mining	Water content raised by breakage from drainage	Washout due to water discharge	Saturation of wall strata (during leakage and draining of water)	Dynamic action of blasts	Washouts during draining water	Thawing of permafrost Hydration and dehydration connected with wetting of walls and slopes
	Rise in hydrostatic gradient due to above reasons	Falls, break and slides as a results of undrainage		Dynamic loading due to haulage units and operation of equipment	Displacements along fissures resulting from seismic action of blasting Carryout of sands, slides and falls produced by vibrations resulting from operation of haulage units and equipment slides and underground workings	Leaching due to same reasons as above
	Water flows down slope of wall	Falls and breaks as a result of improper blasting methods				Disturbances of structural cohesion due to nearby blasting

(continued)

Table 4.2 (continued)

Conditions in pit	Ground water conditions	1st Category Increase in disturbing forces		2nd Category Lowering of resistive forces		
		Class I. Increase in steepness of stope	Class II. Increase in weight of burden	Class III. Additional forces of disturbance	Class IV. Lowering volume and weight of counterbalancing mass	Class V. Lowering of resistive forces of strata
Underground water (seepage)	Pressure in fissures	Flow and carry out off sands and rise in hydrostatic gradient	Saturation of strata by water in zone of active earth pressure		Appearance of up lifting forces in zone of restraint	Conversion of sands into fluid mass by shocks and vibration wetting of soft strata to over saturation

4.2 Stability of Pit Benches and Faces

In practice, distinction is made between the short-term and long-term stability of pit benches and faces. On benches where faces are being worked it is short-term stability, i.e. stability during the period a certain face is being worked that is of practical interest. In cases where final extraction and reclamation operations are carried out in a pit, or where the work is to be discontinued in some given direction or at a certain elevation, the question of long-term slope stability arises, meaning the stability of unworked slopes until the entire pit goes out of operation. The degree of stability of a pit wall bench, or face is characterized by the margin of safety or stability factor (or safety factor) introduced in calculating the limiting conditions of equilibrium. In civil engineering practice, this factor is given a value of 1.0 to 1.5 or even 2 (Yu. N. i\Malyushitsky proposed the use of such values):

For short-term stability: $\eta = 1.1$ to 1.2

For long-term stability: $\eta = 1.5$ to 2.

Both short-term and long-term stability are governed by the interaction of the existing natural and the artificial factors, i.e. under certain given natural conditions stability will be ensured by observing the corresponding mining requirements.

Among these are: maintenance of safe slope angles, face heights and working bench widths; limiting the external loads carried by the benches; selecting the safest possible positions for the benches relative to the strata being worked, and using such methods of breaking down and removing the strata that do not lower the stability of tile faces exposed by the mining operations. The stability of a bench may be quite different, both as to the forces that come into play and the measures to be taken to prevent slides and falls, depending on whether rock or unconsolidated material is being worked. Consider, for instance, the interaction of the forces acting on the side slope of a given bench, neglecting the external dynamic forces due to blasting, the operation of the machines and mechanisms on the bench, etc. Any particle on the slope of the bench, Fig. 4.3, will lend to move under the action of a force that is a component of the force of gravity and is dependent upon the volume weight of the rock being worked and the slope angle of the bench, Eq. 4.1:

$$\boldsymbol{F} = \boldsymbol{q} \sin \boldsymbol{\alpha} \tag{4.1}$$

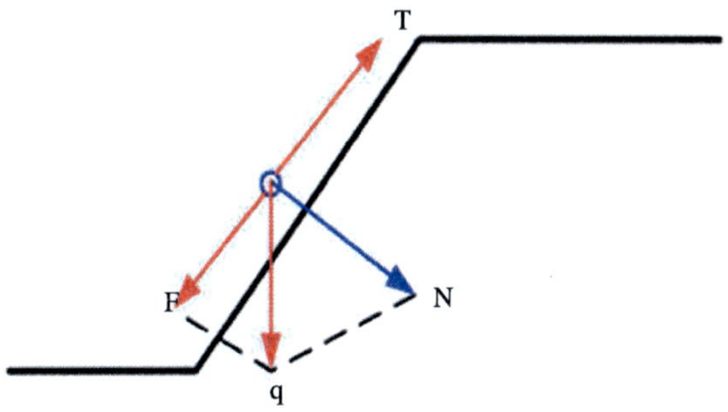

Fig. 4.3 Slope stability determination in unconsolidated material

where:

F force tending to displace the particle, kg
q force of gravity, kg
α angle of bench slope, deg.

If the material of the slope is unconsolidated and loose the movement of the free particle will be resisted by the force of friction, and the condition of equilibrium can be expressed through the Eq. 4.2.

$$\mathrm{a}\sin\alpha = \mathrm{f\,q}\cos\alpha \tag{4.2}$$

where:

f coefficient of internal friction; f = tan φ

(Angle p here is the angle of repose of the given material).

Dividing both sides of the equation by q cos α and replacing f by tan φ, we obtain Eq. 4.3:

$$\tan\alpha = \tan\emptyset \,\mathrm{or}\, \alpha = \emptyset \tag{4.3}$$

Thus, to maintain the stability of a bench in working unconsolidated strata, the basic requirement is that the slope angle of the bench is not allowed to exceed the angle of repose, this being determined by the value of the internal coefficient of friction. The above condition is valid for dry materials. If they are subject to the action of water, the stability of the bench will be considerably lowered. This occurs because the weight of the material increases as it becomes soaked with water, and the coefficient of friction drops to a lower value; moreover, the hydrodynamic pressure of the seeping streams of water also comes into play. For further reading: Jaeger [11], John [12], Salmi et al. [13].

In cases where water seeps outward, the slope angle of a bench of unconsolidated material must be less than the angle of repose depending upon the hydrodynamic gradient of the seepage stream, Fig. 4.4. It is essential that the harmful effect of water on the stability of pit benches and walls be always taken into account. This may manifest itself in the form of a break-through of the water under hydrostatic pressure, a decrease in the frictional resistance forces holding the material in equilibrium, breakdown of the stratum owing to its interaction with the water, loss of mechanical strength, and piping or carry-off of particles of the stratum by the water. The possibility of damage by water necessitates a significant reduction of the slope angles, sometimes by as much as 10–20° and more. However, even this will not always ensure sufficient slope stability. It is therefore better practice to drain the pit and discharge tile water to a point well beyond its boundaries; as such a measure is much cheaper than maintaining small bench and pit-wall slope angles. When cohesive or consolidated strata must be mined, determination of the safe slope angle is more difficult because of the existence of forces of cohesion between the particles of the material and the lack of uniformity in the structure of the strata.

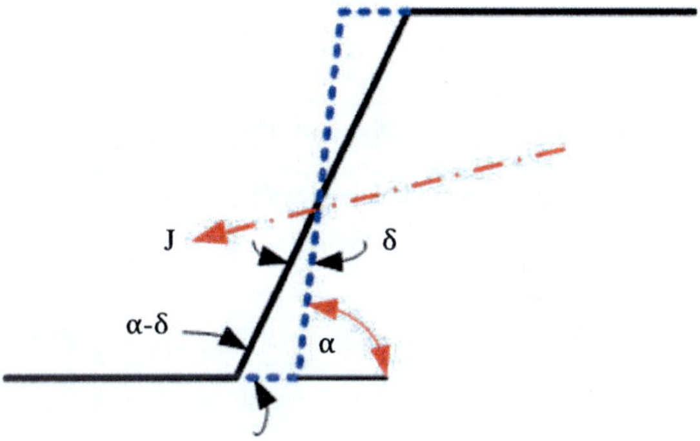

Fig. 4.4 Slope stability calculation in case of water seeping into the material

For determining the permissible slope angles and the maximum height of a face to be used in mining consolidated strata, a number of analytical, graphical and grapho-analytical methods have been developed by Sokolovsky, Tsytovich, Senkov, Clushke-vich, Malytishitsky, Fisenko and others. In consolidated strata it is possible to work with vertical faces, provided the height of the faces is limited. If the height of such a face is gradually increased to a ~ certain value called the critical height, stability will be lost. Considering the case of failure in a face of cohesive material and assuming for simplification that the mass becomes detached along a surface in the form of an inclined plane, Fig. 4.5, we may set up an equation expressing equilibrium between the disturbing forces and the resisting forces for each meter along the face. See Hendron et al. [14].

$$\frac{\gamma\ \mathrm{h}^2 \cot\alpha \sin\alpha}{2} + \frac{\gamma\ \mathrm{h}^2 \cot\alpha(\mathrm{f}\ \sin\alpha)}{2} - \frac{\mathrm{ch}}{\sin\alpha} = 0 \tag{4.4}$$

where:

γ Volume weight of material, t/m^3.

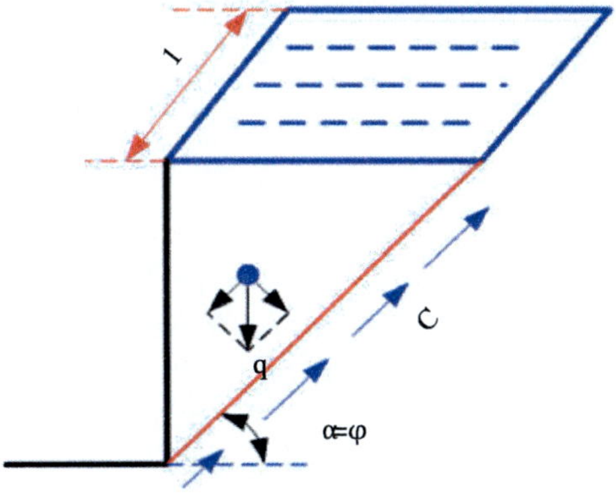

Fig. 4.5 Slope stability analysis in a face of cohesive material

α Slope angle of slip plane, deg.
h Vertical face height, m.
f Coefficient of friction.
c Specific force of cohesion, t/m^2.

If it is assumed that face stability is maintained solely by the force of cohesion, the force of frictional resistance may be neglected and the equation written in the form Eqs. 4.5–4.6:

$$\frac{c\,h}{\sin\alpha} = \frac{\gamma\,h^2}{2}\cot\alpha\,\sin\alpha \tag{4.5}$$

Whence:

$$h = \frac{4\,c}{\gamma\,\sin 2\alpha} \tag{4.6}$$

At the greatest possible value of sin 2α, i.e., at α = 45°, the maximum height the vertical face may have will be:

$$h_v = \frac{4\,c}{\gamma} \tag{4.7}$$

This equation was proposed by N. A. Tsytovich. Since a cohesive mass of material separates from the parent material, not along a plane but along a cylindrical surface, a correction factor should be introduced into Eq. 4.7 and the formula for determining the permissible height of a face takes the form (Fellenius formula, Eq. 4.8):

$$h_v = 0.958\,\frac{4\,c}{\gamma} \tag{4.8}$$

According to V. V. Sokolovsky, the limiting height of a vertical face can be determined from Eq. 4.9:

$$h_v = \frac{4c\cos\varphi}{\gamma(1-\sin\varphi)} \tag{4.9}$$

where:

φ angle of repose.

For comparison, and using the data of S. L. Iofin, Table 4.3 gives the limiting heights of vertical faces in different kinds of strata as determined from the equations given above.

As can be seen from Table 4.3, the heights calculated from these equations do not differ significantly. In actual practice one has to deal with non-uniform strata having local weaknesses that are not taken into account by the equations. The forces

Table 4.3 Limiting heights of vertical faces in various strata

Strata	V, t/m^3	Tan f	C, t/m^2	hv	
				From Eq. 4.8	From Eq. 4.9
Dry clay	1.65	0.839	0.9	2.2	2.3
Wet clay	1.70	0.532	0.3	0.7	0.6
Weak strata	1.95	1.000	40.0	82.0	97.0
Strong strata	2.40	1.000	80.0	133.0	158.0

of cohesion and frictional resistance are also inconstant in value and depend upon the depth of the deposit, the water content, local faults, etc. Besides the internal forces that effect the behavior of a face, various external forces are involved, these comprising the loads applied to the bench by the excavators, drilling rigs and other equipment that are also not taken into account by the equations given above.

To prevent falls and slides, it is not sufficient to ensure that the face has a safe height and angle of slope. It is very essential that the face should be correctly positioned with respect to the layout of the strata, the fissures in the beds, the faults, etc. The shape and the correct slope of the bench in relation to the general slope line is also important fact ores. When benches are cut in strata dipping towards the pit at angles greater than 25–30°, Fig. 4.6, it is usual for slides to occur along the footwall rock; therefore, wherever possible, the faces should be cut so that the strata are inclined away from the pit, Fig. 4.7. For further reading: DiBiagio and Kjekstad [15], Einstein [16], Einstein and Sousa [17], He et al. [18], Sousa et al. [19], Xu et al. [20], Zahiri et al. [21].

Where fissures with an angle of dip greater than 45°. exists in the strata and a face is bared near them, Fig. 4.8, there arises a danger of a slide at the face. To avoid this, the surface of the face and bench above must be thoroughly inspected to detect these fissures in order to take timely safety measures.

While being worked, a face may sometimes acquire an incorrect shape. For example, if the digging height of the power shovel is less than that of the face,

Fig. 4.6 Undercutting of strata inclined in the pit floor direction

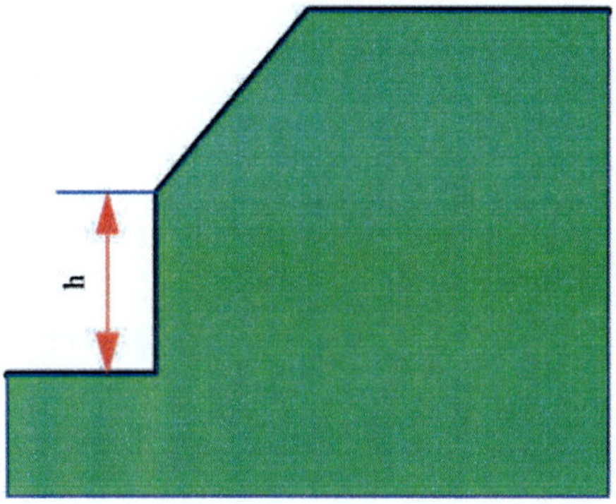

Fig. 4.7 Face cut in strata inclined away pit floor from pit floor

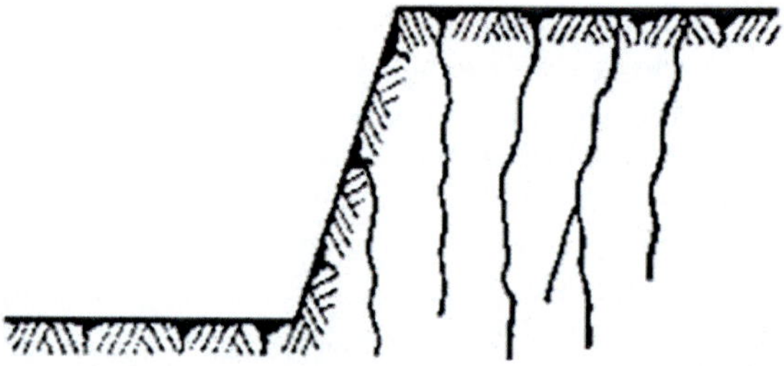

Fig. 4.8 Dangerous position of face relatively dipping fissures

the latter may acquire a hanging or reverse slope, or an overhang, Fig. 4.9. It is evident that such a shape gives rise to the direct danger of a slide or fall and must never be permitted. It is likewise impermissible to allow any form of undercutting, as this also leads to the danger of a slide or fall. For further reading: Fredj et al. [22], Geotechnical Considerations in Open Pit Mines [23], Hutchinson [24], Villegas et al. [25].

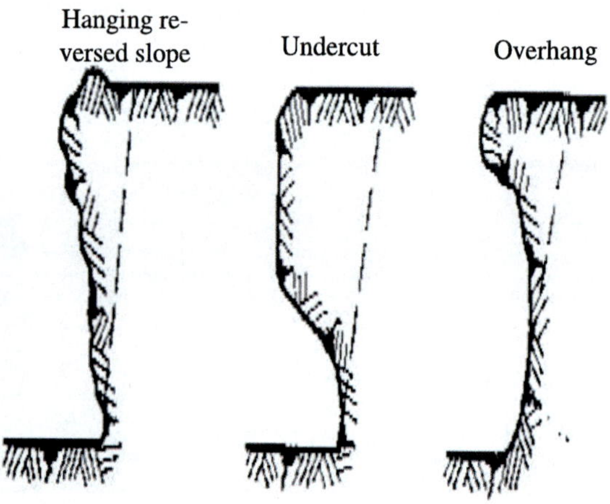

Fig. 4.9 Wrong shapes of faces

These requirements covering the maintenance of pit faces in a proper shape comprise a very important element in the conditions needed for safety in mining and arc therefore included in both the regulations for safety and the regulations for correct operation. According to safety regulations, the height of a face worked with a power shovel and without the use of blasting must not exceed the maximum digging height of the power shovel, a measure aimed at preventing the formation of overhangs in the upper zone of the face. Where a hard rock face is worked with a power shovel after preliminary loosening with explosives, the height of the face may be 1.5times the maximum digging height of the power shovels because the height of the heap of shattered rock produced by the blasting is 1/2–2/3 that of the face. For further reading: Duncan [26].

When the operations are carried out with draglines or chain-and bucket and bucket wheel excavators, the height of the face should not exceed the maximum digging height or depth of the excavator. In the case of manual labor, the regulations permit the face to be 3 m high in loose and running ground (sand, gravel and like materials), 6 m high in soft but stable ground of the stiff clay type, and 10 m high in hard rock. The safety regulations on correct operation permit coal strata to be worked with power shovels, the faces being as high as 30 m and, in individual cases, by special permission by the State mining inspection authority, is up to 40 m. However, special measures must be taken then to preserve the stability of the face and prevent slides and drop-out of coal fragments from the face. With this aim in view, blast holes are not drilled vertically but at an angle of 65° to the horizontal. Besides this, an inspection service is set up to monitor systematically the condition of the slopes and see that all overhangs are scaled off. As a rule the slope of working faces at which power shovels operate should not exceed 80°, while in mining coal at faces up to 10 m highs and interbedded with strong strata it may be as great as 90°. The limiting slope angles of non-working faces and benches intended for long-term stability must be established by a design project or from available mines surveyor's data.

The face height, from the point of view of workplace efficiency, depends upon the performance characteristics of the power shovel and also upon the method of blasting and the probable width of the broken-down sock and coal heap at the given face. Knowing these factors, the height of the face can be determined from the equation proposed by Melnikov, Eq. 4.10:

$$\mathbf{H} = 0.7\,\mathbf{A}\,\frac{\sin\alpha\sin\beta}{\sqrt{\mathbf{k}}\,\eta'(1+\eta'')\sin(\beta-\alpha)},\ \mathrm{m} \tag{4.10}$$

where:

a 0.8 (Rd + R1) = width of broken-down heap of material formed after b1asting, m.

α slope angle of broken-down material, deg.

β slope angle of the face, deg.

k loosening factor of the face material ratio of length of least resistance line of first row.

η' ratio of length of least resistance line of first raw of blast holes to face height, usually equal to 0.55–0.70.
η'' ratio of distance between rows of blast holes to length of line of least resistance, usually equal to 0.75–0.85.
Rd digging radius of the power shovel, m.
Rl loading radius of the power shovel, m.

For a dipper of a given volume, the productivity of a power shovel will be the greater, the shorter its dipper sticks and boom is. The digging-loading cycle will then be carried out at the highest speed and in the shortest time. An increase in these two parameters leads to a greater cycle time and a corresponding decrease in shovel productivity. This circumstance, some 30 years ago, led gradually to lowering of the face height from 30–20 m to 10–12 m. On the other hand, from the point of view of haulage and work organization, it is more economical to have a high face. Taking all these circumstances into account, V. S. Plygunov established that the most expedient face heights were: (a) in workings where no preliminary loosening of the strata is carried out: 9.5–10.5 m when using a type 3kT-4 power shovel; 12.5–13.5 m when using a type 3kT-8 power shovel; (b) in workings where the strata are first loosened by blasting: 15–16 m when using a type 3kT-4 power shovel; 19–21 m when using a type 3kT-8 power shovel. The face heights actually used in opencast coal milling pits correspond to those recommended above and also comply with the safety regulations.

Where a chain-and-bucket excavator is employed, the height of tire face should be such that the length of its slope will be sufficient to enable the buckets to become completely filled as they travel at a speed usually not exceeding 1.25–1.5 m/s. Consequently, the length of the slope will depend upon the capacity of the buckets, the latter, in turn, being selected in accordance with the required digging capacity of the machine, the traffic handling capacity of the rail haulage system, etc. In determining the height of the face to be used with a chain—and-bucket excavator, it is necessary to calculate the following:

The length of the bucket digging path, Eq. 4.11.

$$L_{dp} = \frac{L_v}{\sin \gamma} \tag{4.11}$$

where:

L_v bucket path, travel of excavator neglected, m.

The angle γ is found from the expression, Eq. 4.12:

$$\tan \gamma = \frac{V_e}{V_c} \tag{4.12}$$

where:

V_e speed of travel of the excavator, m/min.

V_c speed of the excavator chain, m/min, usually equal to about 1.25 m/s or 72 m/min.

The thickness of the cut, Eq. 4.13:

$$\delta = \frac{Q_t \sin \alpha}{60\ V_e\ H} \tag{4.13}$$

where:

Q_t Theoretical digging capacity of the excavator, m^3/h.
α slope angle of the face, deg.
H height of the face, m.

For conditions existing in cutting clay, δ = 0.05 m; for sand and gravel, δ = 0.10 m. The width of the cut taken by the bucket, Eq. 4.14:

$$\boldsymbol{b} = \boldsymbol{T} \tan \boldsymbol{\gamma} = \frac{\boldsymbol{V_e}}{\boldsymbol{n}} \tag{4.14}$$

where:

T pitch of the excavator chain, m
n number of buckets emptied per minute, usually about 30–34.

The cross-sectional area of the cut taken by the bucket, Eq. 4.15:

$$\mathrm{S} = \mathrm{b} \times \delta = \frac{Q_t \sin \alpha}{60\ V_e\ H\ n} \tag{4.15}$$

Whence:

$$H = \frac{Q_t \sin \alpha}{60\ b\ \delta\ n} \tag{4.16}$$

Assuming b = 0.15 m and δ = 0.1 m, we obtain, Eqs. 4.16 and 4.17:

$$H = \frac{Q_t \sin \alpha}{0.9\ n} \tag{4.17}$$

The digging capacity of a chain-and-bucket excavator must also be compatible with the capacity of the dump car it loads into, Eq. 4.18:

$$Q_{dc} = \frac{n'\ Q_t\ L_{dc}}{60\ V_e} \tag{4.18}$$

where:

Table 4.4 Limit values of time slope angle and height correspond to stable working conditions in different kinds of strata

Kind of bench and face, and type of material	Height of bench, run	Slope angle, deg
Monolithic rock	Practically no limit	Up to 90
Igneous rock strata in usual conditions	Practically no limit	70–80
Sedimentary strata in usual conditions	Practically no limit	50–60
Semihard rock and dry sandstone	25–30	40–50
Sandy-clay and clay cock strata	25–30	30–40
Waste bank benches (fills) on plough-formed banks of		
Rock and sandstones	20–25	30–35
Sandy-clay rock	10–15	35–40
Clayey materials	8–10	35–40
Excavator-cast banks of		
Rock	40–60	30–33
Sandy-clay materials	30–45	33–36
Clays	20–30	38–40
Moist sandy clays	10–15	18–25

n′ Number of cuts made by the excavator in loading time car td c length of dump car body.

It must be pointed out that face height selection is a complex engineering and economic problem whose solution must also comply with the conditions of safe operation. Based on generalized operating data and theoretical considerations, Table 4.4 gives limo values of time slope angle and height that correspond to stable working conditions in different kinds of strata and materials.

4.3 Stability of Pit Wall

In cases where a pit is worked with a single face, the factors deciding the stability of the pit wall are the same as those covered in discussing the stability of bench faces and bench slopes. Since such cases of working are rare, the problem of pit wall stability requires separate study.

According to the position of the pit wall, relative to tire elements of the strata in which it is situated, Yu. N. Malynshnitsky divides the walls into the following structural forms, Fig. 4.10:

1. Wall in homogeneous strata.
2. Wall in horizontally stratified ground.
3. Wall on hanging-wall side of deposit.
4. Wall on lying-wall side of deposit.

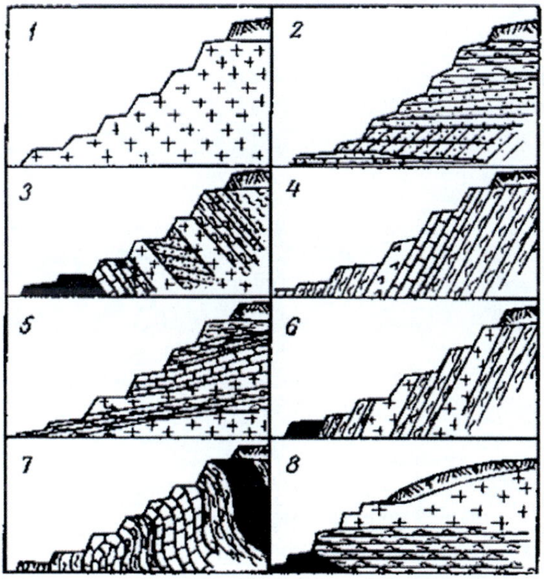

Fig. 4.10 Structures of pit walls

5. Wall with sill-cut strata.
6. Wail in steeply dipping strata.
7. Wall in strata with involved folds.
8. Wall on a slope with a heavy wash cover.

For one and the same kind of ground, the stability of the pit wall will be different, depending upon the structure of the strata, It is a known fact that a wall on the lying-wall side of a deposit will be much less stable than that on the hanging-wall side for the same type of stratum. Accordingly, the slope of a wall must be less on the lying-wall side of a deposit than on the hanging-wall side. Pit wall stability is also strongly influenced by the depth of the workings. Both theoretical study and practical data indicate that the stability of the walls, the internal coefficient of friction and the forces of cohesion drop with considerable increase in the depth of the workings, and it is therefore necessary to take special measures to ensure wall stability. For further reading: Hendron [14].

Time stability of any given pit wall will directly depend upon time stability of time faces arranged along its slope, their slope angles, the relative positions and heights of the bench faces, and time widths of the benches. Figure 4.11 schematically shows the dependence of the pit wall general slope angle upon time dimensions of the benches. This dependence is analytically expressed by the equation proposed by Bogolyubov, Eq. 4.19:

$$\tan\beta = \frac{B}{H} = \frac{\sum h\ f}{\sum(b\ f + h\ f \cot\alpha\ f) + h\ n \cot\alpha\ n} \tag{4.19}$$

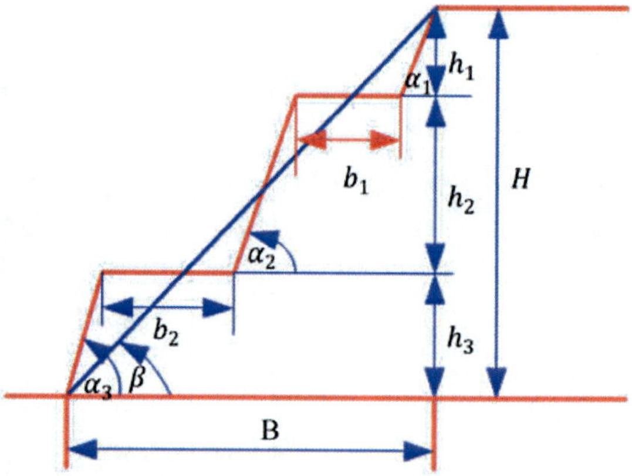

Fig. 4.11 General slope angle of pit wall depends on dimensions of benches

where:

β pit wall general slope angle, deg.
H total height of bench faces, m.
B horizontal projection of wall, m.
hf bench face height, m.
αf bench slope angle, deg.
n number of bench faces.
bf width of bench, m.

In designing the workings for a given opencast pit, the shape of the wall will be selected in accordance with the following considerations: (a) the need to ensure the stability of the separate benches and the wall of the pit as a whole; (b) the need to protect the lower-level benches against the danger of fragments of rock, minerals and other objects falling or rolling down from some upper level; (c) the working conditions to be provided in accommodating the face equipment (excavators, stone-cutting machines, etc.) and the haulage units on the working benches; (d) the configuration of the faces worked by different types of excavators or stone-cutting machines, and also worked hydraulically or with the aid of blasting.

As in the case of bench faces, distinction should be made between the short-term stability applying to the working-side pit wa1l and the long-term stability of the non-working-side wall of the given pit. The attainment of a stable working-side wall is facilitated by the possibility of providing sufficiently wide working benches. As a rule, due to the need to provide space for the broken-down rock and mineral, as well as for the face equipment, such benches will be much wider than would be required solely for stability. Thus, the stability of a working-side wall in a pit can be ensured by increasing the working bench width as required. As to the long-term stability of pit walls, including the nonworking-side wall, it is economically unjustified to make the safety berms as wide as the working benches.

The long-term stability problem in most cases arises when the workings reach considerable depths, i.e. when the mining conditions become more unfavorable. All

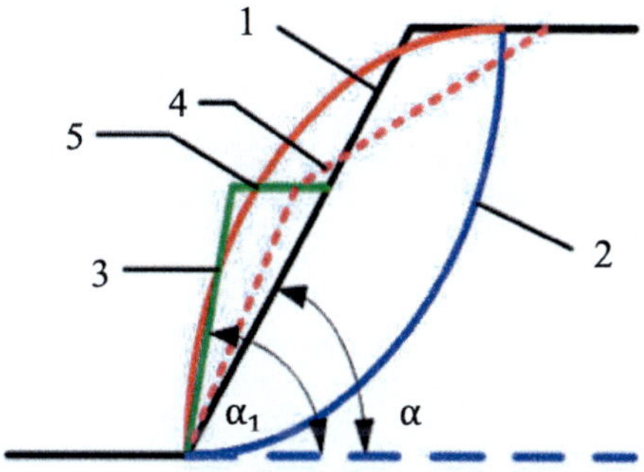

Fig. 4.12 Pit wall shapes: 1-straight; 2-concave; 3-convex; 4-broken-line; 5-stepped

this calls for the selection of lower wall slopes and also imposes certain limitations on the method used in breaking down the mineral and enclosing rock from the parent mass when approaching the boundaries of the pit: limiting the explosive charge in true blast holes, short-delay firing, leaving a cushioning layer at the boundary of the pit, and other measures that will be covered in more detail in Sect. 4.3 of the present chapter. Pit walls may be given different shapes; their slopes may be straight, concave, convex, broken-line and stepped, Fig. 4.12. Simplest of all is a straight slope which is easy to maintain in homogeneous ground and shallow workings. With increase in strength of the strata on deeper sinking of the workings, the wall slope will acquire a broken-line profile that will approach a convex curve in shape. Some specialists recommend leaving a convex wall profile in homogeneous strata because this will require the least amount of overburden removal, at the same time preserving the required stability of the pit wall. A decrease in strengths of the strata with increase in pit depth leads to selecting a concave shape of the slope, which involves a considerable increase in the volume of the overburden to be removed. A more economical method is to give the wall a stepped shape, since this will reduce the volume of overburden to be removed without disturbing the stability of the wall. For further reading: Duncan [26].

At the present stage of knowledge of the conditions governing pit wall stability, perhaps the most suitable approach to the problem of determining tire optimum profile of the pit wall as proposed by Yu. N. Malyushitsky is the following: (a) on the basis of the known conditions of stratification of the deposit and the physical and mechanical properties of the strata making up the wall (with due consideration for their change with increase in the depth of the workings), a profile of the wall is drawn so that its slope sections and surfaces are equally and fully stressed to the limiting profile of pit wall within value from top to bottom; (b) the profile determined as outlined above is now corrected by safety margin factors that serve to flatten the slope sections somewhat but do not change the general profile of the wall; (c) using the corrected wall profile as a base, the positions of the benches compatible with wall stability and the method of mining the pit are laid out (in accordance with the required size of the working benches and the equipment they accommodate, the amount of the original stripping of the overburden).

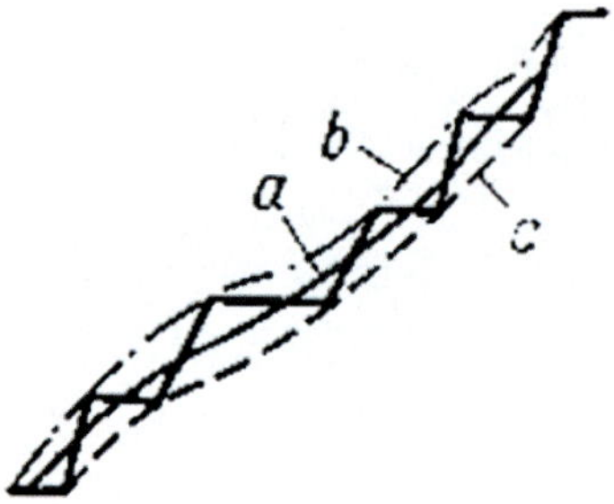

Fig. 4.13 Design profile of pit wall equal stress along slope

Where:

a. Final general slope of pit wall.
b. Initial design profile for equal stress along slope.
c. Resultant profile obtained after correction of design profile with safety factors.

Using the combined profiles, one can now see how the profile answering to the required mining conditions suits the condition imposed by the equal-stress slope profile found by calculation; then corresponding corrections are made so that the design profile of the wall will pass through the midpoint of each of the bench floors, Fig. 4.13. In setting up the profile for an equally stressed wall in a pit where sand, gravel and like deposits are worked, the slope of marginal equilibrium is taken to be a plane surface with a slope angle $\alpha = \varphi$. Since the value of φ varies with depth, the slope of the pit wall will change accordingly, becoming less with increase in pit depth. For consolidated ground, the profile of a pit wall can be considered to consist of three elements, Fig. 4.14.

The upper part, in the form of a vertical wall with a depth given by the Eq. 4.20:

$$H_{90} = \frac{2\,c\cos\emptyset}{\gamma\,(1 - \sin\emptyset)} \tag{4.20}$$

The middle part, in the form of a curve that is a function of the coefficient of internal friction, the force of cohesion, the volume weight, the depth of the working, and [the external loading]. The lower part, in the form of a straight line is drawn

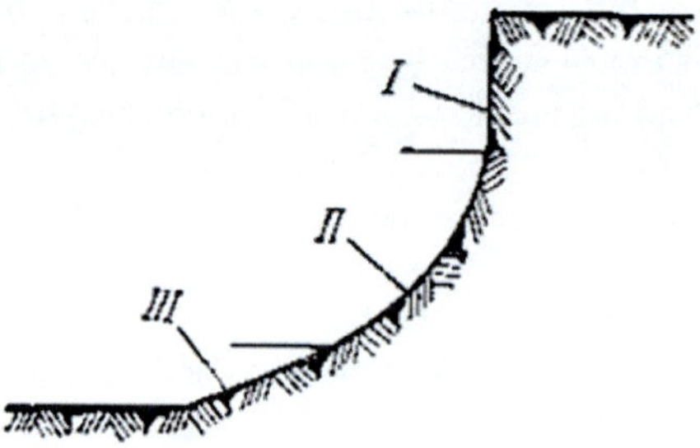

Fig. 4.14 Fully stressed pit wall in consolidated ground

at an angle φ to horizontal. Giving the walls of a pit the profile corresponding to the stable conditions for the strata involved, is an essential condition for ensuring safety of opencast mining operations. The selection of a safe wall profile is for this reason a mandatory part of the preparatory study, exploratory boring, and pit design work that must be done before beginning the exploitation of a deposit. Of paramount importance at this stage is the correct determination of the values of the cohesion force and the coefficients of internal friction, these varying will change in external loading and in depth of the workings. It is understood that [the problem of safety does not end with designing stable pit walls. Care must be taken to see that the design is correctly implemented and is also adjusted as the operations progress and fuller data on the characteristics of the ground and lay of the strata become available.

Although the selection of favorable profiles and contours is important in ensuring wall stability, it is only passive in character in that it adapts itself to the natural conditions existing in a given deposit. In creating safe conditions for operation, it is also necessary to take active measures. Among the very first to be mentioned is drainage. As a rule, drainage of the deposit will increase the strength of the material in the strata and make for greater stability. It should be started well in advance of the opening of the faces. Drainage of a deposit may include a surface water diversion system (using ditches, troughing, flumes, sumps and pump installations) and an underground system. The latter may take the form of a horizontal (gravity) or vertical drain system, a combination of both of these, or a well-point installation. If gravity drainage is employed, the water is diverted by means of a system of open ditches that intercept the ground water channels, or by a system of underground drainage sumps. Gravity drainage is usually employed where the water is freely discharged by a self-draining horizon. Vertical drainage comprises a system of underground workings and pumps or deep-well pumps that are lowered into wells of large diameter, sometimes to a great depth (over 100 in).

In individual cases advantage can be taken of the presence of an underlying bed of water-pervious material such as limestone, etc. Combined drainage uses a horizontal or gravity system in conjunction with means for vertical drainage. Drainage by means of well points is employed mainly in sands arid comprises a system of vertical pipes fitted with needle filters and driven into the sands and connected to vacuum pumps. In the series of measures taken to provide safe working conditions and prevent development of a dangerous emergency, the importance of adequate drainage cannot be overstressed.

Among the other active measures are cementations or freezing of the ground, erection of retaining walls, and so forth. However, these measures are less effective and sometimes do not justify the expenditures. Therefore, they are rarely used. These problems will be covered in more detail in Sect. 5 of the present chapter. Data on actual wall slope angles maintained in several opencast mines are given in Table 4.5.

Table 4.5 Actual wall slope angles in several opencast mines

Opencast pit	Depth of pit, m	Wall slope angle, deg	Remarks
Korkinsk coal pit (USSR)	200	16–18	Lying-wall side
Ilogoslovsk coal pit (USSR)	100	6–12	Lying-wall side
Baidakov coal pit (USSR)	50	25	–
Goroblagodatsk iron ore pit (USSR)	150	48–52.5	–
Vysokogorsk iron one pit (USSR)	152	51	–
Bakal iron ore pit (USSR)	154	43	Lying-wall side
Magnitogorsk iroim ore pit (USSR)	–	32–37	
Kounrad copper mining pit (USSR)	125	31–60	
Sheleinsk nickel mining pit (USSR)	38	17–25	
Morenci copper mining pit (USA)	245	45	–
Utah Copper pit (USA)	490	26	–
Mesabi Mountain iron mine (USA)	138	45–63	–
Menlo iron ore pit (USA)	77	45	–

4.4 Stability of Waste Banks

Waste bank stability in opencast mining conditions depends on the stability of the material forming the slope of the bank and also on the stability of the ground on which the waste is dumped. Distinction must therefore be made between wastes dumped on hard or soft ground. As in the case of a slope of loose ground in a pit face, the particles of material in a waste bank are under the force of gravity tending to move the particles downward and the force of friction resisting this movement. As a result of the interaction of these forces, the material in a waste bank will at first form a slope whose angle will be somewhat greater than the angle of internal friction for the existing conditions. As the height of a waste bank increases, the material in the bottom layers will experience increasing pressure. However, the height of a waste bank of dry material is not actually limited by the strength of the material of which the bank is composed, but is dependent upon two factors: the slope angle and the coefficient of internal friction.

With time, the material in a waste heap becomes more packed at the expense of the voids, and the mass gradually passes over into a two-phase condition of ground and water. Army further packing of the material leads to squeeze-out of the water and sets up a hydrodynamic pressure head. Since the material in time waste bank slope is maintained in position by forces of friction that only exceed the disturbing force by 3–4%, the hydrodynamic pressure head may become sufficient to destroy the condition of equilibrium of the material. This may then result in a slide or fall of the bank.

Thus, to the harmful action of water upon ground stability, resulting from increased weight of the ground, decrease in its resistance to shearing forces, and formation of surfaces of weakness, in the case of waste banks, must be added the harmful effect of hydrodynamic pressure. A typical example of the loss in stability, that may occur in waste banks are the brown clay waste banks in the Urals coal pits. When heaped dry, these banks remain stable up to heights of 10–20 in and at slope angles of 35–37°. These same clays, on acquiring a volumetric water content ratio of 0.45–0.55, will slide down the slope when the height of the heap exceeds 2 m. They can therefore produce dangerous slides.

Another danger of slides and falls of waste banks arises when the height of a bank exceeds the limit considered permissible for the given dump conditions. In practice, many factors are taken into account in establishing the permissible height of excavator cast waste banks. In the Vakhrushev coal trust pits (USSR) they have been limited as follows: for the Veselovsk pit to 12–13 m, for the Northern pit to 18–20 m; for the Southern pit to 18–20 m, and for the Lapchinsk pit to 18–20 m.

A study of the practice of direct waste casting into a worked-out area in the Volchansk and Bogoslovsk lignite deposits has shown that [the maximum height of waste banks under the existing conditions was 35 m, even when the floor on which the; Waste was cast was horizontal, while when the floor of the stripped coal had an inclination towards the seam being worked, the conditions become more unfavorable and the possibility of a slide increases. Under conditions where the overburden is cast into a worked-out space, the limiting height, Fig. 4.15, can be approximated from time equation, Eq. 4.21:

$$H_0 = h - h \cot \beta \tan \alpha \tag{4.21}$$

The angles are being those shown in Fig. 4.15. When the floor is horizontal, $\alpha = 0$ and, accordingly, $H_0 = h$, while for $\alpha = \beta$, the entire bank will slide toward the face because, Eq. 4.22:

$$h_0 = h\,(1 - \cot \beta \tan \alpha) = h\,(1 - 1) = 0 \tag{4.22}$$

Where the floor has an inclination in a direction away from the face, the conditions become much more favorable for casting the waste because in this case an additional force is created and helps in preventing a slide of the material. For this condition,

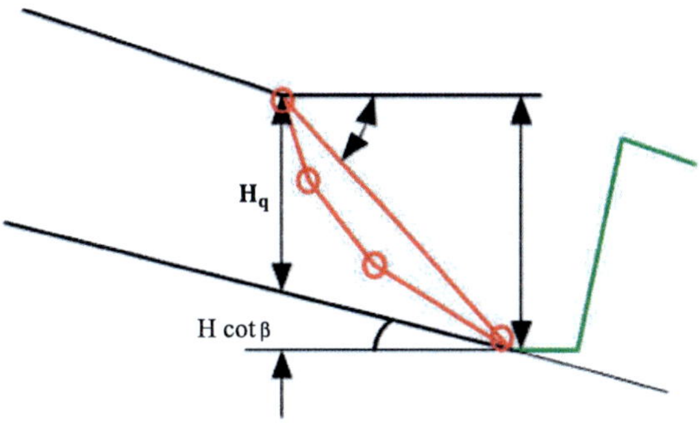

Fig. 4.15 Height of waste bank cast directly in stripping pit

Table 4.6 Height of waste banks at different conditions of waste disposal

Means used for waste bank formation	Kind of waste material		
	Rock	Hard and stable	Loose clayey
Terracing ploughs	30	15	10
Power-shovels	40	30	15
Chain-and-bucket overburden excavators	–	50	–
Overburden conveyor bridges and overburden stackers overburden stripping excavators (power shovels, draglines)	Within limits of stacking capacity of equipment		

Eq. 4.23:

$$H_0 = h(1 + \cot\beta \tan\alpha); \; H_0 > h \tag{4.23}$$

The examples given above show that if a waste bank is to be safe from slides it must have a height established in accordance with the conditions of waste disposal existing in the given pit. The regulations on mine operation set forth a requirement that the height of the waste banks must be fixed by the design project for the workings and be selected in accordance with the physical and mechanical propel-ties of the waste material, the method of dumping or casting, and time topography of the waste disposal area.

Table 4.6 contains data on the height of waste banks for different conditions of waste disposal. These have been obtained by generalizing the practice of a number of opencast pits. For further reading: Salmi et al. [13].

Substantial influence on the stability of a waste bank is exerted by the sequence in which the material is deposited on the bank. If materials with different characteristics are dumped on the same bank, it will not settle and consolidate uniformly. The most unfavorable condition is created when soft waste, especially clayey material, is deposited at the bottom of the bank. As a result of the non-uniform settling of the hank, and the presence of layers of clayey material in contact with the surfaces, a possibility of bank slides arises. In all cases where the operating conditions are favorable and suitable equipment is available, the bottom layers of the waste banks should be built up of the hardest and most stable waste that will facilitate drainage of water from the bank. Finally, waste bank slides and falls may be due to a soft floor on which time bank has been deposited. When the weight of the bank dumped on such a floor becomes too high, it will begin to heave. This is what occurred, for example, in the internal waste banks in the Bogoslovsk pit. A matter deserving due consideration is waste bank drainage. Wherever it is necessary to drain the area on which the banks are to be deposited, as well as the banks themselves, the following means should be used:

A system of drainage ditches to divert all surface waters from the waste disposal area.

Drainage ditches to lower [the ground water level and discharge all the water that percolates through the waste banks from the surface.
A system of wells is lowering the ground water level by means of deep-well pumps.

Waste disposal practice in the mines has shown that wherever-it is possible to keep water from wetting the deposited waste the stability of the banks remains firm. This is why it is imperative to incorporate a system of drainage directly into the banks proper. With this aim in mind, the bottom layers of the banks should consist of sands containing the least amount of water, or of crushed rock or ham-d rock waste. Wherever a water-permeable material is not available, perforated pipes should be laid out on the floor of the waste bank and be covered, first with brushwood, and then with a layer of hard-rock waste. This will allow time clayey and water-bearing waste to drain the water off through the pipes away from the banks.

Waste dump surfaces should be given a shape that is favorable for good drainage of surface waters. It is not permissible to leave isolated low areas or hollows, or a bumpy sum-face that will allow water to gather and thus lead to the danger of bank slides, and falls. Thus, the basic measures in ensuring adequate waste bank stability are: drainage of the deposit and the waste banks, limiting of the bank height to a safe value, and graded dumping of the waste materials in accordance with their physical and mechanical properties. Timely and comprehensive use of the above means for slide prevention will permit the waste banks to retain their stability and make for safe working conditions in the pits.

References

1. Álvarez-Fernández MI, González-Nicieza C, Menéndez-Díaz A, Álvarez-Vigil AE (2005) Generalization of the n-k influence function to predict mining subsidence. Eng Geol 80(1–2):1–36
2. Call RD, Savely JP (1990) Open pit rock mechanics. In: Kennedy BA (ed) Surface mining, 2nd edn. Society for Mining, Metallurgy and Exploration, Inc., pp 860–882
3. Lie G, Yongli G, Junjie L (2001) A preliminary discussion on the slope sliding mass in the bottom wall in Nanfen Iron Surface Mine and its control measures. Min Res Div 21(6):4–6.
4. Wyllie D, Mah C (2007) Rock slope engineering, 4th edn. Spoon Press, London and New York
5. Karam KS (2005) Landslide risk assessment and uncertainties. Massachusetts Institute of Technology. Ph.D. Thesis, 768 pp
6. Zharikov SN (2019) Pit wall stability and drilling-and-blasting. IOP Conf Ser Earth Environ Sci 262:012084
7. Xia K, Chen C, Fu H, Pan Y, Deng Y (2016) Mining-induced ground deformation in tectonic stress metal mines: a case study. Eng Geol 210(3):212–230
8. He MC, Sousa LR, Jili F, Zhigang T (2014) Monitoring of sliding forces for opens pits in China. In: Rock mechanics for natural resources and infrastructure, SBMR 2014 specialized conference 09–13 Sept, Goiania, Brazil
9. Hoek E, Bray JW, Boyd JM (1973) The stability of a rock slope containing a wedge resting on two intersecting discontinuities. Q J Eng Geol 6(1):1–55

10. Marschalko M, Yilmaz I, Bednárik M, Kubečka K (2012) Influence of underground mining activities on the slope deformation genesis: Doubrava Vrchovec, Doubrava Ujala and Staric case studies from Czech Republic. Eng Geol 147–148(15):37–51
11. Jaeger JC (1971) Friction of rocks and the stability of rock slopes. Geotechnique, London 21(2):97–134
12. John KW (1968) Graphical stability analysis of slopes in jointed rock. J Soil Mech Found Div, ASCE 94(2):497–526
13. Salmi EF, Nazem M, Karakus M (2017) Numerical analysis of a large landslide induced by coal mining subsidence. Eng Geol 217:141–152
14. Hendron AJ, Cording EJ, Aiyer AK (1971) Analytical and graphical methods for the analysis of slopes in rock masses. U.S. Army Engineering Nuclear Cratering Group Tech. Rep. No. 36. U.S. Army Engineers Waterways Experiment Station, Vicksburg, Miss
15. DiBiagio E, Kjekstad O (2007) Early warning, instrumentation and monitoring landslides. 2nd Regional Training Course, RECLAIM II, 29 Jan–3 Feb 2007
16. Einstein HH (1997) Landslide risk: systematic approaches to assessment and management. In: Cruden A, Fell R (eds) Landslide risk assessment. Balkema, Rotterdam, pp 25–50
17. Einstein HH, Sousa RL (2007) Warning systems for natural threats. Georisk 1(1):3–20
18. He MC, Tao ZG, Zhang B (2009) Application of remote monitoring technology in landslides in the Luoshan mining area. Min Sci Technol 19(5):609–614
19. Sousa F, Akciz SO, Harvey JC et al (2014) Field and LiDAR observations of the hector mine California 1999 surface rupture, AGU Fall Meeting, San Francisco, CA, USA
20. Xu N, Kulatilake PHSW, Tian H, Wu X, Nan Y, Wei T (2013) Surface subsidence prediction for the WUTONG mine using a 3-D finite difference method. Comput Geotech 48(3):134–145
21. Zahiri H, Palamara DR, Flentje P, Brassington GM, Baafi E (2006) A GIS-based weights-of-evidence model for mapping cliff instabilities associated with mine subsidence. Environ Geol 51(3):377–386
22. Fredj M et al (2018) Study of slope stability (Open Pit Mining, Algeria). In: Frikha W, Varaksin S, Viana da Fonseca A (eds) Soil testing, soil stability and ground improvement. GeoMEast 2017. Sustainable civil infrastructures. Springer, Cham. https://doi.org/10.1007/978-3-319-61902-6_1
23. Geotechnical Considerations in Open Pit Mines: guidelines. Department of Minerals and Energy, Western Australia (1999). ISBN 0730978079
24. Hutchinson J (2008) Rock slide hazards: detection, assessment and warning. Presentation at the 33rd International Geological Congress, Oslo, Norway
25. Villegas T, Nordlund E, Dahnér-Lindqvist C (2011) Hangingwall surface subsidence at the Kiirunavaara mine, Sweden. Eng Geol 121(1–2):18–27
26. Duncan JM (2000) Factors of safety and reliability in geotechnical engineering. J Geotech Geoenviron Eng 126:307–316

Chapter 5
Surface Mine Development

5.1 Order of Development of Opencast Mining Work

The order of development of opencast mining work cannot be established arbitrarily. It depends, first of all, on the kind of deposit to be mined, surface relief, shape of the deposit, position of the deposit relative to the prevailing surface level, angle of dip, capacity, structure, quality distribution of minerals, and kind of overburden rock. The next logical consequence is the choice of the method of opencast mining: surface, deep, on-slope, on-slope-deep or underwater. For further reading: Dutta et al. [1].

Another stage of reasoning is the principal (preliminary) decision on the open-pit field: its probable depth, dimensions at the floor and surface level, slope angles of flanks, and the total reserve of the mined rock and of the mineral in particular. It is also essential to establish the points of location of consumers of the mined mineral, waste dumps, tailing dumps and their approximate capacities, which will make it possible to outline the probable directions and ways of transportation of pit loads. On the basis of this reasoning the probable dimensions of the open-pit field, its position in conformity with the surface relief, and the approximate contours of the mining allotment can be selected. Only after that and considering the planned capacity of the quarry, one can start working out the order of development of mining work within the limits of a given open-pit field.

Schemes of development of mining work and quarry benches in profile and plan are shown in Fig. 5.1. Arrows indicate the directions of advance of the mining work for deposits of various shapes on flat-land surface. In order to put a quarry in exploitation as soon as possible and minimize the capital costs, mining work should be started in a portion of the field where the mineral deposit is nearest to the surface and the scope of mining-construction work will be at a minimum; this decision should also take into account the probable solutions on stripping the mining levels for the future periods and the choice of a mining system that will allow a high degree of integrated mechanization of mining work.

The principal purpose of opencast mining-extraction of minerals from the earth with simultaneous removal of a large volume of capping and enclosing overburden

M. M. Ali Elbeblawi et al., *Surface Mining Technology*, Topics in Mining, Metallurgy and Materials Engineering, https://doi.org/10.1007/978-981-16-3568-7_5

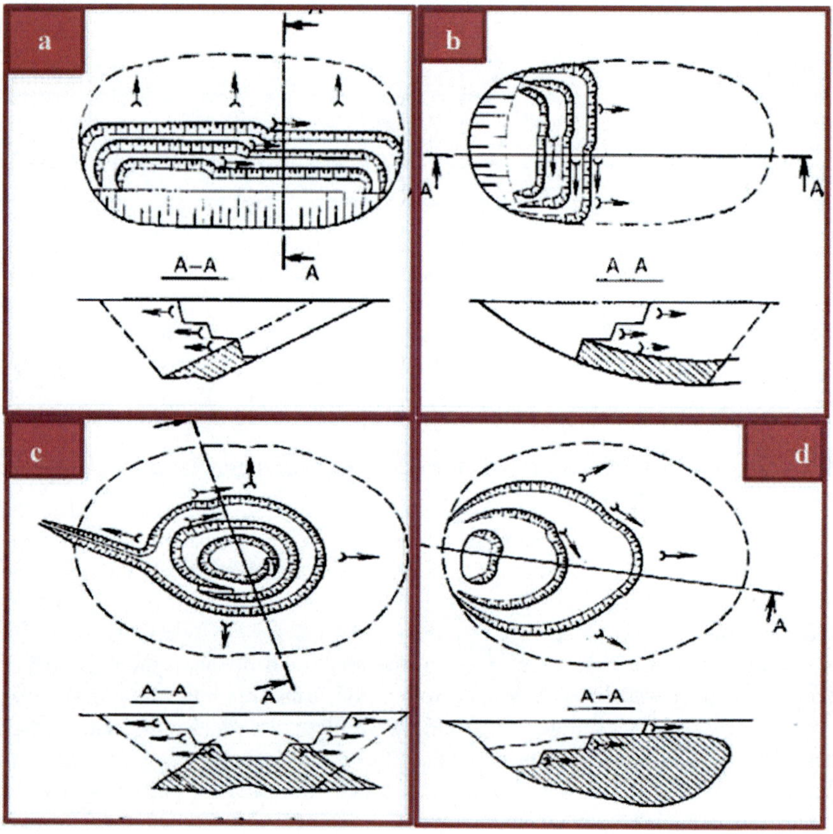

Fig. 5.1 Mining work development **a** mining oriented along the longer axis of quarry **b** along the shorter axis **c** concentric mining line **d** elliptical mining line

rock-can only be achieved with a clear-cut and highly efficient organization of the leading and most expensive process of opencast mining-transfer of the mined rock from quarry faces to reception points at storages and waste dumps. Ensuring stable operation of the freight traffic lines for mineral and overburden attains the effective transfer of pit loads. Various problems of stripping the mining levels of the open-pit field and selecting the capacities of transport means to be employed are in turn solved in connection with the effective operation of the traffic lines. For further reading: Jeong and Phillips [2], Moris [3], Rzhevsky [4].

5.2 The Concepts of Regimes and Stages of Mining Work

The technical solutions and economic results of opencast mining are determined primarily by the ratio of the volumes of stripping and mining work as a whole and in various periods of the quarry activity. Using what is called the overburden ratio does the quantitative estimation of these relationships. The average overburden ratio K_{av} (m^3/m^3) is the ratio of the volume of overburden rock V_{ob} within the limits of a quarry to the total assured reserve volume of the mineral V_m within these limits, Eq. 5.1:

$$K_{av} = \frac{V_{ob}}{V_m} \tag{5.1}$$

The average operational overburden ratio $K_{av.op}$ (m^3/m^3) is the average overburden ratio during the period of mining exploitation work in a quarry. It is defined at the total volume of overburden to be mined, V_{ob}, in a quarry, minus the volume of overburden V_{ov} that has been removed during the quarry construction, related to the total reserve volume of the mineral V_m minus that portion, V_{mc} which has been mined out during quarry construction, Eq. 5.2:

$$K_{av.op} = \left(\frac{V_{ob} - V_{ov}}{V_m - V_{mc}}\right) \tag{5.2}$$

The current overburden ratio K_{cur} (m^3/m^3) is the ratio of the volume of the overburden rock ($V_{ob\ cur}$) that has been actually transferred from the rock massif to waste dumps in a particular period (month, quarter or year) to the volume of mined-out mineral $V_{m\ cur}$ during that period, Eq. 5.3:

$$K_{cur} = \frac{V_{ob\ cur}}{V_{m\ cur}} \tag{5.3}$$

The boundary overburden ratio K_b determines the volume of overburden rock to be stripped per unit volume of the mineral that is permitted to be transferred from the rock massif to waste dumps by the condition of profitable opencast mining. The planned overburden ratio K_p is used for planning the current production cost of the mineral, C_{cur} (roubles/t); it characterizes the volume of stripping work whose expenses will be paid back in the process of current performance of opencast mining work, Eq. 5.4:

$$C_{cur} = C_{m\ cur} + K_p C_{ov\ cur} \tag{5.4}$$

where:

C_{cur} and $C_{ov\ cur}$ are current expenses on mining of 1 m^3 of the mineral and 1 m^3 of the overburden.

In many quarries, overburden ratios are measured as ratios of the volume or mass of stripped overburden to ton of the mineral. The ratio of the current volumes of stripping and mining work determines, in the first place, the production utilization (capacity) of a quarry in terms of the mined rock which is not constant, first of all, because of variations of the annual volumes of stripping work in individual periods. These variations are the result of variable capacity of the overburden and deposit of the mineral, varying conditions of mineral bedding, various geological disturbances, and uneven concentration of mineral components in the deposit. These variations can also be caused by some economic factors. On the other hand, the plants for processing the mined minerals are designed for a particular production capacity and must receive the mineral in strictly definite volumes and of specified quality. These circumstances are fundamental for selection of the regimes of the mining work in quarries. For further reading: Lesin et al. [5], Siskind [6].

The regime of mining work is understood as the time sequence of accomplishment of stripping and mining work which ensures planned, safe and economic development of a deposit during the quarry age. This sequence is specified in the quarry design or established on the basis of investigation. The regime of mining work is expressed as graphs showing year variations in the volumes of mining and stripping work during the quarry age, Fig. 5.2.

For relatively short periods (up to 5 years), the regime of mining work for existing quarries is established during planning of mining work for that period (five years). The regime of mining work is considered economically effective if it ensures the highest profit from development of the deposit and production of the mineral of the specified quality. With the period of operation of a quarry up to 8–12 years (which corresponds to the time of depreciation of the main quarry equipment), economical effectiveness of the mining work is achieved if the annual volumes of stripping work are constant for as long a period as possible, Fig. 5.2a. With a longer period of operation of a quarry, it is advisable in the general case to divide the whole time of exploitation into individual periods each of which is characterized by a constant annual volume

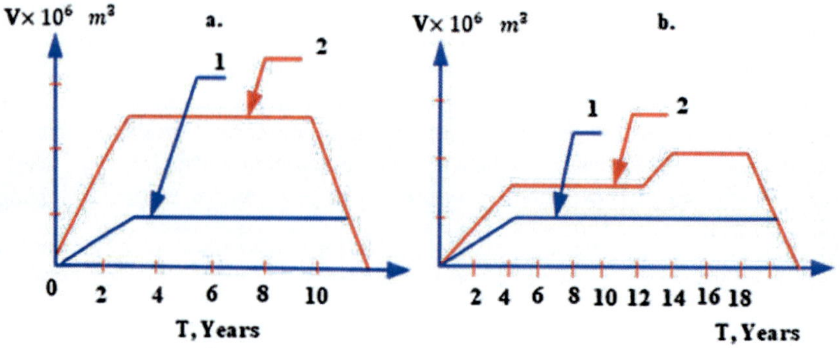

Fig. 5.2 Graphs of variation of volume V of mining (1) and stripping (2) during a period of T years: **a** with the quarry age of 10 years; **b** 20 years

of stripping work; in each next period~ these volumes are increased or decreased as required, Fig. 5.2. The periods of exploitation of a quarry with substantially different volumes of stripping work are called development stages. With a short age of a quarry, the exploitation period is not divided into stages. With a longer age, several stages of development are preferable. For Further reading: Agrawal et al. [7], Beyglou [8], Tzu-Hsien [9].

In the former case, it is recommended that the mining work be carried out with a constant current overburden ratio which should be close to the average operational overburden ratio. In the latter-case, the graph of mining regime slopes down stepwise; Fig. 5.2. The time of each stage should be matched with the depreciation periods of the main equipment. A change from one stage to another is usually timed to the moment when a quarry requires reconstruction with replacement of the obsolete mining and hauling equipment. A non-uniform regime of mining work during a stage can result in 'peak' volumes of stripping work being performed in certain years. This can worsen the indices of economic efficiency of mining, since a large number of units of mining and hauling equipment and of power-generating units must be concentrated in a quarry, more manpower engaged in it and additional shops and amenity rooms must be built. The adverse effects of non-uniform regimes are especially sensible in quarries with a relatively short age and those exploited in less developed regions of the country.

Proper selection of the mining regime for a quarry is of prime importance for increasing the plant economy, and quicker turnover of the working capital. It is also favorable in minimizing prepaid and unproductive expenses during the period of quarry operation when the overburden ratio and the cost of mined mineral are changed owing to changes in natural conditions. According to the calendar ~ time stages, there are definite volume stages of the development of a quarry, i.e. certain intermediate contours of the quarry along the depth and in plan, Fig. 5.3. The problem of establishing a rational regime of mining work consists essentially in finding such stage contours and, within them, annual contours (positions of mining work) on each bench.

5.3 The Theory of Stripping of Mining Levels

5.3.1 The Order of Formation of Freight Traffic

The diversity of forms of deposits and conditions of their bedding, on the one hand, and the main principle of opencast mining, i.e., layer-wise (bench-wise) extraction of the overburden and mineral, on the other, predetermine the necessity for organizing such freight traffic in quarries that will minimize the cost of hauling the mined rock from faces to waste dumps and storages and thus ensure the highest economy of opencast mining. The problem is resolved by the formation of freight traffic in a quarry as the basis for stripping the mining levels in it.

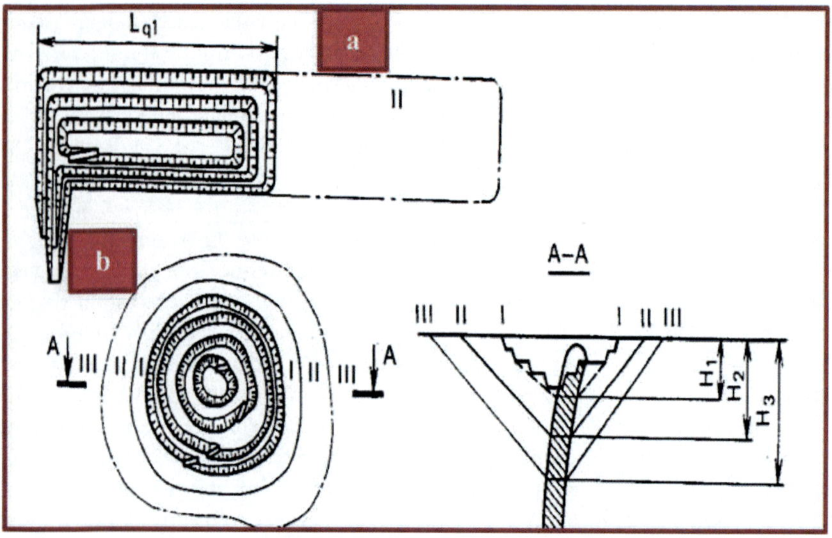

Fig. 5.3 Schemes of stage contours of quarry development: **a** Elongated quarry **b** Rounded quarry

The order of formation of freight traffic will be discussed below by considering as an example an elongated open-pit field with two deposits I and II of different quality of the mineral, Fig. 5.4a, and relative elevations of mining levels ranging from—1-20 m to − 60 m. Mining work is started from stage I which is closer to deposit I, on height elevations − 10 m and ± 0 at which mined rock is extracted in the quarry faces and where freight traffic flows begin. Mining is intended to be done on both flanks (right and left); in each of the flanks and in each slice being worked, the volumes and quality of rock are different and vary both in various stages (1–6) of mining work and during the entire period of exploitation. For further reading: Gartner [10], Phlevani et al. [11], Vasil'ev and Kolesnikov [12], Valery et al. [13].

With the order of development of mining work shown in Fig. 5.4, the volumes of the overburden and mineral are first estimated (calculated) by their grades in each mining level and separately for the right and left flank and are plotted as a stage schedule of regimes of mining work, Fig. 5.4b. When plotting the stage schedule of mining work regimes, it is essential to envisage the earliest possible data of starting the work. It is also reasonable to postpone the transfer of the main mass of overburden to later periods of extraction. The stage schedule offers the possibility for estimating the economic efficiency of the adopted version of development of mining work by comparing it with other probable versions. If the given order of mining development is adopted as the basis, analysis and formation of freight traffic can be commenced. For this, a summary table is compiled, Fig. 5.4c, which shows the volumes of various loads supplied from each working level at each stage of the development (1–6) and for each flank of the quarry. For further reading: Giri and Yun [14], Patti and Watson [15], Rappold and Yoho [16], Samanta et al. [17].

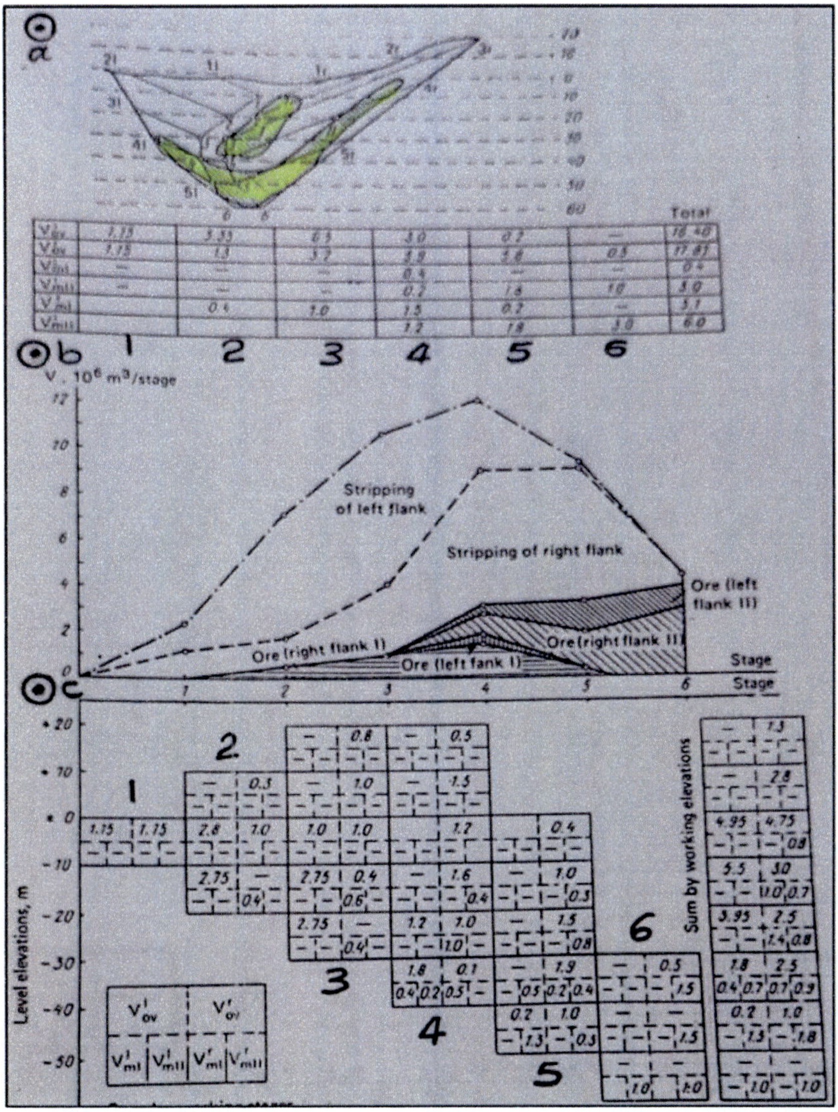

Fig. 5.4 Stage freight traffic a Graphs of mining work regimes b Stage-wise c Distribution of freight traffic flows

The data summarized in the table can be used as the basis for the formation of freight traffic. In order to estimate the time schedule of the mining work for the adopted capacity of a quarry (in terms of the mineral), the stage schedules and tables must be transformed into calendar ones, Fig. 5.5a and b, in which the abscissae are

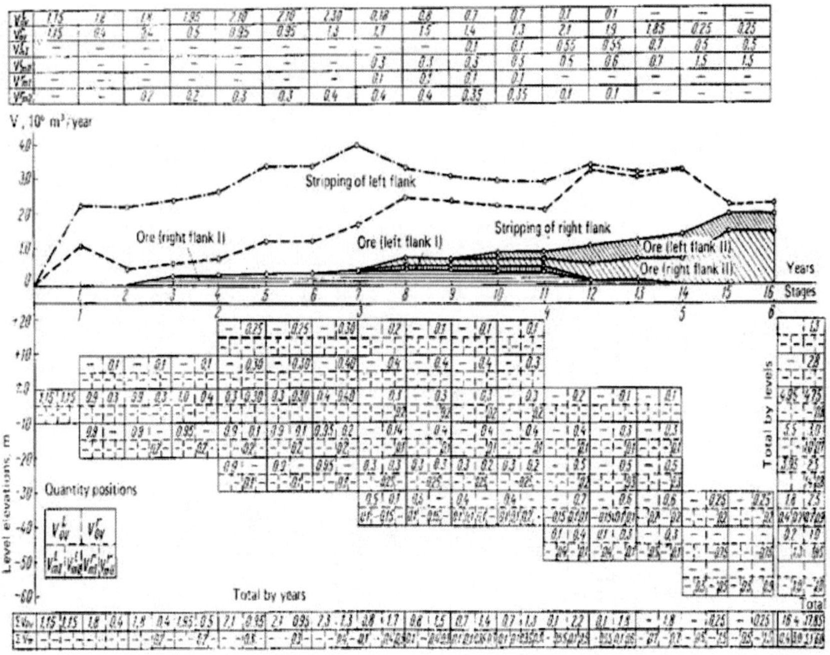

Fig. 5.5 Annual calendar schedule of freight traffic on levels (a) and diagram of its distribution among development stages

years of the quarry age. The order of transformation of schedules will be shown below. For further reading: Chen et al. [18].

The above-given example of constructing schedules demonstrates how the required volumes of extracted and transferred quarry loads can be determined for the stages of mining work and the years of quarry age in accordance with the plans of production development. Using the method of variants, the stage and calendar schedules can be perfected in order to optimize the economic results of opencast mining of a given deposit. On the other hand, even rough calculations made by this method make it possible to substantiate the formation of freight traffic in a quarry at all stages of mining work, and therefore, to prove the economic efficiency of the adopted method of stripping.

Schedules of formation of freight traffic for all kinds of deposits must be constructed by considering the surface relief. If needed, the volumes of stripping work should be divided by their kinds and the mineral should be divided into quality grades so as to find the best solution as regards the mining and hauling complexes and the time of functioning of each freight traffic flow. In that case, the general elevation of the surface and the ascending and descending portions of the open-pit fields are fixed on the schedules.

5.3.2 *Kinds of Freight Traffic*

In the general case, each layer (slice) extracted in a quarry can be represented by:

- Overburden rock (compact or weathered igneous and metamorphic rock, dense or soft rock.
- Off-quality and off-balance minerals which are stored in separate dumps to be utilized in later periods.
- Minerals which are divided into types and grades for separate transportation according to planned production tasks.

The flow of loads of a particular quality, characterized by a relatively constant (in time) direction and a definite volume per unit time (shift or day) is called an elementary freight traffic flow. If the rock in a face is homogeneous (a simple face), a single elementary freight traffic flow will start from that face. In complex faces (with different rocks and separate extraction), two or three elementary flows may commence. Thus, the number of elementary freight traffic flows on a bench depends on the number of faces and the method of rock extraction in them and is usually greater than the number of operating faces. For further reading: Li et al. [19], Manyele [20], Schultz [21].

Elementary freight traffic flows may differ from one another by their directions, Fig. 5.6a and b, the kind of transport, Fig. 5.6b, and transport lines, Fig. 5.6c, or by the models of quarry transport of the same kind. For instance, elementary rock and

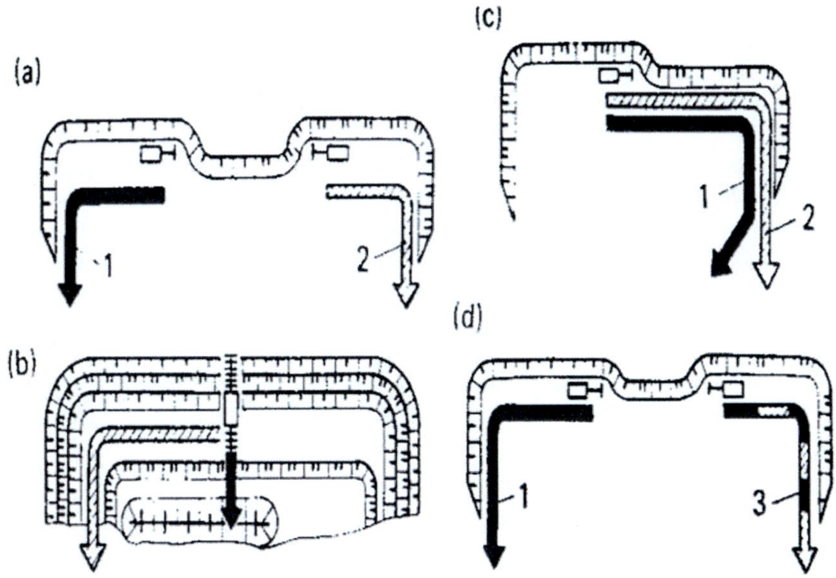

Fig. 5.6 Scheme of elementary freight traffic flows: (1) overburden (barren rock) (2) mineral (3) traffic flow with alteration of barren rock and mineral

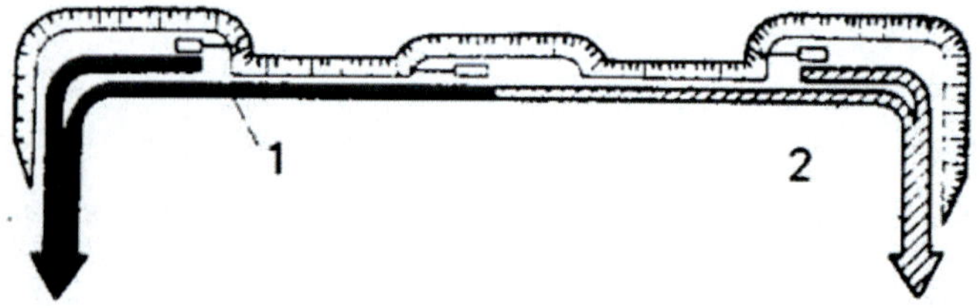

Fig. 5.7 Freight traffic scheme flows from a bench (1) overburden (2) mineral

mineral traffic flows from a complex ore face, with the use of automobile transport and a single road for both, often differ only in that the rock and mineral are hauled by different dump trucks of the same type and capacity, Fig. 5.6d. If conveyer transport is employed under the same conditions, separate conveyers are needed, i.e., the elementary freight traffic flows differ by transport lines and transport means, Fig. 5.6.

In quarries where the rock is homogeneous in several faces, the number of elementary flows is usually reduced by combining them into a single freight traffic flow from a bench (Fig. 5.7). By the same principle, freight traffic flows from several benches can be combined into a homogeneous flow from a group of benches or from all benches of a quarry, Fig. 5.8a and d. For further reading: Kolesnikov [22], Lesin et al. [5]. A group of connected elementary flows having common transport lines forms a converging freight traffic flow, Figs. 5.7 and 5.8a. A common freight traffic flow of a quarry or a quarry section which is then divided into individual flows is called the diverging freight traffic flow, Fig. 5.8b. Diverging flows are mostly typical for the transport of overburden rock and mineral, less frequently for different kinds of rock and even less frequently for homogeneous rock.

A common freight traffic flow which is formed by initially converging elementary flows and which then (more often on the surface) becomes diverging is called a complex freight traffic flow (Fig. 5.8c). The reloading or sorting points are provided on the route of loads, the freight traffic flow is called combined. Complex and combined freight traffic flows are predominant in the practice of opencast mining. If freight traffic flows consist of different kinds of rock, they are called heterogeneous freight traffic flows. A common freight traffic flow is called concentrated if its component freight traffic flows are hauled from the quarry on the same transport lines, Fig. 5.8a; if they are moved on different transport lines, the freight traffic flow is called dis-concentrated, Fig. 5.8d. Reduction of the number of freight traffic flows in a quarry makes it possible to utilize mining equipment more efficiently, improve the quality of roads, and reduce the number of stripping workings and the expenses on their construction.

A number of freight traffic flows in a quarry may be:

- Independent of one another if the operation of the complex of machines serving a given flow (from beginning to end) does not depend on the operation of the equipment that serves the other freight flows, i.e., the equipment is allotted strictly to a particular flow.

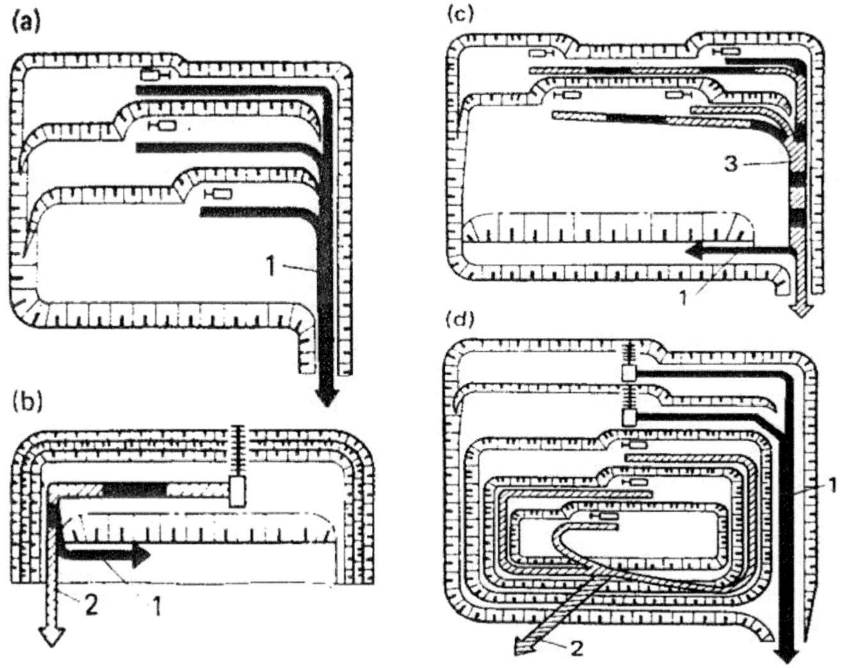

Fig. 5.8 Schemes of freight traffic flows from a quarry (1) Overburden (2) mineral (3) overburden and mineral alternately

- Dependent on one another if the serving equipment, in particular, the hauling means, must be rearranged periodically among the adjacent freight traffic flows for better utilization; rearrangement (redistribution) of transport means is carried out by traffic control service.
- Heavily dependent if the traffic control service continuously (according to a schedule) changes the utilization of the equipment, redistributes equipment units and controls the volumes of elementary freight traffic flows (for instance, for blending the extracted mineral before sending it to a concentrating plant).

Dependent freight traffic flows are the most typical type. Freight traffic flows are organizational links which combine all processes of mining: preparation of rock for extraction, mining proper, loading, hauling, dumping, and storage. Proper functioning of freight traffic is essential for economical mining and effective utilization of the equipment.

5.3.3 Prerequisites for the Formation of Freight Traffic

Transportation of quarry loads is defined by the planned volumes of stripping and mining work. The volume of a load (in tons or cubic meters) transferred per unit time (hour, shift, day, year) constitutes the load turnover of a quarry. The largest fraction of the quarry turnover usually falls on the transport of overburden rock to waste dumps. In many cases, especially when several kinds of rock are transported separately, it may be reasonable, from the engineering and economical standpoints, to organize several freight traffic flows in a quarry; this can simplify division of loads destined to different reception points and reduce the total length of transportation. It is a trend to divide, first of all, the freight traffic flows of overburden rock and mineral, especially if these flows are conveyed by different transport means, Fig. 5.9a.

The transport of overburden rock is divided into individual freight traffic flows in the following cases:

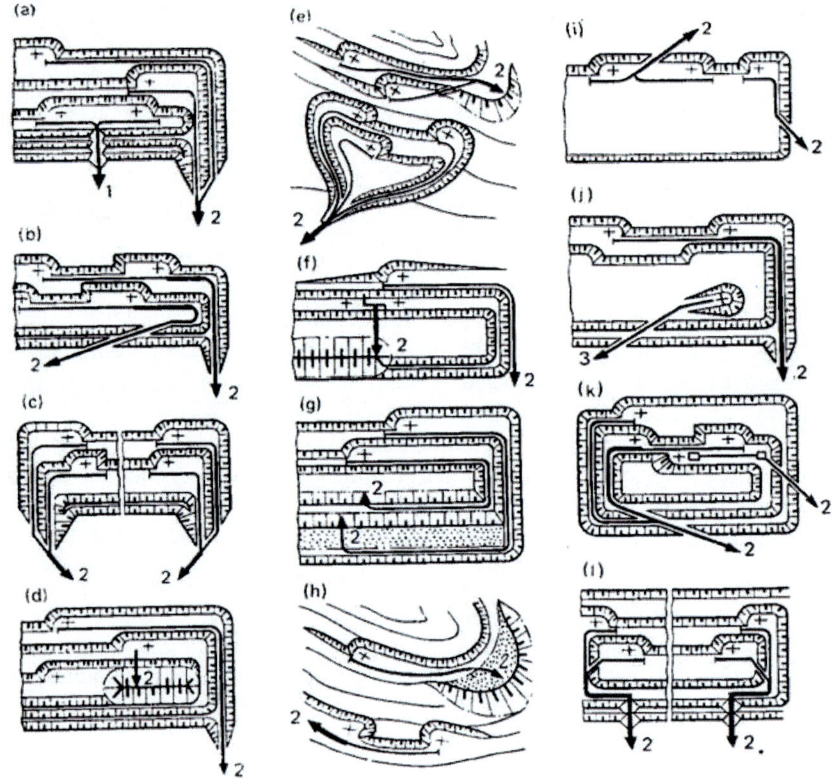

Fig. 5.9 Schemes of dis-concentrated freight traffic burden

- If the volume of stripping work is very large, two or three freight traffic flows of overburden rock are organized and the stripping workings are grouped accordingly. Under common conditions, a single-track railway can pass through 20–30 × 103 m^3 of mined rock per day, a double-track railway, up to 50–60 × 103 m^3, and a single-line automobile road up to 40–50 × 103t per day. If the freight traffic exceeds these figures it must be divided into two or three flows, Fig. 5.9b.
- If the open-pit field is very vast, Fig. 5.9c and l, it is expedient to form two or three transport exits from a group of stripping benches. The formation of single common freight traffic of overburden rock under such conditions may be associated with a reduction of the possible number of excavators arranged on a mining level, increased total run of transport means, and increased length of haulage lines in the quarry and on the surface.
- If the overburden rock must be hauled from upper levels to external or internal waste dumps and part of the rock must be reloaded to internal waste dumps in order to reduce the transport expenses, Fig. 5.9d and f.
- If dis-concentrated waste dumps are employed and their capacity and receiving ability are insufficient and also in the cases when it is essential to shorten the transport distances in on-slope and on-slope-deep quarries, Fig. 5.9e and h.
- If overburden rock is transferred to internal waste dumps on mining levels and stored in individual stages, Fig. 5.9g.
- If mobile transport means (dump trucks, scrapers) are employed, which move the rock through a system of temporary trenches to waste dumps nearby, Fig. 5.9i. The construction of temporary trenches and access tracks is especially effective during the period of quarry construction, since it favors intensification of mining-construction work and improves its economic indices.
- Freight traffic flows from individual benches and groups of benches are, as a rule, separated when various kinds of transport are employed for rock hauling, Fig. 5.9j and h.

Freight traffic flows of the mineral are divided mostly in the cases when it must be extracted and hauled separately by types and quality grades. For instance, if the mineral flow is directed to several crushing-grading and concentrating mills which accept different types and grades of the mineral. Individual freight traffic flows divide a quarry into process zones each of which includes the portion of the working zone that is serviced by a particular flow and a non-working portion of the quarry where the transport lines of this freight traffic flow are arranged. Each of such zones has its own complex of loading and hauling equipment. Usually the range of variations of the quality and properties of mined rock in a process zone is restricted, which makes it possible to select a proper complex of equipment best suitable for this rock within the boundaries of the freight traffic flow. For further reading: Chen et al. [18], Ta et al. [23].

As the mining work is being developed, the total turnover of the quarry and individual freight traffic flows are changed periodically. The formation of individual elementary freight traffic flows is substantiated as necessary and effective beginning from the formation of freight traffic flows of benches or their groups and is the first complicated problem, since it must be solved in combination with the problems of stripping the mining levels and locating the waste dumps and together with the selection of technical means for extraction loading and hauling operations. With converging freight traffic flows, some sections of trench and main routes may be loaded most heavily and often pass through the entire turnover of a quarry. The section among them that has the most intricate shape in plan and the heaviest track profile (for railways, also the greatest length) is called the limiting section, since it is exactly this section that limits the traffic capacity. Organization of hauling is calculated for a limiting section (leg), since the volume of the load to be conveyed by the route is determined by the throughput capacity of that section. The transport systems of shops for processing the extracted minerals have their own freight traffic flows independent of the quarry, as required by the technology of mineral processing and removal of wastes.

5.3.4 Initial Stages of Mining Work Development

Stripping of mining levels is performed by constructing mining workings specially intended for the purpose. In order to provide for hauling of the mined rock, each mining level must be stripped by a main trench, Fig. 5.10a, usually inclined (dipping), and since it connects the elevation of the level being stripped with the elevations of the existing levels and of the surface.

Mining work on a working level is started from the formation of the initial work line by cutting a working trench, Fig. 5.10b or a working pit, Fig. 5.10c. If the

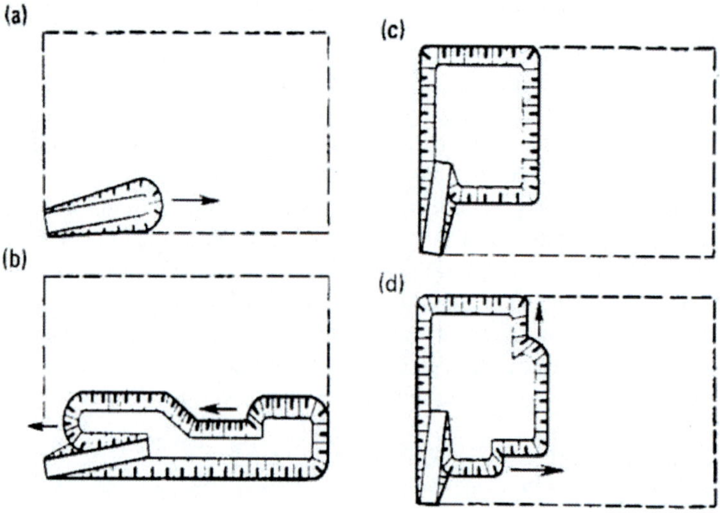

Fig. 5.10 Schemes of the initial period of mining development on a level

characteristics of the excavators employed permit it, the mined rock is sometimes unloaded into a dump near the pit flank, but more often it is transported to an external dump. Further, spacing of one or both flanks of a working trench, Fig. 5.10b, or working pit, Fig. 5.10d, is performed. As the upper bench has been properly advanced, the next lower working level can be stripped and extended within the limits of the working. The longitudinal slope of the working levels should be established so as to ensure safe operation of transport means during loading.

The choice of places for cutting the working trenches is mainly governed by the relief of the surface and deposit roof and by the need to minimize the scope of mining-construction work for putting the quarry into exploitation as soon as possible. In mining horizontal and gently dipping deposits, working trenches are usually cut along the strike of the deposit.

This ensures a sufficiently wide spread of work for high-capacity mining machines and makes it possible to open at once an appreciable reserve of the mineral. In small quarries, the preparatory work may be carried out successively in a number of relatively short sections. This order is often employed for mining building construction minerals, which makes it possible to reduce the initial expenses on stripping work and on the equipment used.

In mining suites of gently dipping strata and deposits of intricate structure, the direction of mining work development must ensure separate extraction of the mineral and gangue. In the mining of sheet deposits, mining work is developed in time descending sense and only rarely, in the ascending sense. If the conditions of stripping are such that the working line must be oriented along the shorter axis or diagonal of the open-pit field, the required capacity of the quarry is achieved by increasing the rate of advance of the working line. With any orientation of the working line and any direction of mining work development, the whole bulk of the rock in a bench of an area F (m^2) (measured on the surface) and of the average capacity H (in) must be worked off in T months according to the calendar schedule. The comparable index of intensity of mining may be taken as the average monthly stripped area, Eq. 5.5:

$$F_m = \frac{F}{T} = L_{b1}\, V_1 \tag{5.5}$$

where:

L_{b1} is the adopted average length of working front on a bench, m.
V_1 is the average monthly rate of advance of the working line, m/month.

The rate of line advance is primarily determined by the intensity of extraction of the mined rock.

Table 5.1 Division of main trenches (Sheshko)

Division feature	Principle characteristics	Type of trench
Position of trenches relative to quarry con tour	Outside the quarry contour	External
	Inside the quarry contour	Internal
Number of benches	One bench	Individual
Serviced by trench system	Several (a group) of benches	Group
	All benches of the quarry up to the ultimate depth	Common
Purpose of trenches	For motion of loaded and empty trains (pendulum motion)	Single
	For motion of only loaded or only empty trains (stream motion)	Doubled
Position of trenches in time	Constant position outside the quarry contour or on flanks	Permanent
	Temporary position inside finite contour on flanks to be mined	Temporary (sliding)

5.3.5 *Stripping Workings*

The division of main trenches is given in Table 5.1. External and internal permanent main trenches are utilized during a long period. Their parameters (initial and ultimate depth, longitudinal slope, length, and flank slope angles) are strictly specified depending on the particular conditions, the properties of enclosing rock, and the specifications for the design of transport lines. Main trenches have a trapezoidal or triangular cross section. They have a step like shape if transport and safety berms are arranged on their flanks. The depth of main trenches usually varies from zero to a value equal to the height of one or more benches. The lifts (slopes) of main trenches depend on the kind of transport being employed, Table 5.2.

The slope angles of flanks of main trenches are determined by the service life of trenches and the properties of the rock, in particular, by water content. The flanks of a trench cut for a long life should have a sufficient long-term stability; for sandy, soft, dense and weathered igneous and metamorphic rock it is taken not more than the angle of repose and for compact rock, up to 50–60°. In external main trenches, both flanks are permanent, i.e. have a constant position. With an internal main trench, only one of its flanks has a constant position. The minimum width of the floor of main trenches is chosen according to the sum of overall dimensions of the transport means, so as to allow safe clearance between them and the transverse dimensions of strips and ditches arranged in the floor.

The width of the floor of a main trench as found from the conditions of arrangement of transport tracks is then checked for the possibility of trench cutting. The cross-sectional area of underground stripping workings is determined by the overall dimensions of transport means and by the schemes of track development (with the required clearances taken into account). For wide-gauge track railway transport

Table 5.2 Typical gradients of main trenches

Kind of quarry transport	Gradient in direction of motion of transport vessels, ‰	
	Loaded	Empty
Inclined trenches		
Railway transport		
steam locomotives	0.02–0.03	0.025–0.035
diesel and electric locomotives	0.025–0.04	0.025–0.06
powered cars	0.04–0.05	0.06–0 0.08
Automobile transport	0.05–0.1	0.08–0.12
Steep trenches		
Cageless hoisting with tractors	0.12–0.25	–
Belt conveyers	0.25–0.33	–
Cage hoists	0.25–0.5	–
Skip hoists	0.50–1.0	–

(dump cars, gondola-cars and industrial locomotives), the cross-sectional dimensions of underground workings (tunnels) are specified by the State standards.

5.3.6 Methods of Stripping of Working Levels in a Quarry

Stripping, or opening, of working levels is done in order to provide transport lines for the freight traffic flows formed on working benches, i.e., to move these loads from the working levels to reception points on the surface or on intermediate levels. Stripping workings begin on the surface or on a level opened earlier and end on the elevation of the working berm of the level being opened.

The method of stripping is defined by a number of features, in the first place, by the kind of stripping workings. In some cases (with the application of tower excavators and cable cranes), mining of the entire deposit and hauling of quarry loads are carried out without making stripping workings. Transport access to particular working levels of a quarry can be provided without stripping workings, for instance, when overburden in quarries of the on-slope or on-slope-deep type is transferred to waste dumps arranged at intermediate levels, when belt conveyers mounted on the non-working pit flank are employed. This method is called trenchless stripping. In most cases, however, working levels of a quarry are opened by cutting main trenches and hillside trenches (or half-trenches). Less frequently, stripping is done by means of underground workings (inclined and vertical headings, drifts and tunnels) or by a combined method.

Trenches intended for motion of wheeled transport vehicles (railway and automobile transport) should be inclined and those for hoisting means may be steep. Depending on the number of benches (one, a group or all benches of a quarry) serviced by trenches with a common hauling route, it may be distinguished between single or individual trenches, group trenches and common trenches, Table 5.1. External trenches may be permanent or semi-permanent. Internal trenches may be permanent (located on non-working flanks of a quarry), semi-permanent, temporary and sliding. Temporary and semi-permanent trenches on working flanks of a quarry are employed in order to decrease the scope of permanent mining work or when the scope of stripping work must be redistributed in time. On working levels opened by a single (individual) trench, pendulum (reciprocating) motion of hauling means is employed most frequently. If a working level is opened by two workings (for loaded and empty vehicles), this makes it possible to organize through motion of hauling means on the benches and thus improve the utilization of mining equipment in time, which will compensate for larger expenses on the construction of opening workings. Such workings are called double; they may be cut from the inside or outside and consist of a pair of single, group or common trenches or half-trenches. Accordingly, the hauling routes laid in the working are called single or double. Double trenches and routes are mainly employed in quarries of a small depth and with intensive freight traffic. In accordance with the indicated main features of division of main trenches, Table 5.3 gives the classification of the main methods of stripping based on the classification proposed by E. F. Sheshko.

When stripping the levels located below the prevailing level of the earth surface, the longitudinal profile of main trenches is given a rising gradient (slope) in the direction of motion of loaded hauling vehicles. Trenches cut on the levels above

Table 5.3 Classification of stripping methods

Characteristic of stripping method	Method of stripping		
	By open workings (trenches)	By underground workings	By Combination of open and u.g. workings
Position of opening workings relative to final quarry contour	External, internal and combined trenches and half trenches	External, internal and combined workings	External, internal and combined workings
Stationary of workings	Permanent, semi-permanent, temporary trenches and half-trenches	Stationary underground workings	Stationary workings or combinations of stationary and semi-stationary (temporary) workings
Inclination of workings	Steep or inclined trenches and half trenches	Vertical, steep, inclined or horizontal	Combinations of vertical, steep, inclined or horizontal workings
Number of serviced levels	Individual, group or common trenches and half trenches	Individual, group or common workings	Individual, group or common workings
Pattern of motion of hauling-means on benches (stream or pendulum)	Single and double workings trenches and half trenches	Single and double workings	Single and double workings

the prevailing surface level have a rising slope in the direction of motion of empty hauling vessels. By the position of stripping workings relative to the open-pit field and deposit it may be distinguished between opening with flank and central trenches (or underground workings), opening from the lying or hanging side of a deposit, and opening from the end of a quarry.

5.3.7 Routes of Stripping Workings

The route of a trench or another working is a line whose spatial position determines the plan, shape and profile of the roadbed of a hauling track. The horizontal projection of the route is the plan of the track and its vertical projection is what is called the track profile. In plan, a track consists of straight and curvilinear portions and in profile, of horizontal and inclined portions and smooth transition curves which connect them.

Route lying consists in establishing the axis of a transport track in plan and profile. The points through which a route must pass are determined by the combination of topographic, geological, building and other factors. According to the position of a route relative to the quarry contour, external, internal and combined routes may be distinguished. As regards their service life, there are stationary, semi-stationary and temporary sliding routes. Stationary routes are mostly arranged on non-working flanks of a quarry; semi-stationary routes are laid on portions of the working flanks of a quarry which have been out of operation for a certain period; sliding (temporary) routes are built on portions of the working flanks being mined.

The basis for laying main trenches is the intermediate or final position of the quarry flanks, which is depicted in a plan by lines of equal height elevations with intervals between them equal to the height of bench. The route of external trenches is drawn from the surface to the horizontal that determines the position of the bench being opened. The route of internal trenches passes on a flank and intersects the horizontals which define the benches, Fig. 5.11. The route is introduced into the contour of a quarry from the quarry end with lower levels of the surface relief; this simplifies route lying inside the open-pit field contours and reduces the volume of mining-construction work. Selection of the position of a route may be governed by some other considerations, for instance, the stability of the flank portions on which main trenches will be cut, the possibilities for increasing the life of these trenches, convenient location of stations and waste dumps on the surface and convenient arrangement of access tracks to waste dumps, the length of tracks on the surface, the length of connecting lines between trenches and face tracks in the quarry.

The principal parameters of a route are the ruling gradient, the difference of height elevations between the beginning and end of route, radii of curved sections of tracks, the theoretical and actual route length, and the number and design of adjoining points between horizontal and inclined sections of tracks. The theoretical length of a route, Lt (m), is determined by the difference of extreme height elevations H_0 and H_x through which the route passes and the angle of incline, I (degrees) of the route to the horizontal, Eq. 5.6:

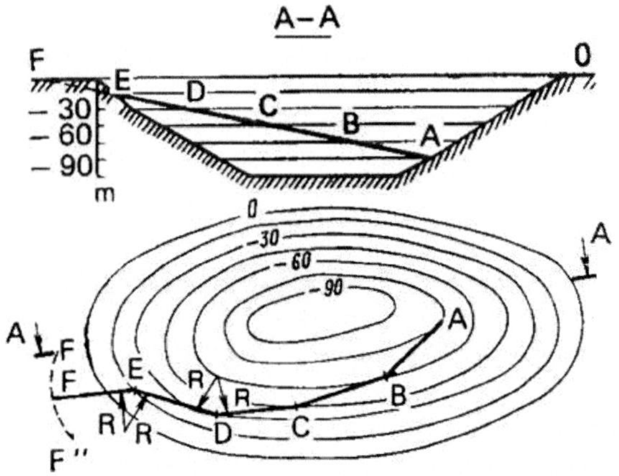

Fig. 5.11 Scheme of route lying of permanent (main) trenches: A, B, C, D, and F, points of junction of the route to levels; F, beginning of the route

$$L_t = \left(\frac{H_0 - H_x}{\tan i}\right) = \frac{H}{i_x} \tag{5.6}$$

where:

i_x is the ruling gradient (upgrade or downgrade) of the route.

The actual length of a route, La (m), is greater than the theoretical length, since curved portions of the route and the track sections connecting the trenches with working levels are necessarily laid with a smaller gradient. For that reason, La = K_{el} LT, where K_{el} is the coefficient of route elongation.

In curved sections of tracks for wheeled transport, the resistance to motion increases by ωc (N/t), and therefore, the gradient (upgrade) of trenches should be diminished to La = ir – ωc/g ‰. Where: La is the actual length of the route in meters, ir is the ruling gradient (upgrade or downgrade) of the route, ωc is the resistance coefficient to motion, and g is the gravity force. The magnitude of ωc depends on the radius R of curves. The smallest radius Rmin is established depending on the curve-in radius of rolling stock. The smallest radius Rmin influences the volume of spacing of quarry flanks as required for laying curved tracks, because of which it is desirable in the general case to employ rolling stock units allowing the least radius of curvature. With railway transport, the shortest length of an element of profile (a track section with a constant upgrade) is determined by the conditions of safe motion of trains. Uninterrupted motion of trains can be ensured if a train passes at any instant of time not more than one change of gradient of the track profile. Thus, the length of a profile element must be not more than the length of a train.

5.3.8 *Route Forms of Permanent Workings*

The form in plan of the route of a permanent working is simple if the route is located on a single quarry flank and does not change its direction essentially along the entire length. A route is said to be complex if it consists of two or more sections of different direction, connected with each other, or if it passes over all flanks of a quarry. The routes of external trenches are always simple and those of internal ones are usually complex. The shape of a route in plan is established according to the dimensions of the open-pit field, the ruling gradient, and the typical elements of the profile. If the actual length of the route of internal trenches does not exceed the length of quarry along the strike on a particular level, Lq, a simple route can be located completely on one flank. The condition that: $La = \left(\frac{(K_{el}\,Hq)}{i_r}\right) \leq L_q$ can, however, be fulfilled only with a favorable relation between the quarry length Lq and quarry depth Hq at a given ruling gradient, ir and given route elongation coefficient (K_{el}).

If it turns out that $La = \left(\frac{(K_{el}\,Hq)}{i_r}\right)$ is greater than L_q two cases of route lying will be possible. The route is laid on one quarry flank and its direction is changed from direct to reverse as many times n1 as needed for route location, Eq. 5.7:

$$La = \left(\frac{(K_{el}\,Hq)}{i_r}\right) = n_1 L_q \tag{5.7}$$

The factor n_1 may be integer or fractional.

Straight sections of a route are connected at the ends by dead-end reversing sidings or curved sections (loops) of a small radius. Loop connection, Fig. 5.12a, is usually employed with automobile transport and dead-end connection, Fig. 5.12b, with railway transport.

Arrangement of the entire route on one flank of a quarry is preferable when the deposit is worked from the lying to the hanging side and with a parallel advance of the working line. Dead-end sections, however, drastically decrease the traffic-carrying capacity of the route, since trains must brake and stop in dead-end sidings where the direction of their movement is reversed. Besides, this scheme requires more intricate traffic control. For that reason, dead-end routes are not recommended for application at least at the group of upper levels of a quarry. The route is drawn from one flank of a quarry to another as many times n_2 as required for its positioning on the corresponding levels of the quarry flanks, Eq. 5.8:

$$n_2\,P = \frac{K_{el}\,H_q}{i_r} \tag{5.8}$$

where:

P (m) is the average total length of the perimeters of flanks.

In that case the route encircles the quarry in the form of a spiral, Fig. 5.12c. A spiral route includes curvilinear sections which are arranged at the end flanks of a quarry and usually have a large radius. The laying of curves involves no difficulties and usually

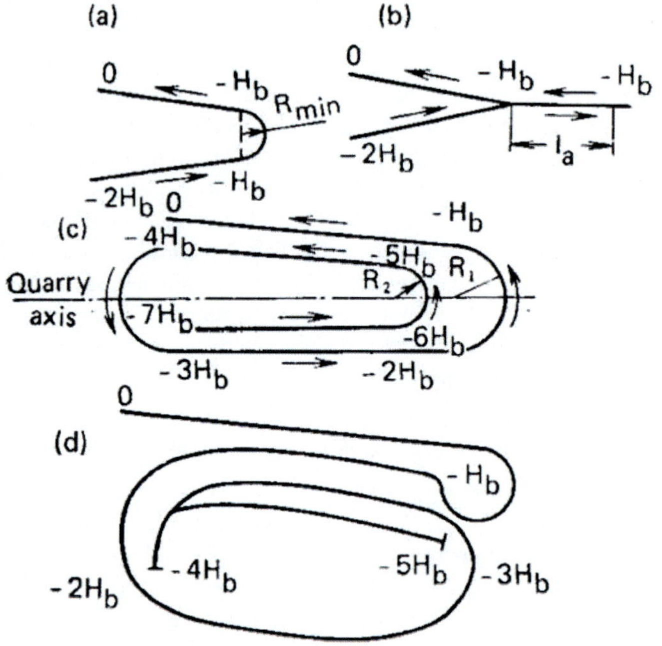

Fig. 5.12 Plans of routes: La, length of adjoining berm

does not require a half-embankment or half-pit. In many cases, the internal route may contain straight, spiral and dead-end (or loop) sections, Fig. 5.12d. Such complicated routes can, however, improve the conditions of opening of individual levels, increase the efficiency of operation of the quarry transport, and permit application of more rational working systems. An internal route is an immediate continuation of the external one. Such a combined route is usually employed for opening in deep quarries: a number of upper levels are opened with rock hauling by the external route, after which an internal route is laid to lower levels of the quarry. The lowering (descent) of a route of internal main trenches is determined in terms of the average route gradient and the actual length of the route. For further reading: Ji [24], Krzyzanowska [25], Kusar et al. [26].

A simple route can be employed in working of deposits of an appreciable extent and a relatively small quarry depth. Dead-end routes are applicable with deposits having relatively small dimensions along the strike, especially with steeply dipping deposits and with the quarry having a small size across the strike. Loop-end routes are used in cases when a deposit is opened from internal trenches and rock is hauled by automobile transport or, where loop radius permits, by railway transport. A spiral route is laid when the application of a dead-end or loop-end route is impossible or unadvisable from the conditions of bedding of the mineral body, spacing of quarry flanks, the required traffic-carrying capacity and efficiency of operation of the quarry transport. On spiral routes, the railway tracks must be stationary where possible, since track shifting (relaying)—incurs large difficulties. With automobile transport, road sections of a spiral route can be re-laid comparatively easily.

5.3.9 Volumes of Main Trenches and Half-Trenches

The volume off an individual inclined trench, V_{tr} cut from the horizontal surface to a final depth H, with the bottom width b and the angles of slope of trench sides, cc, is determined as the sum of volumes of regular geometrical figures which constitute the trench, Fig. 5.13a:

- The volume of figure A which is the mid portion of the inclined trench.
- The volumes of figures 2B which represent the spacing of trench sides.
- The volumes of figures D and 2F which constitute the end (face) portion of the trench after cutting.

As follows from Fig. 5.13a, A is a rectangular semi-prism with the base being the rectangle with sides' b and H. Its height is H/tan I or H/i and its volume (m^3) is, Eq. 5.9:

$$A = \left(\frac{b\,H^2}{2\,i}\right) \tag{5.9}$$

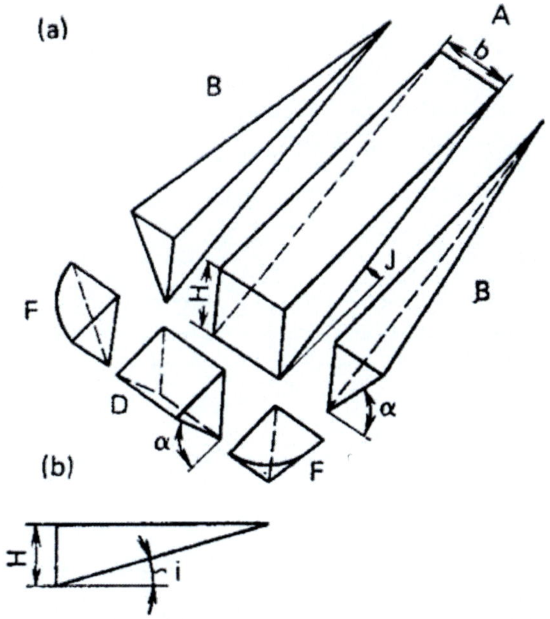

Fig. 5.13 Diagram to calculate the volume of an inclined trench

where:

i the angle of slope of the trench toe, degrees, and i is the trench bottom gradient.

Figure B is a pyramid with the base formed by a triangle of an area $\frac{H^2}{2\tan\alpha}$; its height is $\frac{H}{i}$ and its volume (m^3) is, Eq. 5.10:

$$2\,B = \frac{H^3 \cot\alpha}{3\,i} \quad (5.10)$$

Theoretical principles for calculating the volumes of maim trenches have been developed by Sheshko.

Figure D is a rectangular semi-prism with the base bH and height H/tan cc; its volume (in^3) is, Eq. 5.11:

$$D = \frac{b\,H^2 \cot\alpha}{2} \quad (5.11)$$

Figure F is part of a cone whose base is a quarter of a circle of radius H/tan α and the height is H; its volume (m^3) is, Eq. 5.12:

$$2\,F = \frac{\pi\,H^3 \cot\alpha}{6} \quad (5.12)$$

Thus, the volume of the inclined trench (m^3) is, Eq. 5.13:

$$V_{tr} = A + 2B + D + 2F \quad (5.13)$$

If an inclined trench has only a slight slope, the volumes D and 2F can be neglected. Then we have, Eq. 5.14:

$$V_{tr} = A + 2B = \frac{H^2}{i\left(0.5\,b + \frac{H\cot\alpha}{3}\right)} \quad (5.14)$$

If the volume of a trench is determined by considering the spacing of the end portion (with slopes more than 40‰), one has to sum up (Fig. 5.14).

The volumes: A + 2B + D + 2F. In that case we obtain, Eq. 5.15:

$$V_{tr''} = \frac{H^2}{i\left(\frac{b}{2} + \frac{H\cot\alpha}{3}\right) + H^2 \cot\alpha\left(0.5\,b + \pi\frac{H\cot\alpha)}{6}\right.} \quad (5.15)$$

The construction volume (m^3) of an individual inclined half-trench, Vhtr, with a depth (height) H, gradient i, angle of slope of the hillside γ, and the angle of slope of half-trench side α is equal to the volume of an inclined prism with the base formed by triangle ΔABC and with the height equal to h, Fig. 5.15 and Eq. 5.16:

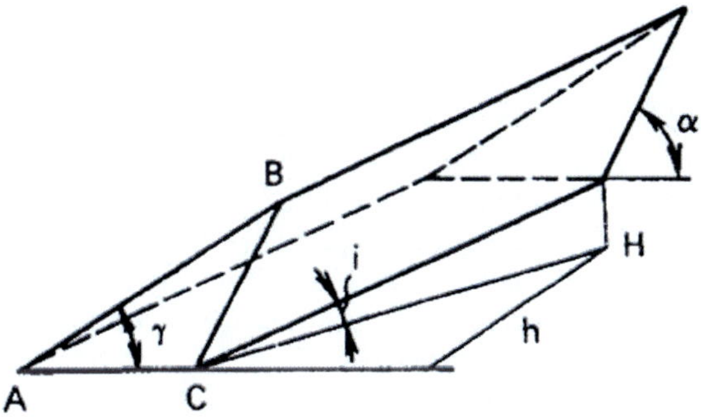

Fig. 5.14 Diagram to calculate the a volume of an inclined half-trench

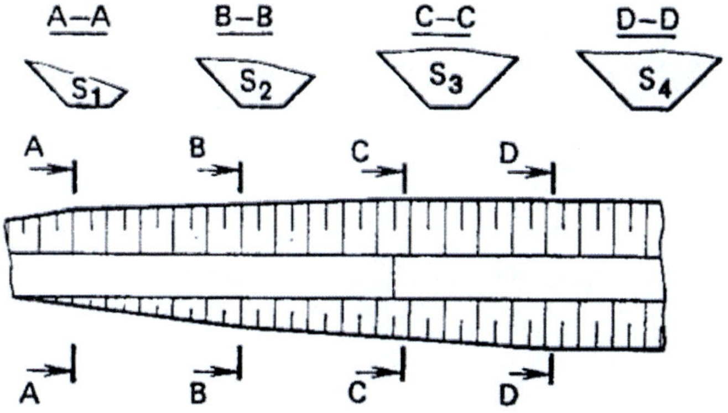

Fig. 5.15 Diagram to calculate the volume of a main trench of intricate shape of these sections

$$V_{htr} = \frac{H\,b^2 \sin\alpha \sin\gamma \sqrt{\left(\frac{1}{i^2}\right) - \left(\frac{1}{\tan^2\gamma}\right)}}{2\sin(\alpha - \gamma)} \tag{5.16}$$

If $\gamma \geq 10°$, the volume of a half-trench (m^3) can be found from a simpler formula which is quite accurate, Eq. 5.17:

$$V_{htr} = \frac{H\,b^2 \sin\alpha \sin\gamma}{2\sin(\alpha - \gamma)} \tag{5.17}$$

Inclined trenches usually cut in hillsides and worked-off flanks of quarries. In the latter case they called inclined access ways or simply access ways. With an intricate surface relief of the deposit and curvilinear form of external trenches in plan, their volumes are determined by making a number of parallel vertical cross sections in the typical points of the longitudinal profile of a trench, Fig. 5.15. The areas of these sections are determined by means of a planimeter, after which the volume of the trench is found as the sum of these blocks.

$$V_{tr} = 0.5\left[\frac{(S_1 + S_2)}{l_1} + \frac{(S_2 + S_3)}{l_2} + \cdots + \frac{(S_{n-1} + S_n)}{l_{n-1}}\right] \quad (5.18)$$

where:

$S_1, S_2, \ldots S_n$ are the cross-sectional areas of the trench, m_2, and $m_{l1}, m_{l2}, \ldots$, ln are the lengths of the blocks into which the trench has been divided, m. This method gives greater accuracy with a smaller distance between parallel cross sections. With an intricate shape of a trench in plan and sharply uneven surface relief, it is essential to divide the length of a trench into a larger number of blocks. The construction volumes of common and group internal trenches are determined as the sum of volumes of the individual trenches and half-trenches. Under comparable conditions, the volumes of common and group external trenches depend on their cross-sectional form, type of junction points, number of benches being stripped and number of transport exits from a trench. A group or common trench may be made in one of two versions. In the first version, Fig. 5.16a, the exit from a trench is common for all levels and in the second version, Fig. 5.16b, an independent exit is made from each bench. With two benches being opened, the volume of an external trench (m^3) with a common exit is determined by the formula:

With one-sided junction of the roads of working levels, Eq. 5.19:

$$V_{tr} = \frac{4H^2b\left(\frac{b}{2} + \frac{2Hb}{3\tan\alpha}\right)}{i} + \frac{(b_t + b_s)bH^2}{i} \quad (5.19)$$

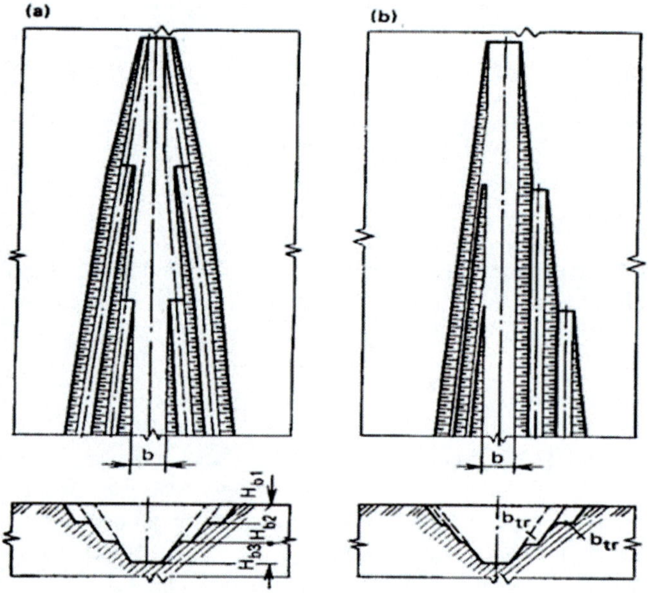

Fig. 5.16 Schemes external trenches

With two-sided junction, Eq. 5.20:

$$V_{tr} = \frac{4\,b\,H^2\left(b + \frac{2\,H\,b}{3\tan\alpha}\right)}{2} + \frac{2\,b_t\,H^2\,b}{i} \tag{5.20}$$

where:

bt and bs are the width of the transport berm and safety berm respectively, m.

With independent exit onto the surface from each bench, the volume of an external trench will be smaller and can be determined for instance, as for the above conditions of one-sided junction, by the formula, Eq. 5.21:

$$V_{tr} = \frac{4\,H^2 b}{i}(\left(\frac{b}{2} + \frac{2\,H\,b}{3\tan\alpha}\right) + \left(\frac{bt\;b_s}{2\,i}\right)H^2 b \tag{5.21}$$

With the use of conveyer transport, the volume (m^3) of an external trench is limited by the design position of the slope of the non-working flank of the quarry and can be given by the formula, Eq. 5.22:

$$V_{tr} = \left(\frac{H^2}{i}\right)\left(\frac{b}{2} + \frac{H}{3\tan\alpha}\right) - \frac{H^2}{\tan\alpha}\left(\frac{b}{2} + \frac{H}{2\tan\alpha}\right) \tag{5.22}$$

In modern mining practice and in projects, deep external trenches with a common transport exit are employed widely. This is explained by the fact that such trenches are, as a rule, cut in water-saturated soft and loose rock, so that the first version of opening simplifies drainage of horizontal rock layers. Besides, the length of railway tracks and electric contact system can be reduced. On the other hand, a trench with an independent exit to the surface can decrease the volume of mining-construction work and allows intensification of quarry development. Trenches of this type are promising for opening deeply bedding deposits.

The total mining-construction volume of an external trench for railway transport should always be determined by considering the location of junction curves, Fig. 5.17a. An additional volume of work V_0 (m^3) for the construction of a junction curve in a single side of trench, for opening a bench, can be determined approximately as the difference between the volume of the rectangle, Fig. 5.17b, with the basis in plan OCEB and ODE'A and the volume of the portion of truncated cone with the bases in plan OCB and ODA, with their heights being equal to Hb.

In the general case, the volume of opening of n benches by an external trench is found as given per Eq. 5.23:

$$V_0 = K\sum H\,b\,h(R^2K - RK\,Hb\,K\cot\alpha) \tag{5.23}$$

where:

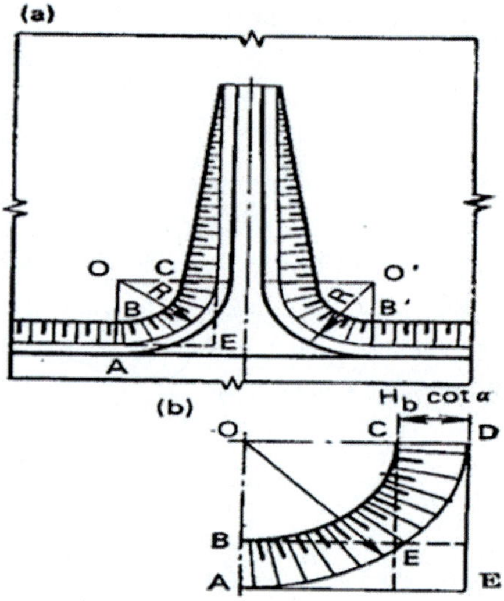

Fig. 5.17 Diagrams for determining the volume of junction of external trenches with the use of railway transport

K is a coefficient depending on the number of sides of junction (with junction curves on one or two sides respectively.

K = 0.215 and K 0.43); Rk is the radius of junction curve of the Kth bench, m ($R_{min.}$ = 250 m).

The total volume of an external trench will then be found as:

$$V_{ex.tr} = V_{tr} + V_0$$

Steep trenches in deep-type quarries are usually cut internally. By their position relative to the quarry flank, they may be divided into transverse and diagonal. Transverse steep trenches, Fig. 5.18b, are employed in the cases where the total angle of slope of quarry flank does not exceed the maximum angle of rise for transport vehicles, which is typical for:

- Skip and cage hoists. Diagonal steep trenches, Fig. 5.18a, are commonly used for accommodating conveyers or automobile lifts. These trenches are typical for the cases when transport hems of the width b t > 12–15-in are left on the non-working flank of a quarry. If the flank has only relatively narrow safety hems or doubled (tripled) benches, a conveyer lift is arranged in a steep half-trench or an opening working which is a combination of steep trenches and a half-trench.

The mining-construction volume of an internal steep trench (m^3) can be given by the formula, Eq. 5.24:

$$V_{tr.} = H^2(\cot I - \cot \gamma_{nw})\left[\frac{b}{2} + \frac{h}{3}(\cot I - \cot \gamma_{nw})\frac{\cot \alpha}{\cot I}\right] \quad (5.24)$$

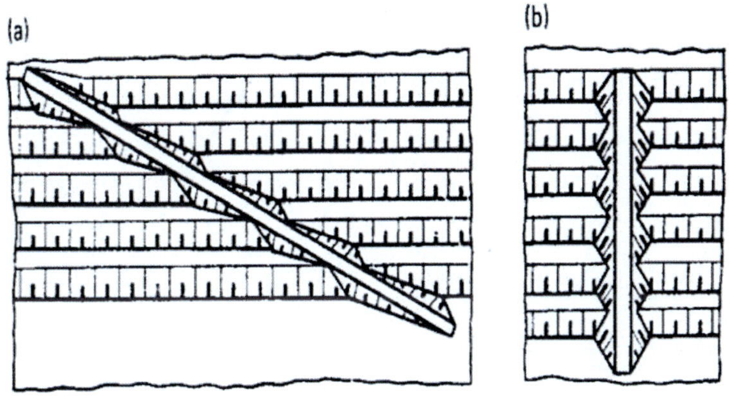

Fig. 5.18 Schemes of steep trenches

where:

H	is the trench depth, m.
I	is the slope angle of the trench, degrees.
γ_{nw}	is the angle of slope of non-working flank of the quarry, degrees.
b	is the width of the trench bottom, m, and.
α	is the slope angle of trench sides, degrees.

The volume of a steep half-trench is determined similarly to that of an inclined half-trench.

5.3.10 Working Trenches and Pits

Working trenches and pits are made on each bench in order to form the initial spread of mining work. They usually are the continuation of stripping inclined trenches. Working trenches have a small (3–5%) longitudinal gradient for water drainage from a bench. Their cross-sectional form is usually trapezoidal. The length and width of cutting pits are usually of the same order of magnitude.

Planned working of a bench is started by spacing one or both sides of a working trench in the direction of the boundaries of a working level. In pits, two, three or even four flanks may be spaced simultaneously. Working trenches may be made in the deposit, in the rock of the hanging or lying side of the deposit, for the whole length of a bench or only for part of that length. The depth of working trenches and pits is equal to the adopted height of a bench or sub-bench. Deep trenches in contact with a dipping or gently dipping deposit are rather frequently cut in slices of the height H_{s1}, in Fig. 5.19.

In order to diminish losses and avoid dilution of the mineral, Eq. 5.25:

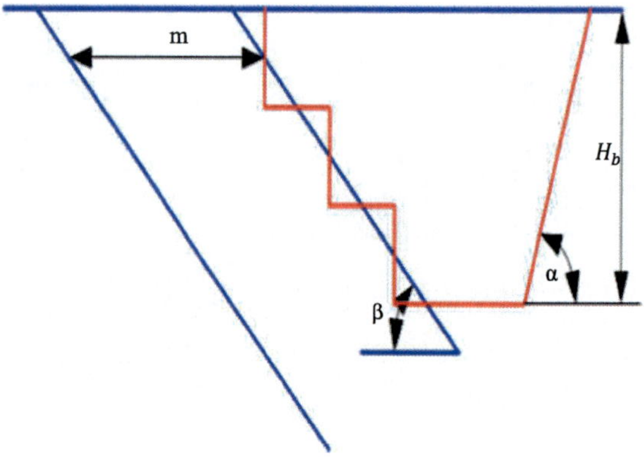

Fig. 5.19 Selection of cross-sectional form of working trenches

$$\mathbf{H_{s1} \leq 2\frac{Rm}{[100(\cot\beta - \cot\alpha)]}} \tag{5.25}$$

where:

R is the permissible coefficient of dilution of the mineral.
m is the horizontal capacity of the deposit, m.
β is the angle of dip of the deposit, degrees.

The width of a working trench at the bottom, b, should be such as to ensure safe motion of hauling means, accommodate mining equipment, and provide the possibility for an excavator to make the first cut. In some cases, the cross sections of some portions of cutting trenches are widened for laying crossings and for other purposes. The slope angles of the sides of working trenches and pits are usually equal to the angles of slope of working benches (60°–85°). The sides of preparatory workings which coincide with the final contour of a quarry are given a slope angle that ensures their long-term stability.

The volume (m^3) of a working trench is found approximately as the volume of a rectangular prism with a trapezium as the base, Fig. 5.20a and Eq. 5.26:

$$V_{c.tr} = (b + H_b \cot\alpha) H\, b\, L \tag{5.26}$$

where:

L is the trench length, m.

The volume of a working half-trench (m^3) made in a hillside or quarry flank, Fig. 5.20b, is determined as follows, Eq. 5.27:

$$V_{c.htr} = \frac{b^2 \sin\alpha \sin\gamma\, L}{2\sin(\alpha - \gamma)} \tag{5.27}$$

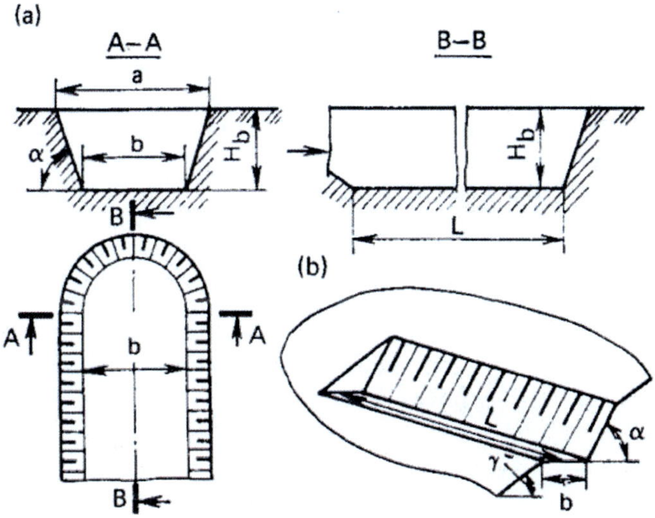

Fig. 5.20 Volumes of main trenches (**a**) and half trenches (**b**)

With an appreciable length and narrow width, working trenches often have an irregular geometrical shape. Their volume (m^3) is determined by neglecting the transverse slope of the terrain, but considering the longitudinal slope, Eq. 5.28:

$$V_{c.tr} = \left[S_0 + \frac{\frac{U}{12}}{(H_2 + H_t)^2} \right] \tag{5.28}$$

where:

S_0	$\left(\frac{S_1+S_2}{2}\right)$ is the average area of adjacent parallel cross sections, m^2.
U	is the slope coefficient numerically equal to the cotangent of slope angle,
H_1 and H_2	are the average heights of the cross sections, m.
1	is the spacing between adjacent sections, m.

The second term in the equation is called the slope correction; it considers the volume of the elements of the body (working) which have the shape of a pyramid. Under the conditions of hilly terrain relief, the volumes are calculated by considering additionally the effect of transverse slope, using the formula, Eq. 5.29:

$$V_{c.tr} = \left(S_0 - \frac{y_1 - y_2}{6} \right) \tag{5.29}$$

where:

y_1 and y_2 are the areas of the bases of pyramids in the volume of the mining body (to be determined graphically).

The preparation of working levels by cutting pits can only be employed when excavating machines operate with mobile kinds of transport, most often with automobile transport. This preparation favors quick development of mining work on a level in different directions. Pits are usually located within the limits of a mineral deposit. The plan dimensions of a pit are determined from the conditions of proper delivery of hauling means to an excavator and by the possibility of placing a second excavator on the level; they usually measure from 40×40 m to 100×100 m.

5.4 The Nature of Surface Mining

Surface mines come in many different forms and sizes. However, most of our surface-mined mineral products come from large mines that practice large-scale (mass production) methods. The sheer magnitude of the volume or tonnage of material broken and handled in surface mining is staggering. Table 5.4 outlines some of the statistics on production of ore and waste in surface and underground mining in the United States for the year 1997. Note that the average tonnage of waste required to produce a ton of ore is about 15 for surface coal and about 2.6 overall for surface mining, but only 0.1 for coal and 0.07 overall for underground mining. Because of the waste associated with surface mining, the cost to mine a ton of material in surface mines must be much lower than that for underground mines.

The proportions of ore and total material tonnage produced in surface and underground mines are provided in Table 5.2. The tonnage of ore and coal being mined on the surface keeps increasing. When the first edition of this book appeared in 1987, the proportion of ore and coal mined in surface mines in the United States was 85%; it is now more than 88%. Note also that the percentage of all material mined in surface mines is greater than 96% on a tonnage basis. This heavy emphasis on surface mining is due to two factors: the increasing difficulty of finding deposits that can be economically mined underground and the ever-increasing efficiency of mining in surface mines. This is true even though the cost of reclamation for a surface

Table 5.4 Ore and waste production in the United States

	Surface			Underground			All Mining		
	Ore	Waste	Total	Ore	Waste	Total	Ore	Waste	Total
	Million tons			Million tons			Million tons		
Metals	1290	1863	3153	64	3	67	1354	1866	3220
Non-metals	2778	449	3227	123	0	123	2901	449	3350
Coal	669	10,303	10,972	421	45*	466	1090	10,348	11,438
Total	4737	12,615	17,352	608	48	656	5345	12,663	18,008

Table 5.5 Proportions of ore, coal tonnage and total tonnage

	Ore tonnage		Total tonnage	
	Surface (%)	Underground (%)	Surface (%)	Underground (%)
Metals	95.3	4.7	97.9	2.1
Non-metals	95.8	4.3	96.3	3.7
Coal	61.4	38.6	95.9	4.1
Total	88.6	11.3	96.4	3.6

mine is increasing steadily. The success of surface mining operations is thus highly dependent on efficient drilling, blasting, and haulage.

Table 5.4 gives the estimates for both ore and waste for different mineral-commodity classes in surface and underground Mining. These values based on ore/waste ratios (Table 5.5).

Mine planning and development are crucial steps in the operation of a surface mine. Certain factors (see Sect. 1.8) may require special attention in preparation for surface mining. Of the location factors, climate is of more critical concern in surface operations than in underground mines. Today, harsh climates at high altitudes or in northern latitudes rarely preclude surface mining, but they can be detrimental to efficient and cost-effective operation. Among the natural and geologic factors, terrain, depth, spatial characteristics of the deposit, and presence of water are very important variables. The environmental concerns and the costs of overcoming environmental problems rank among the most important considerations in the planning and development of a surface mine. The steps in the sequence of mine development are enumerated in Sect. 1.8; three of these are unique to surface mining:

- Initiation of a land reclamation plan as part of the environmental impact statement (EIS).
- Provision of topsoil stockpiles and waste disposal dumps.
- Performing advanced stripping of overburden to gain access to the deposit.

Other steps in mine development must be carried out as well, but these three steps require significant resource allocation and planning for a surface mine. Figure 5.21 shows the mine development schedule for a typical mine. Note that the environmental studies and the permitting process are a significant part of the overall mine development, and that environmental concerns are continuous throughout the development process. Thus, the first of the aforementioned three steps in development will normally be a major consideration in the development of the mine. Land reclamation, waste disposal, and advanced stripping are scheduled during Stage 3 of Fig. 5.21 (Notice that it is customary terminology to say we mine ore, stone, or rock but that we strip overburden or waste.). We will consider the three tasks, as well as plant layout, in more detail in the sections that follow. The scheduling diagram, Fig. 5.21, is reported with the permission from American Institute of Professional Geologists, Colorado Section, Golden, CO.

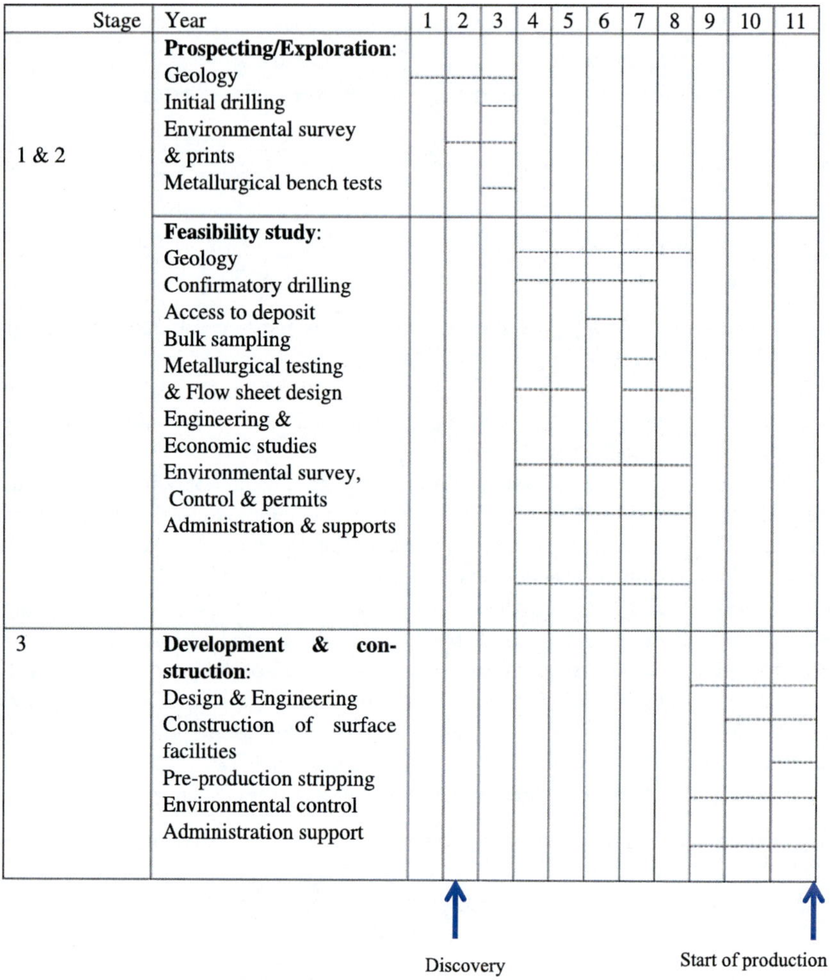

Fig. 5.21 Scheduling diagram for surface metal mine

5.4.1 Land Reclamation

The federal Surface Mining Control and Reclamation Act of 1977 require that land disturbed by mining be reclaimed and restored to its pre-mining condition or better. Some of the many requirements are summarized in Chap. 8. During development, the first steps are taken to ensure that the EIS filed by the company is fully implemented. The company must post a sizable bond to cover the expected cost of reclaiming the land. The bond on each section of land is released only after the land has been reclaimed and suitable vegetation has been established. Provisions of the law, which are primarily applicable to surface coal mines, are quite stringent and costly. Restoration of land to "approximate original contour" following mining is a requirement that necessitates careful planning, surveying, and mapping. Preserving surface drainage may require stream relocation or diversion. Careful control of the runoff from the mine is also necessary to prevent stream siltation. Maintaining wildlife and fisheries is of special concern on lands to be surface mined. Provisions to protect fish and game resources and to provide acceptable habitat must be initiated early in the development process and carefully maintained throughout the life of the mine. Finally, archaeological sites, both known and uncovered during mining, must be protected. State or international codes may also enter significantly into the planning picture. For provisions of a typical state mining code.

5.4.2 Topsoil Stockpiles and Waste Disposal

During the development stage of any surface mine, topsoil stockpiles and waste disposal areas are located. Separate areas may be required for topsoil, subsurface soil, rock, low-grade or leachable ore, and tailings. Separating these materials enhances the opportunities to better utilize the materials in the extraction of valuable components and in the reclamation of the mined area. Site selection must ensure convenient disposal and retrieval but must avoid interference with production and its related auxiliary operations. Advanced planning is necessary to ensure that the materials in storage never conflict with mining activities. For further reading: Roberson [27].

5.4.3 Advanced Stripping

The geometry of the mineral deposit and the overburden and the planned production rate dictate the minimum amount of stripping that must be done to maintain the desired rate of ore or coal extraction. Economic considerations would normally suggest that the stripping be performed only as needed because the expenses associated with stripping are not matched by an economic return.

However, researchers outline three general plans for stripping in a surface mine: the increasing stripping ratio method, the decreasing stripping ratio method, and the constant stripping ratio method. The increasing stripping ratio method strictly follows the rule of stripping only as much overburden as required for production; this procedure is optimal in terms of the cash flow if other variables do not significantly enter the picture. The other two methods attempt to level out the stripping requirements and spread them more evenly over time. In a deposit that does not outcrop, advanced stripping is required before ore production can begin. In addition, it is a general rule that a certain amount of ore should always be available for mining so that production scheduling is not constrained by the lack of available ore. A rule of thumb, common in truck-shovel operations, is to maintain at least a 30-day supply of broken ore available for the loading equipment. For further reading: Chanda and Gardner [28].

A major consideration in stripping decisions may be the climate. Severe cold weather may favor stripping in the summer months when the ground is thawed, with mining being conducted during the winter. In the iron ranges of the Lake Superior district and in regions of Alaska, shipping may be restricted by ice formation on the waterways. Thus, stripping may be emphasized in winter and mining during the summer. A final consideration may be the decision as to whether the stripping should be performed in-house or by a contractor. Large mining companies generally prefer to do their own stripping, but a smaller company may find it expeditious to contract it out. Contract stripping is often more expensive, but it may be quicker. In addition, the mining company is relieved of the capital expense of purchasing stripping equipment.

5.4.4 Plant Layout

Some of the factors that must be considered in selecting the mine plant size and layout were identified. For a surface mine, the task is complicated by the special considerations just discussed: land reclamation, topsoil stockpiling, waste disposal, and overburden stripping. These must be carefully planned to minimize the cost of mining. In addition, the support activities will become an important additional mine plant layout task. The plant layout of the Black Thunder mine in Wyoming, among the largest coal mines in the world, provides an interesting case study and is shown in Fig. 5.22. The pit, waste dumps, and topsoil stockpiles are off the diagram to the lower

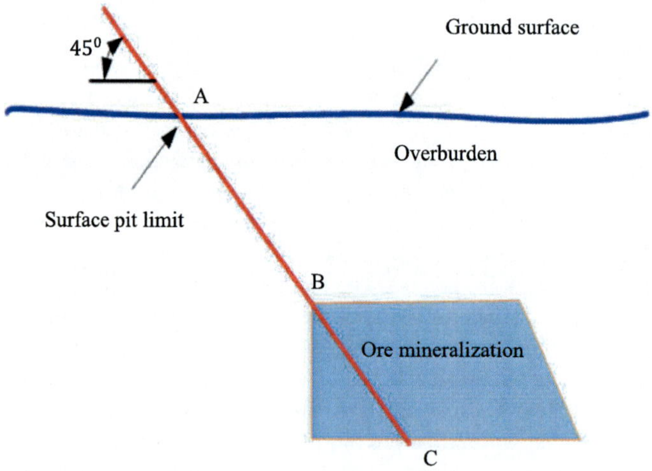

Fig. 5.22 Strip ratio and pit slope relation

right. The coal haul road and crusher, mineral processing plant, slot storage facility, and clean coal storage facility are all connected by single-flight belt conveyors for efficient materials handling. The storage silo is on a rail loop that facilitates unit train loading. The maintenance shops, administration building, change house, warehouse, fuel storage, and other necessary facilities are then arranged in close proximity to the processing facilities. When laid out in a logical manner, the physical plant enhances the ability to mine and contributes to the efficiency of the operation.

5.5 Pit Planning and Design

5.5.1 Introduction

Open pit mining is a method of operating a surface mine that is simple in concept but complex in its cost and efficiency requirements. Recall the earlier discussion in Sect. 5.4 about the amount of waste mined; it is evident that open pit mining must be carefully planned and executed to keep unit Costs to a minimum. Accordingly, the average open pit mine is heavily engineered even though it is simple in configuration. There are a number of factors that must be considered in the initial planning:

- Natural and geologic factors: geologic conditions, ore types and grades, hydrologic conditions, topography, metallurgical characteristics, climate, and environmental variables of the site.
- Economic factors: Ore grade, ore tonnage, stripping ratio, cutoff grade, operating cost, investment cost, desired profit margin, production rate, processing and/or smelting costs, and market conditions.
- Technological factors: equipment, pit slope, bench height, road grade, property lines, transportation options, and pit limits.

The pit planning team will most likely strive to optimize the pit design in respect to the technological factors. Most of the other factors are beyond their control and become part of the constraints. The overall plan for the pit is then studied, including both the overall pit limit and the sequence of extraction. Many variables must be considered in this exercise so that the ore is brought into production as early as possible and the sequence is conducted without disrupting production or cash flow. The initial cash flow is very important, as the income generated during the first five or ten years of exploitation is more apt to make or break the mine than the economics of the long-term mine plan. In this regard, he lists a number of objectives that apply to pit planning in most open pit operations:

- Mine the ore body so that the production cost per lb (kg) of metal is a minimum (i.e., mine the "next best ore" to generate income as early as possible).
- Maintain proper operating parameters (adequate bench width and haul roads).
- Maintain sufficient exposure of ore to overcome miscalculations or delays in drilling and blasting.

- Defer stripping as long as possible without constraining equipment, manpower, or the production schedule.
- Follow a logical and achievable start-up schedule (for training, equipment procurement and deployment, etc.) that minimizes the risk of delays in the initial cash flow.
- Maximize pit slopes, while maintaining reasonably low likelihood of slope failure provide safe berms, employ good rock mechanics, implement good slope monitoring systems, etc.).
- Examine the economic merits of various production rates and cutoff grades.
- Subject the favored choice of method, equipment, and pit sequence to exhaustive contingency planning before proceeding with development.

In open pit quarrying/mining, planning is of considerable importance because the failure to maintain adequate 'blocked out' reserves of mineral or stone will result in an inability of the mine to maintain production, which ultimately may need an extensive re-development program to restore a full production position. This can occur where overburden is not stripped off fast enough, especially where overburden thickness increases. The concentration of output in a few production units can also present problems, if large variations result from lack of forward planning. For multi-bench operations the correct balance between benches is necessary to achieve synchronized advance and to maintain access roads and auxiliary works, which must keep pace with this advance. Planning must also ensure that high capital cost (expensive) equipment is fully utilized in production and not subjected to unnecessary movements or delays. It is important to realize that imposed on mine planning are many financial implications along with legal constraints. The approach utilized here is to concentrate on operational (physical aspects) with limited economic comparisons. At the operation level, therefore, it is unlikely that operation personnel actually make a capital cost equipment decision such as the purchase of a type of truck. However, a consensus on relative suitability will be sought.

In planning there are a number of recognized levels which constitute an overall long-term framework. Some of the procedures and objectives associated with them are as follows:

- Long-term planning; exploiting a deposit profitability to obtain extraction of the whole of the reserves or up to the cut-off point.
- Medium-term planning; programs which are in greater detail and are related to, say annual production requirements.
- Short-term planning; detailed control of day-to-day production.

Short-term planning can never be static and must operate flexibly within the longer term framework in order to meet the current geological information, changes in production requirements, changes in ore grades and other prevailing conditions. Where the mine is operational, the design decisions have already been made and hence day-to-day supervision control must aim for:

- Economy
- Safety

- Efficiency
- Practicality.

However, overall the major aspects of:

- Access;
- Services;
- Production including ore stocks within the pit and in stockpiles, must be attended to maintain continuity of operation.

5.5.1.1 Practices and Approaches for Effective Open-Pit Mine Operation

There are a great many items considered to influence the efficiency in open-pit mining. Those recognized to be of major consequence are:

- Human element—the availability and retention of good management, supervisory and operating personnel.
- The mine location and topography.
- The climatic conditions.
- The material characteristics of the ore body.
- Equipment selection and use.
- The mining practices employed.

5.5.1.2 Material Characteristic

In the planning of a new mining operation or in the expansion of an existing operation, the first task is to determine the tonnage and characteristics of the material to be mined. Tests are made to ascertain if the material is:

- Heavy mineral sand (either unconsolidated, or soil with heavy induration.
- Hard rock requiring drilling and blasting.
- Dense, for both broken and in situ ore.

Other factors are the hardness, moisture content, strength and load bearing properties of the materials. From this type of data the projected angle of pit slope both working and final may be estimated. Properly designed pit slopes will reduce the stripping ratio and increase the amount of ore which can be economically mined, thereby increasing the ore reserve and the potential life of the operation. A decrease in the waste removal has a direct effect on the efficiency of a pit operation, especially if achieved realistically without catastrophic slope failure. The requirement for capital equipment is reduced, along with manpower. Long production life for a mine influences the likely equipment selection. Long life and high daily production tonnage will permit the use of larger, more efficient equipment. Pit stability, along with the location and grade of the final ramp system, will play an important part in achieving effective pit operation. The working pit slope, usually at a lower grade than the final pit slope, should be maintained at a maximum within safety and space.

5.5.1.3 Operating Practices and Equipment Selection

In the initial equipment selection there are several factors to consider. One of the first would be the total volume and type of overburden to be removed. Where the volume is considerable, unconsolidated or soft, scraper or conveyor type handling systems may be justified. If the volume is small or the characteristics of material require drilling and blasting, then shovel and truck or front—end loader systems are the logical economical choice. Where little stripping is involved, the equipment can be standardized to handle ore and waste. The degree of ore control necessary is also important. An ore body of uniform grade may be mined more efficiently and with lower operating costs than one with varying ore grades. The principal factors to be considered in selecting mining equipment along with the above are:

- Life of the property.
- The daily production rate desired.
- The strip ratio.
- The capital available.
- Haul distance.
- Bench height.
- The working area available.
- The road width available.
- Climatic conditions.

In the case of conventional hard rock operations, the largest size shovel possible can accrue certain advantages:

- Lower operational costs per ton of material;
- Fewer operating shovels per shift;
- Overall, possibly lower capital expenditure;
- Fewer operating faces, which results in fewer haul roads to maintain.

However, these may be off-set by a lower opportunity for ore blending (grade control) and larger rocks being fed to the primary crusher.

For maximum truck-shovel efficiency or productivity it is necessary to match the haulage unit to the excavator and to the haul distance and haul grade. The abilities of various truck sizes and types would be compared for total cycle times and total cost per tonne while being loaded by various size excavators. Also, in obtaining a proper balance between excavator bucket size and truck size, it is necessary to look at the number of full buckets (passes) required to fill the truck to its rated capacity. This is usually in the order of 4–6 passes. Partial bucket loads reduce shovel efficiency. Also, trucks may be either under-loaded or over-loaded. Overloading and over-speeding rapidly result in excessive wear and tear on most equipment components, culminating, in the case of continuous abuse, in premature failure and escalating costs.

The body of the truck therefore must match the density characteristics of the load; for example, 30 cu. meters of coal weighs approximately 36 tonnes, 30 cu. meters of iron ore weighs approximately 135 tonnes. Hence, a coal hauler is likely to be totally unsuited for iron ore.

In selection of blasthole drills the most important factor is the hardness of ground which will determine the type of drilling method and drill bit used. The daily mine production tonnage determines the number of holes required, hence, basically, the size in relation to ground hardness along with the power required. Drills range from surface-mounted, pneumatic, and hydraulic crawler-mounted to down-the-hole hammer. Large blasthole rotaries may be either diesel powered or electric powered. The latter offer lower maintenance costs.

An important factor in efficient pit operation is good blasting practice. Actual analysis of ground-rock quality by borehole analysis, seismic testing, can be important to determine blastability of various types of rocks. To achieve good rock fragmentation, factors of hole size and proper powder ratio and drill patterns are essential. Bulk explosives permit greater safety in handling, charging and lower total cost due to lower consumption and less manpower required by using efficient bulk handling and placement equipment. For large excavators, the larger blasting rounds are required to give sufficiently deep material pile for efficient digging. Where off-highway trucks are used for haulage, tyres can constitute something of the order of 20% of the haulage unit's total operating cost. Hence, to optimize this cost proportion it is necessary to match tyre size to the unit as well as to operating conditions.

After equipment selection which appears to offer overall optimization of projected costs against productivity, it is necessary to look at mining practices. The construction and maintenance of good haulage roads is essential. This provides access and rapid routing of haulage out. A good road layout has closely engineered grades and curves. Passing room is essential so that a slow unit does not set the pace for all. Direct radio communications between pit foreman, shovels, trucks, maintenance and shovel and dump clean-up equipment assists in efficient dispatching and supervision control.

Equipment records are an essential part of anticipating replacement requirements, scheduling maintenance and modifying practices if necessary. In selection of equipment there is usually a range of sizes of equipment and differing methods to do a job. The final choice needs to be based on equipment and methods which provide safety, and the greatest production for the lowest cost. A very important factor is supervision, that is, the pit foreman. To have a good safety record along with the lowest possible operating costs, the pit/quarry foreman must have a good working relationship with his men.

5.5.2 Long-Term Mine Planning

The initial step in open-pit mine design is the compilation of a long-term mining plan or a final pit design. There are two objectives associated with this. First the ore reserves of the mine should be determined, and second, the extent of the orebody and the ultimate pit geometry should be defined. In reality, long-term plans usually change over time to reflect the changing economy, increased knowledge of the orebody and improvement in mining technology. Hence periodic revision of the long-term plans is necessary. To offset the comparatively manpower-intensive task of looking at combinations and varying parameters, use of computers today is necessary. This allows a broader range of variables to be considered in developing a plan. In accomplishing the objective of long-term planning the following conditions need to be met:

- Orebody characteristics such as grade, geological structure and formation need to be established and depicted on vertical and horizontal sections and a mineralization inventory needs to be developed.
- The basic pit design requirements, such as economic and physical parameters' and legal constraints, must be defined.
- The open pit design technique to be used must be determined.

5.5.2.1 Mineralization inventory

The first step in the development of an open pit mine design is the establishment of a mineralization inventory. This is based on results from drilling and surface mapping. In essence it represents a model of the orebody's mineralization, topography and geology. To provide sufficient reliable information for such a model the drilling as well as sampling must be adequate to classify the material and give a definition of geological features. The drill data expanded should hence include assays, geological formation, structure and mineralization type. Results of metallurgical tests also should be tabulated. The assay results of drill hole data can be composited to intervals coinciding with the selected bench height being used (e.g. for a multi-bench pit). This is then used to compile horizontal and vertical sections. The vertical Sections are of particular value when attempting to visualize the orebody and estimate final pit limit, while the horizontal sections are used to evaluate each estimate.

To facilitate the compilation procedure the horizontal sections are divided into a number of blocks or regular shape of some unit volume. Hence the physical characteristics (if necessary) are assigned to each block. The block height reflects the predicted bench height while the actual block width dimension is likely to be determined by:

- General geometry of the orebody.
- Size and shape of the geological features as seen in the model.
- The time interval which the mine plans represent;
- Sample density and spacing (drill-hole frequency).

With the drilling data composited and a block size selected, the next step is the compilation of the horizontal sections by assigning assay values to the blocks. The most common grade prediction technique employed is the principle of nearest points or gradual change. The principle of nearest points consists of assigning to each block in the inventory the value of the nearest sample value. Where there are two or more samples equivalent from a block center, the block is assigned the average of these values. Thus the principle of gradual change assumes a gradually changing condition between sample points. Interpolation is the most common technique used to determine a value between two drill holes (sample points). For widely-spaced holes in a relatively uniformly mineralized orebody, an unsampled block can be assigned a value from values of the surrounding samples.

However, it must be noted that the interpolation technique indicated may not be appropriate to a particular ore body. Mineralization control within an orebody may be the geological structure or favorable strata. Hence a grade prediction should include methods which match the orebody, e.g. to match zonal trends. The method adapted should be one which can be readily adapted to a computer. Once all the ore values are properly set out on a plan map for each level, the project is then ready for pit design.

5.5.2.2 Basic Pit Design Requirements

To determine the final pit limits of an orebody and its associated. Quality (grade) and quantity (tonnage) it is necessary to consider certain economic and physical aspects.

5.5.2.3 Economic aspects

The development of a pit design to establish the final pit limits (unless defined by some artificial boundary, e.g. property boundary) requires that criteria be decided upon to which the pit design is to conform. These criteria are often a function of the corporate objectives of company/s involved and can vary with the type of mineralization - involved. The most common objective is to develop a pit design that will maximize the present value of an orebody or, alternatively, to design such that the total net value of the orebody be maximized, excluding the time value of money (diminishing). The latter criterion gives rise to a 'break-even pit' which infers that in order to maximize total net value of the orebody, the final pit must be expanded to a point where the economic value of the cost cut mined down the final slope approaches zero or break-even.

Hence, the break-even stripping ratio can be defined as follows:

$$\frac{\text{Recoverable value per tonne of ore} - \text{production cost per tonne of ore}}{\text{Stripping cost per tonne of waste}}$$

The production cost is the total of all Costs, through to the saleable product exclusive of stripping cost. The break-even stripping ratio is usually based on variations of ore grade, metallurgical characteristics (anticipated recovery) and the projected market prices. Of course, it is essential initially to ascertain whether the orebody and grade would support a viable open-pit operation. To do so it is necessary that the return revenue will pay for all production and stripping (overburden) removal costs as well as return a margin of profit. Hence, for a completely defined orebody it may be possible to aim for complete extraction. However where the orebody is not fully defined in depth, then economics, that is, the stripping ratio, is the determining factor. The initial viability is checked by comparing projected surface mining costs and underground costs, for example, break-even cost differential between open-pit methods, as against underground methods. Break-even stripping ratio:

$$\frac{\text{Underground mining} \frac{\text{cost}}{\text{tonne}} \text{of ore} - \text{open pit mining} \frac{\text{costs}}{\text{tonne}} \text{of ore}}{\text{Open} - \text{pit stripping} \frac{\text{cost}}{\text{tonne}} \text{of waste}}$$

Where open-pit methods costs are \$0.50/t and, underground methods costs are \$3.50/t hence, in theory we can strip:

$$\frac{6.5 - 1}{\frac{\text{costs}}{\text{tonne}} \text{stripping waste}} = \frac{5.5}{0.5}$$

Tonnes of overburden before we arrive at a break-even limit. The break-even strip ratio where the revenue/tonne is \$10.00 and overall mining, milling and smelting costs are \$5.00, then the strip ratio is: $10 - 0.5/0.50 = 10$.

Where, the stripping cost is 0.50/tonne.

The break-even stripping ratio concept is based on average grades and costs. It can be applied to orebodies with ore-metal or products being mined.

5.5.2.4 Physical aspects

After determining the allowable stripping ratio, the final slopes must be defined. To minimize the overall stripping ratio, the slope should be as steep as possible and still remain stable. Knowledge of the location, orientation and characters of geological structures, as well as operating experience can serve as a basis of estimating final slope. The application of soil mechanics and rock mechanics techniques are significant where large, deep pits are involved. One aspect which needs to be adequately considered is that of time. Steep slopes may stand for a period of several months or even years. Mining plans in this case need to be designed to recover ore as soon ~ as possible. Ground water pressure can also be a significant factor in pit slope stability. With the establishment of the basic pit design requirement, the geometric configuration of the pit can be developed in three ways.

- Manually.
- With a computer.
- A combination of manual and computer techniques, depending upon the size of the orebody, degree of accuracy desired time limits and availability of computer and suitable programs.

Here we examine a manual approach with the use of vertical and horizontal sections. In the simple case we can assume the all-round (average) ultimate pit slope will be approximately 450 and bench height on equipment selection which, in turn, is based on the scale of operations, for example, a 10 m bench. The design is developed from vertical sections on which the surface topography and ore limits cut to the adopted cut-off grade. The cut-off grade being the point where mineralization is not classed as ore and either goes to a low grade stockpile if stripped off for ore recovery; or goes to waste or is left unmined.

The sections are drawn at regular-spaced intervals parallel along the longitudinal axis of the orebody. The pit limit, at the adopted degree of slope, is placed on trial sections at a point that results in the allowable stripping ratio for the grade of ore in that area. In section the length of pit slope line in waste over the length of the pit slope in ore gives a limiting stripping ratio, Fig. 5.22. However, this approach is only applicable to very simple regular orebodies. In practice it is necessary to calculate out the stripping ratio bench by bench as incremental steps until the limiting position is ascertained. Since orebodies are often irregular in terms of shape as well as depth below the surface, radial sections can be used to determine the pit limits. The trail pit line is so located that the stripping ratio is less than that allowable (for the curved geometry of the sector).Where:

- Limiting stripping ratio AB: BC (simple regular ore bodies only)
- Line ABC is the projected pit limit.

The trail pit limit, at the adopted degree of slope, is placed on the trail vertical section at a point that results in the allowable stripping ratio for the grade of ore in that area. The data can then be transferred onto the horizontal plan and evened out by averaging out. The allowable limiting stripping ratios are determined by measuring the respective ore and waste areas in plan (with a planimeter). To determine whether or not the actual stripping ratio for a sector satisfies the allowable stripping-ratio criteria, it is necessary to determine the grade of ore at the pit limits. The grade of ore shown in plan at the pit limit is obtained from vertical sections or from horizontal sections corresponding to each bench level. Hence, a break-even stripping ratio can be determined against the ore grade by sectors. It is important to realize that the ratios so determined for each sector are at the final pit surface and do not reflect the overall stripping ratio of the orebody. Averaging of ore grades and stripping ratios is not, a sound technique for pit design. Each sector needs to meet the economic criteria on its own merits. Where mineralization extends to some depth it may be necessary to determine whether it is to be classified as open pit or underground ore. Project profitability should be the criteria used.

5.5.3 Short-Term Mining Planning

Once a long-term mine plan has been established it is essential to develop a series of short-term mining plans. These will define the intermediate steps required to reach the final pit limits under physical, operating and legal constraints. Here the emphasis is on operation and fundamentally this must be achieved under supervision in order to optimize output for effort input. The short-term plans provide the pit geometry, ore grade, stripping ratio and expected profit information necessary for future production forecasts and equipment needs. Key factors of significance in short-term mine plans are:

- Pit operating procedures relating to productive capacity.
- Existing and projected mining and treatment capacities including refining such as smelting if practiced.
- Orderly mining of the pit.
- Required operating slopes.
- Haulage profiles.
- Water drainage.
- Dump space.
- Equipment maneuverability.
- Equipment availability.
- Corporate goals related.

To ensure that all important factors are taken into account, then a number of alternative short-term plans may have to be examined by the mining engineer. Manually a set of horizontal sections are depicted along with a series of proposed mining cuts, the locations and extent of which will reflect operational factors, that is, covering differing ore types or rock products in quarrying. The mining sequence of the orebody should be analyzed to take into consideration such aspects as metallurgical characteristics, for example, oxide-sulfides, but also varying ore grades, availability of ore, haulage routes, mining capacity etc. Hence, the projected production schedule of both tonnage and grade can vary considerably depending upon objectives adopted. This can vary from mining initial high grades to mining a uniform grade over the life of the mine. The combinations and permutations of such approaches and resultant pit profile outlines are often best done by computer.

An important element in short-term open pit planning is to provide ample operating room to permit more economical mining practices. Tight bench room affects a minimum stripping ratio but results in a costly operation, as well as hampering drill and blast operations. Shovel and haulage operations are facilitated by ample working room. This means flat pit slopes in contrast to the final pit slope that must be as steep as possible to minimize the overall stripping ratio. Obviously it is necessary to avoid high peak ratios for short periods. Poor equipment utilization can occur where high investment is necessary to carry the operation over such peaks. Failure to adequately develop ore stocks ready for mining by adequate stripping can result in significant production shortfalls. Hence, the compromise is to minimize high stripping ratios during early mine life with operation slopes as steep as possible while providing ample bench room for optimum operating efficiency. It is therefore necessary to know the room required for efficient utilization of equipment.

Short-term planning is hence largely related to production personnel who must co-ordinate production, equipment locations and development work within that long-term plan. Fundamentally, this production is solely concerned with men and machines. The sequence of operations involved varies from large-scale operations to small quarry pits.

5.5.3.1 Short-Term Design

The main purpose of short-term planning and operation is to provide for:

- Access.
- Services.
- Production inclusive of stockpiles.

These should be achieved while providing for economy, safety, efficiency and practicality. To achieve open-pit equipment operation efficiency, it is necessary to create adequate space or pit room in which to operate. Hence, when conditions become too tight, then the overburden stripping situation can become critical. To emphasize the need for adequate room, the shovel needs to dig along a face, rather than into it so that its swing circle is approximately 900 not 1800. For conventional truck haulage, the routes need to be across the bench and need to meet the requirements of regulations as well as to achieve acceptable production output. Likewise, there must be adequate room for a drilling pattern to be laid out to prepare the next bench.

In the preparation of plans it is necessary to indicate the shoulder and toe of faces, and hence the berm widths at the final pit limits. Where, say, 450 is accepted as a safe slope with 10 m high benches and, say, a face angle of around 80°, then the resultant safety berm is not 10 m but in actuality approaches 7 m due to the slope of the face itself, Fig. 5.23. While a berm lift at each bench may be acceptable, other alternatives are available. A possible saving in waste stripping can be achieved by providing a double width berm on alternate benches and eliminating the intermediate ore, Fig. 5.24. The wider berms can provide better protection against multi-bench failure. Single berm widths can present problems in the latter stages of the pit where they have crumbled or have become covered with fallen debris, so presenting a smooth

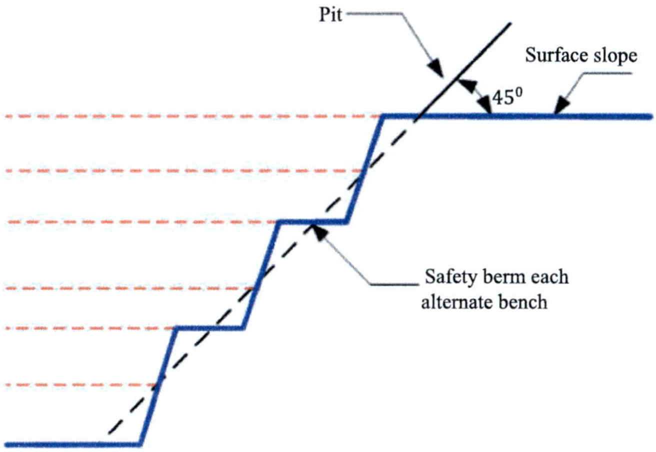

Fig. 5.23 Bench profile – 10 m bench height with safety berm at pit limit

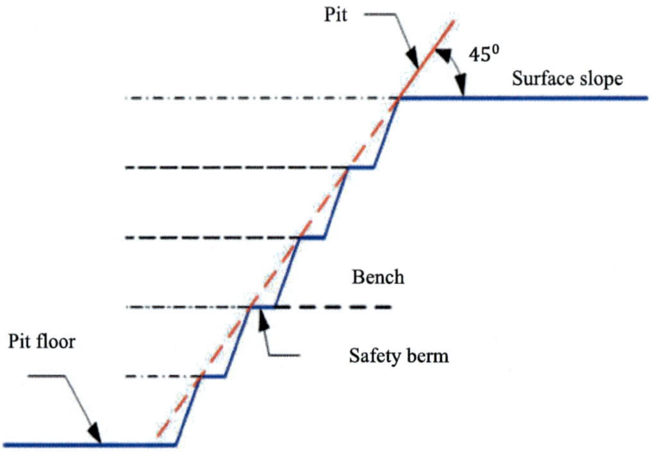

Fig. 5.24 Bench profile −10 m bench height with double safety berm width at pit limit

slide for further falls. In some cases reinforcement of berms may be necessary by anchoring the slope or stabilizing with shotcrete or concrete, but only in extreme cases.

5.5.3.2 Pit Size and Shape

Many factors govern the size and shape of an open pit. These must be properly understood and used in the planning of any open pit operation. The importance of each will depend on the particular project, but the following are the key items affecting the pit design: geology, grade and localization of the mineralization, extent of the deposit, topography, property boundaries, production rates, bench height, pit slopes, road grades, mining costs, processing costs, metal recovery, marketing considerations, strip ratios, and cutoff grades.

5.5.3.3 Bench height

The bench height is the vertical distance between each horizontal level of the pit. The elements of a bench are illustrated in Fig. 5.25. Unless geologic conditions dictate otherwise, all benches should have the same height. The height will depend on the physical characteristics of the deposit; the degree of selectivity required in separating the ore and waste with the loading equipment; the rate of production; the size and type of equipment to meet the production requirements; and the climatic conditions.

The bench height should be set as high as possible within the limits of the size and type of equipment selected for the desired production. The bench should not be so high that it will present safety problems of towering banks of blasted or unblasted

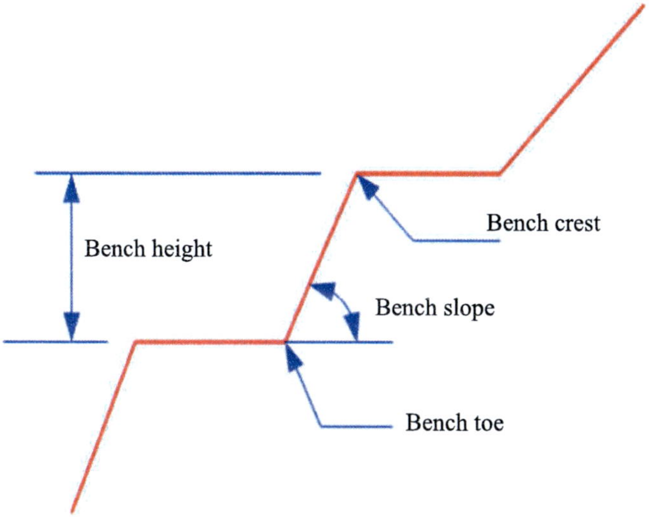

Fig. 5.25 Cross-section of the bench

material or of frost slabs in winter. The bench height in open pit mines will normally range from 15 m (49 ft.) in large copper mines to as little as 1 m (3.3 ft.) in uranium mines.

5.5.3.4 Pit slopes

The slope of the pit wall is one of the major elements affecting the size and shape of the pit. The pit slope helps determine the amount of waste that must be moved to mine the ore. The pit slope is usually expressed in degrees from the horizontal plane. A pit wall needs to remain stable as long as mining activity is in that area. The stability of the pit walls should be analyzed as carefully as possible. Rock strength, faults, joints, presence of water, and other geologic information are key factors in the evaluation of the proper slope angle. The slope may be stated as a simple, overall average for the pit (45°), but a more detailed study may show that the physical characteristics of the deposit cause the pit slope to change with rock type, sector location, elevation, or orientation within the pit. Figure 5.26 illustrates how the pit slopes may vary in the deposit.

A proper slope evaluation will give the slopes that allow the pit walls to remain stable. The pit walls should be set as steep as possible to minimize the strip ratio. The pit slope analysis determines the angle to be used between the roads in the pit. The overall pit slope used for design must be flatter to allow for the road system in the ultimate pit.

Figures 5.27 and 5.28 show the need to design the pit with a lesser slope to allow for roads. The pit in Fig. 5.27 has been designed with a 45° angle for the pit walls.

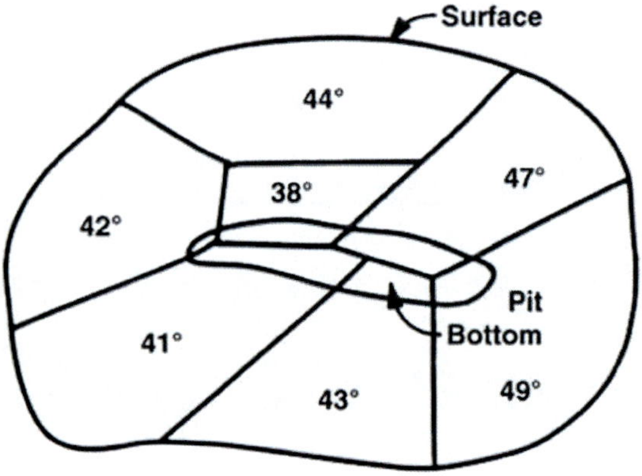

Fig. 5.26 Example of pit slopes varying in a deposit

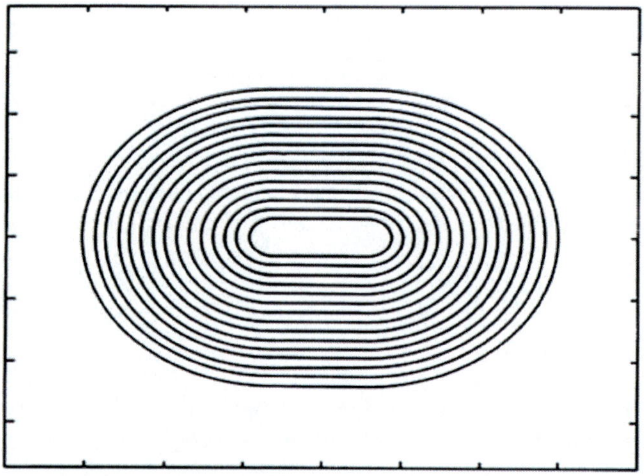

Fig. 5.27 Pit designed with a 45° pit slope

The pit in Fig. 5.28 uses the same pit bottom and the 45° inter-ramp slope between the roads, but, a road has been added. Note the larger pit that results. In the example, almost 50% more tonnage must be moved to mine the same pit bottom.

In the early design of a pit a lesser pit slope can be used to allow for the road system. The pit in Fig. 5.29 was designed with an overall slope of 38°. The overall slope to use will depend on the width, grade, and anticipated placement of the road.

Figure 5.30 shows a vertical section of a pit wall from Fig. 5.28. The inter-ramp angle is projected from the pit bottom upward to the original ground surface at point B. The overall pit slope angle is the angle from the toe of the bottom bench to the crest of the top bench. Point A shows the intercept of the overall pit slope angle with the original ground surface.

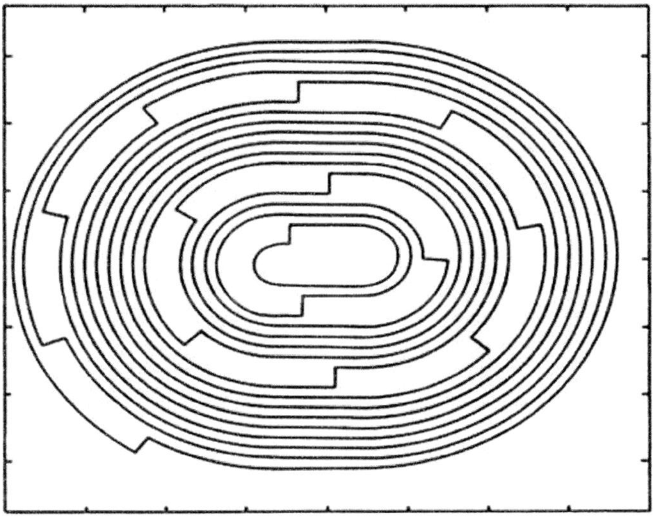

Fig. 5.28 Pit designed with a 45° inter-ramp slope

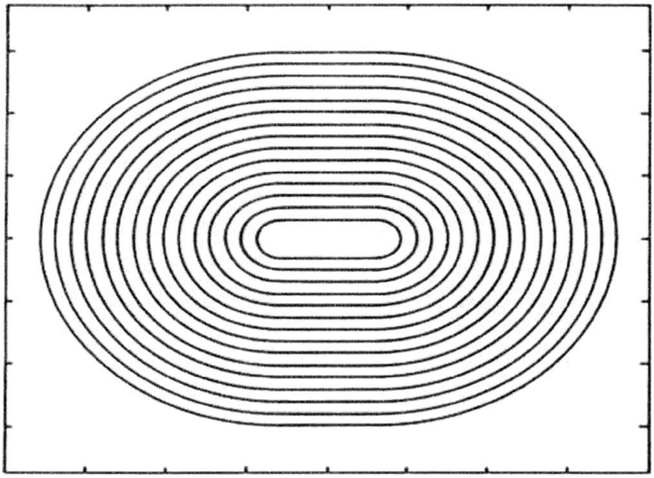

Fig. 5.29 Pit designed with a 38° overall slope to allow for a 45° inter-ramp slope and a road system

5.5.3.5 Cut-off grade

A "cutoff grade is any grade that for any specified reason is used to separate any two courses of action." The reason used in setting a cutoff grade usually incorporates the economic characteristics of the project. When mining, the operator must make a decision as to whether the next block of material should be mined and processed;

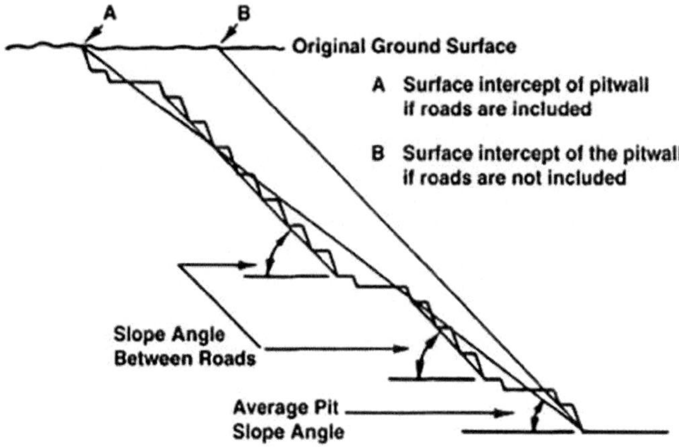

Fig. 5.30 Vertical-section through a pit wall

mined and stockpiled; mined (to expose ore) and sent to the waste dump; or not mined at all. The grade of the block is used to make this decision. For any block to be deliberately mined, it must pay for the costs of mining, processing, and marketing. The grade of material that can pay for this but for no stripping is the breakeven mining cutoff grade. A second cutoff grade can be used for blocks that are below the mining cutoff grade and would not be mined for their-own values. These blocks may be mined as waste by deeper ore blocks. The cost of mining these blocks is paid for by the deeper ore. The final destination of these blocks is then only influenced by costs for the blocks once they have been mined.

The blocks can be processed at this point if they can pay for just the processing and marketing costs. Because the revenue for the block does not need to cover the mining cost, the milling cutoff grade is lower than the mining cutoff grade. The cutoff calculation depends on the point of the cutoff decision in the life of the mine. In deciding whether to mine one more block at the end of the mine life, the only costs used would be the cash operating costs and a minimum profit to reflect the opportunity costs of using the money elsewhere. For a decision to mine one more year, the costs would be the cash operating costs, plus the replacement capital needed, plus all general and administrative costs that would be incurred. For a mine in the planning stage, the costs to be used are more complex and must be carefully considered. All direct costs of mining, processing, and marketing should be used. In the mining phase this would include the drilling, blasting, loading, and hauling costs. The processing costs would cover crushing, conveying, grinding, and concentrating costs. Depending on the final form of the product, the marketing costs could include concentrate handling, smelting, refining, and transportation. Additional direct costs for royalties and taxes would also be included. Overhead costs should also be added to the calculation. The general and administrative costs for the mine, mill, and administrative office staff should be included. Until the size of the pit has been determined and the associated overhead costs developed, the costs to be used for

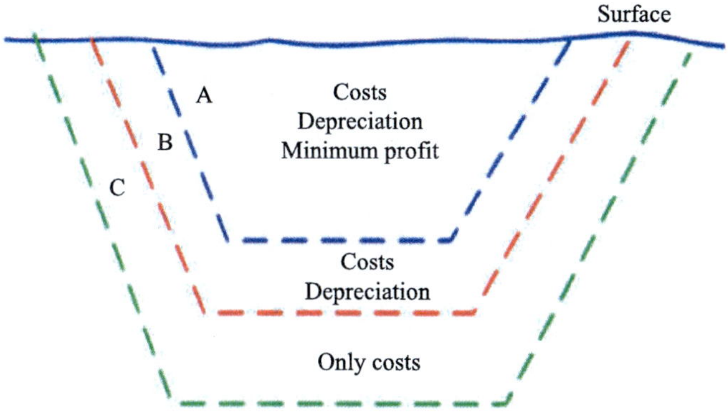

Fig. 5.31 Relative pit sizes using different levels of costs

the calculation can only be estimated and are therefore subject to later refinement. Depreciation is used in the calculation for the purpose of setting the pit size. As shown in Fig. 5.31, the size of the pit will increase if the burden of some costs is removed. The cutoff grade is lower in increment C than in increment B. This is due to the lower costs used in determining the cutoff grade.

The material in increment C can only be economically mined after the plant has been depreciated. A plant built to handle the material in increment C would not be justified because the revenue would not cover the cost of the plant. If the plant was fully depreciated by the time increment C was mined, the ore would be worth processing. A minimum profit can also be used to calculate the cutoff grade. It will further decrease the size of the pit as shown by increment A, in Fig. 5.31. The purpose of adding a minimum profit is twofold: (1) it confirms that a block is ore only if it can be mined and processed at a profit; and (2) it sets an economic limit below which a company would find an alternate investment more attractive.

The amount of minimum profit to be used is a difficult decision. A true profit calculation would include the role of depreciation, depletion, and taxes. At the design stage, these are not known. An approximation can be made by increasing the costs. Other costs and changes in revenue can be included if they are known. These would include recoveries that vary with the ore grade, mining costs that vary with the distance or elevation of haulage, and the time lag between stripping the waste from a block of ore and the mining of the ore. These values should only be added if they are well known and the added degree of sophistication is warranted. The calculation of the mining cut-off grade, for a copper project, is given below with the following parameters:

30	Kt./d. (33,000 st pd) of ore mined for 20 years
$300,000,000	Capital cost (including replacement capital)
$1.00	Mining cost per tonne of ore
$0.95	Mining cost per tonne of waste

(continued)

(continued)

$3.00	Processing cost per tonne of ore
$1.00	General and administrative (G&A) cost per tonne of ore
$0.75	Freight, smelter, and refining (FSR) cost per kilogram of copper
85%	Overall copper recovery

The results are shown graphically in Fig. 5.32. Note that the cutoff grade will increase as the costs increase. The difference between the mining cutoff grade and the milling cut-off grade is shown in Fig. 5.33 (Table 5.6).

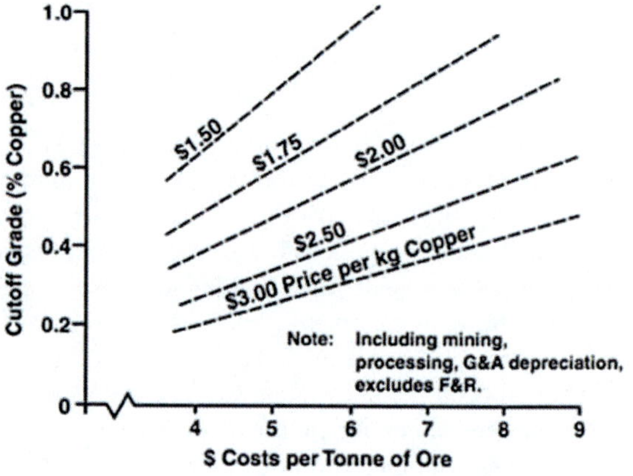

Fig. 5.32 Cut-off grades for different costs and metal prices

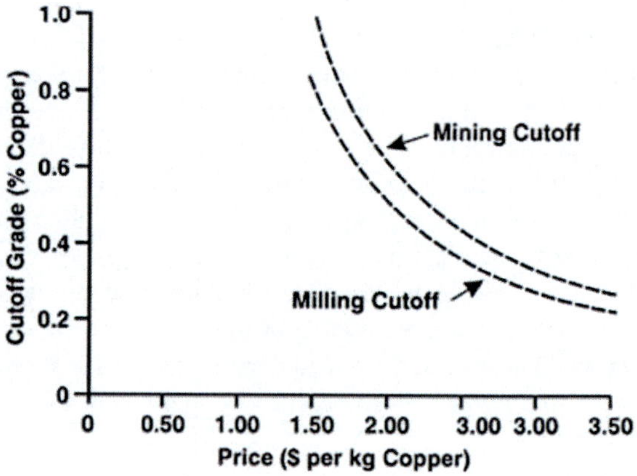

Fig. 5.33 Relationship of mining and milling cut-off grades

Table 5.6 Calculation of break-even cut-off grade

Head grade	0.80% Cu		0.70% Cu	
Recovery	0.85% Cu		0.85% Cu	
Recoverable copper per tonne	6.80% Kg		5.95% Kg	
Costs	Per tonne ore	Per kg Cu	Per tonne ore	Per kg Cu
Mining	$1.00		$1.00	
Processing	3.00		3.00	
General and administrative	1.00		1.00	
Depreciation	1.40		1.40	
Total	$6.40	$0.94	$6.40	$1.08
Freight, smelting				
Refining	5.10	0.75	4.46	0.75
Total	$11.50	$1.69	$10.86	$1.83
Value @ $1.75/Kg	$11.90	1.75	10.41	1.75
Net value	$0.40	$0.06	($0.45)	($0.08)
Cutoff grade		0.753% Cu (by interpolation)		

5.5.4 *Stripping Ratio and Pit Limit*

The strip ratio is the ratio of the number of tonnes of waste that must be moved for one tonne of ore to be mined. The results of a pit design will determine the tonnes of waste and ore that the pit contains. The ratio of waste and ore for the design will give the average strip ratio for that pit. This differs from the breakeven strip ratio used to design the pit.

The breakeven strip ratio refers only to the last increment mined along the pit wall. The strip ratio is calculated for the point at which break even occurs and the necessary stripping is paid for by the net value of the ore removed. The calculation for the breakeven strip ratio (BESR) is, Eq. 5.30:

$$\text{BESR} = \left(\frac{\text{A} - \text{B}}{\text{C}}\right) \tag{5.30}$$

where:

A Revenue per tonne of ore.
B Production cost per tonne of ore (including all costs to the point of sale, excluding stripping).
C Stripping cost per tonne of waste.

In certain studies a minimum profit requirement is included in the formula, Eq. 5.31:

Table 5.7 Typical equivalent yardage ratings

Material	Ratings (e)
Dredged mud, water	0.5
Loose sand	0.7
Common soil (sand, loam, till)	1.0
Hard soil (clay, hardpan)	1.5
Shale rock	1.5–2.5
Sandstone, limestone	2–3
Hard taconite	3–5

$$\mathrm{BESR} = \left[\mathrm{A} - \left(\frac{\mathrm{B}+\mathrm{D}}{\mathrm{C}}\right)\right] \tag{5.31}$$

where:

D minimum profit per tonne of ore.

Table 5.7 contains the information for calculating the strip ratio for the example used in calculating the cutoff grade previously. The results are shown graphically in Fig. 5.34.

5.5.4.1 Maximum Versus Overall Stripping Ratio

It is on the basis of calculating stripping ratios that we are able to locate pit limits and to express volumes of overburden to be moved per unit weight of ore, coal, or stone uncovered. We must distinguish between two stripping ratios (units: yd^3/ton or m^3/tonne):

- Maximum allowable stripping ratio SRmax, = volume of overburden per weight of ore at economic pit limit = $\frac{V}{W}$
- Overall stripping ratio SR0 = volume of overburden per weight of ore for entire ore body or cross-section = $\frac{\mathrm{V}}{\mathrm{W}}$

In that singular case, SRmax misses significance, and the pit limits are located at the property lines. Another distinction is that SR0 is an actual numerical ratio of yd^3/ton (m^3/tonne); whereas; SRmax is expressed in units of equivalent yards (see the next section). Under long-range planning, we discovered that the maximum stripping ratio, while a physical quantity, is determined solely by economics. The overall stripping ratio, on the other hand, has mainly physical significance. It is because of its economic basis that we can employ SRmax to locate the pit limits of a deposit in the general case, that is, an ore body of varying thickness, dip, or grade occurring beneath an inclined or horizontal surface.

In only one instance; SRmax does not establish the pit limit or exceed SR0 in magnitude, and that occurs when:

Head Grade	1.0 8.50		0.9 7.65		0.8 6.80		0.7 5.95		0.6 5.10		0.5 4.25	
(% Cu)	ton	kg	ton	kg	ton	kg	ton	kg	ton	kg	ton	kg
Kg Cu recovered per tonne ore	Ore	Cu	Ore	Cu	Ore	Cu	Ore	Cu	Ore	Cu	Ore	Cu
COSTS:												
Mining* ($)	$1.00											
Milling ($)	3.00											
G&A ($)	1.00											
($)	5.00	0.59	5.00	0.65	5.00	0.74	5.00	0.84	5.00	0.98	5.00	1.18
FSR ($)	6.38	0.75	5.74	0.75	5.10	0.75	4.46	0.75	3.83	0.75	3.19	0.75
	11.38	1.34	10.74	$1.4	10.10	1.49	9.46	1.59	8.83	1.73	8.19	1.93
Depreciation ($)	1.40	0.16	1.40	0.19	1.40	0.20	1.40	0.24	1.40	0.28	1.40	0.33
Total Cost ($)	12.78	1.50	12.14	1.59	11.50	1.69	10.86	1.83	10.23	2.01	9.59	2.26
BREAKEVEN STRIPPING RATIO:												
@ $1.75/kg Cu												
Value ($)	14.88		13.39		11.90		10.41		8.93		7.44	
Net	2.10		1.25		.40		(0.45)		(1.30)		(2.15)	
Ratio†	2.2:1		1.3:1		0.4:1		—		—		—	
@ $2.00/kg Cu												
Value ($)	17.00		15.30		13.60		11.90		10.20		8.50	
Net	4.22		3.16		2.10		1.04		(0.03)		(1.09)	
Ratio†	4.4:1		3.3:1		2.2:1		1.1:1		—		—	
@ $2.25/kg Cu												
Value	19.13		17.21		15.30		13.39		11.48		9.56	
Net	6.35		5.07		3.80		2.53		1.25		(0.03)	
Ratio†	6.7:1		5.3:1		4.0:1		2.7:1		1.3:1		—	
@ $2.50/kg Cu												
Value ($)	21.25		19.13		17.00		14.88		12.75		10.63	
Net	8.47		6.99		5.50		4.02		2.52		1.04	
Ratio†	8.9:1		7.4:1		5.8:1		4.2:1		2.7:1		1.1:1	
* Excludes stripping cost.												
† At stripping cost of $0.95 per tonne of waste.												
() Indicates negative value.												

Fig. 5.34 Breakeven strip ratios calculation

- The surface is flat, and
- The deposit is flat, tabular, and of constant thickness.

5.5.4.2 Equivalent Yardage

In developing a relationship and procedure to locate the pit limits in an ore body or coal deposit, we will employ a unit called the equivalent yard. Equivalent yardage is the volume of overburden, which costs a unit amount to move (expressed in $/yd^3, or $/m^3) and is accepted as a standard for the mine or district in which it occurs. It is a dimensionless unit (Fig. 5.35).

Examples of some standards and typical costs are:

- Lake Superior iron ranges (loaded and hauled),
- Glacial till: $0.25–0.S0/yd^3 ($0.33–0.65/m^3),
- Eastern U.S. coal fields (cast),
- Soil or decomposed rock: $0. 10–0.30/yd^3 (50.13–0.39/rn^3),
- Western U.S. porphyry copper district (blasted, loaded, and hauled), and
- Quartz monzonite porphyry: $0.50–1 0.00/yd^3 ($0.65–1.31/m^3).

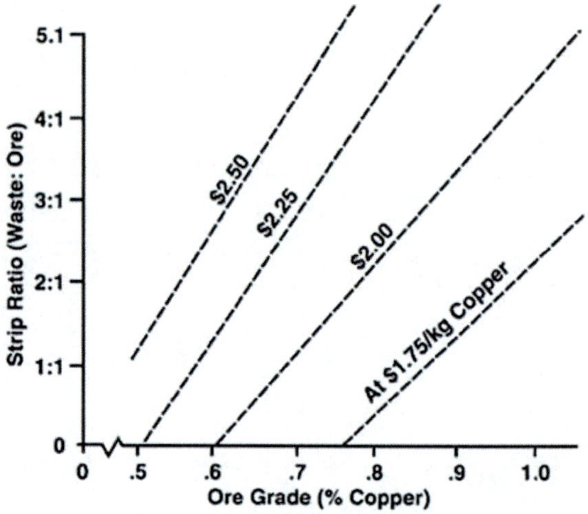

Fig. 5.35 Strip ratios for different ore grades and metal prices

The equivalent yardage rating e of a material is calculated with reference to the above standards, which are assigned a value of unity (e = 1). For example, at an eastern U.S. surface coal mine, if it costs $0.40/yd³ ($0.52/rn³) to cast well-blasted rock, then the rating of the rock is e = 2, based on an average cost to cast soil of e = 1 of $0.20/yd³ ($0.26/rn³), the standard material in that district. Table 5.7 gives typical equivalent yardage ratings.

Equivalent yardage is a colloquial term, borrowed from the construction industry. More precisely, we should say "equivalent cubic yardage," since in reality we are using a Unit of volume measure. The concept of equivalent yardage is useful in dealing with a variety of stripping materials in calculating the maximum allowable stripping ratio and pit limits.

5.5.4.3 Relationships for Maximum Stripping Ratio and Pit Limit

The physical and economic relationships of the maximum allowable stripping ratio at the pit limit enable us to develop a mathematical expression to locate the pit limit. We can do this by using simplified geometric representations of the pit. In Fig. 5.36a, an inclined mineral deposit intersecting a horizontal surface is presented in cross-sectional view. The deposit thickness is t, its dip is a, its tonnage factor is TF, and its inclined length to the pit limit is m. The pit slope is, β its inclined length is, l its vertical height is h, and the horizontal distance from the outcrop to the pit limit is d. Note that d is measured to the crest of the bank and m to the toe. For purposes of simplicity, we assume the thickness of the cross section to be 1.0 ft. Figure 5.36b. The overall volume of the overburden is V, and the overall weight of the ore is W. We can establish a geometric relation for SRmax if we assume that the cost of removing the cross-hatched quadrilateral of overburden in Fig. 5.36a will equal to the net value

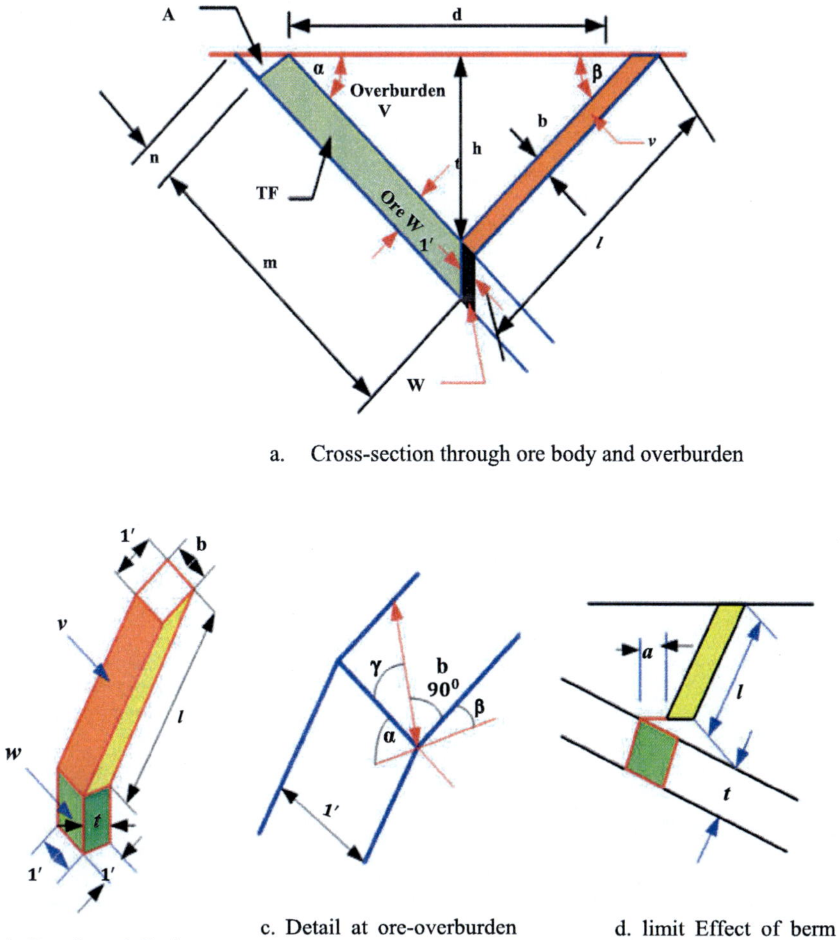

a. Cross-section through ore body and overburden

b. Detail at pit limit

c. Detail at ore-overburden intersection

d. limit Effect of berm at pit limit.

Fig. 5.36 Geometrical-relation of pit parameters and SRmax at pit

of the ore that is uncovered. The volume of overburden removed is designated v, and the tonnage of ore recovered is w, mathematically, the calculation of the overburden can be written as follows, Eq. 5.32:

$$v = \frac{1.0 \times b \times 1}{27} \tag{5.32}$$

where:

b and l are in ft. or (m) and v is expressed in units of yd^3 or (m^3).

Note that the constant 27 must be dropped if meters are used as the unit in the equation. For the ore, Eq. 5.33 the mathematical formula is introduced as:

$$w = \frac{1.0 \times 1.0 \times t}{TF} \tag{5.33}$$

Note that: unit dimensions are indicated by 1'.where:

- t is in ft. (m).
- TF is the tonnage factor in ft.^{3}/ton (m^3/tonne).
- w is expressed in tons (tonnes).

The ratio of the two is the maximum allowable stripping ratio, Eq. 5.34:

$$SR_{max.} = \frac{V}{W} = \left(\frac{\frac{ebl}{27}}{\frac{t}{TF}} \right) \tag{5.34}$$

Again, omit the 27 when using SI units. Insertion of the equivalent yardage e in the equation will permit us to use the equation when different materials are present in the overburden. Finally, it is convenient to determine a geometric expression for b. Referring to Fig. 5.36c, we see that:

$$\alpha + \beta + \gamma + 90^\circ = 180^\circ.$$

Therefore, Eq. 5.35:

$$b = 1.0 \times \cos\gamma = \cos(90^\circ - \alpha - \beta) \tag{5.35}$$

Because SRmax is calculated using the economic relationship in Eq. 5.34, we are in a position to determine l, the inclined slope at the economic pit limit. Rewriting Eq. 5.34, we obtain, Eq. 5.36:

$$l = \frac{27\,t \times SR_{max.}}{e \times b \times TF} \tag{5.36}$$

By trigonometry, we can find the vertical height of pit slope h in ft. (m) as expressed in Eq. 5.37:

$$h = l \sin\beta \tag{5.37}$$

Assuming a berm width a as a safety features, Fig. 5.36d, the horizontal distance from the outcrop m in ft. (m) is given in Eq. 5.38:

$$d = a + \frac{h}{\tan\alpha} + \frac{h}{\tan\beta} \tag{5.38}$$

Finally, the inclined length of ore from the outcrop m in ft. (m) is expressed as per Eq. 5.39:

$$m = \frac{h}{\sin\alpha} \tag{5.39}$$

There are several variations of the previous deposit geometry that affect the determination of SRmax and the pit limits. Two of them are shown in Fig. 5.37. In both, two different overburden formations occur, with equivalent yardage ratings of e1 and e2, respectively. In Fig. 5.37a, the surface is horizontal and the deposit dips. In Fig. 5.37b, the surface is inclined and the deposit is flat. In writing a relation for maximum stripping ratio, we modify Eq. 5.34 to get SR_{max} as given in Eq. 5.40:

$$SR_{max.} = \frac{V}{W} = \left(\frac{\frac{e_1 b_1 l_1 + e_2 b_2 l_2}{27}}{\frac{t}{TF}}\right) \tag{5.40}$$

Figure 5.37a depicts typical dipping tabular non-coal deposit with different overburden formations e_1 and e_2. While Fig. 5.37b shows the typical flat tabular coal deposit with different overburden formations e_1 and e_2.

If the bank angle is constant in different overburdens, then $b_1 = b_2$. After calculating a numerical value for l_2 using trigonometry, the equation can be solved for l1, the only remaining unknown. Then the inclined pit slope $l = l_1 + l_2$ and values of h, d, and m can be found using Eqs. 5.37, 5.38, and 5.39, modified trigonometrically if the surface slopes. One caution in calculating the value of b using Eq. 5.35: both α

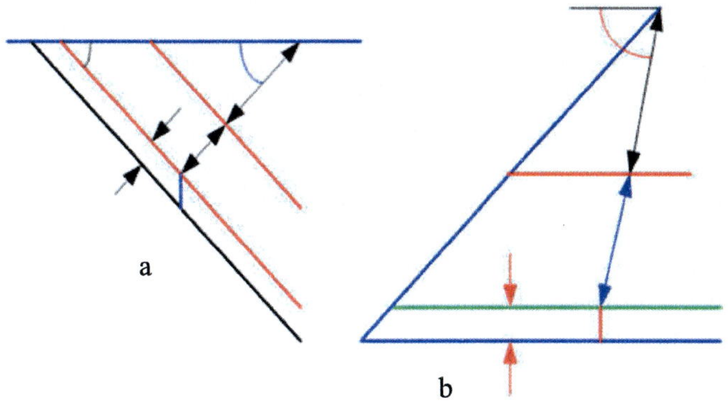

Fig. 5.37 Variations in deposit geometry and overburden composition

and β should be measured from the horizontal. An error in b will result otherwise. If there are two variables in the deposit geometry (i.e., both t and I vary), then Eq. 5.36 must be solved by trial and error. In this case, a graphical solution may be quicker than an algebraic one.

5.5.4.4 Determination of Overall Stripping Ratio

In calculating the potential return on a surface mining venture, the overall stripping ratio is more important than the maximum stripping ratio. We must therefore also calculate the overall stripping ratio for a given cross section or for the entire mine. In our case, we may now want to determine SR0 for a cross section where we have already calculated SRmax. Referring to the general case in Fig. 5.37, the total volume of overburden V in yd^3 is found trigonometrically as, Eq. 5.41:

$$V = \frac{1.0 \times 0.5 \times h \times d}{27} \tag{5.41}$$

Similarly, the total weight of ore W in tons (tonnes) is:

$$W = \frac{1.0 \times m \times t}{TF} \tag{5.42}$$

Therefore, the resulting expression for overall stripping ratio in yd^3/ton is:

$$SR_0 = \frac{V}{W} = \left(\frac{\frac{0.5\,h\,d}{27}}{\frac{mt}{TF}} \right) \tag{5.43}$$

Note that Eqs. 5.41 through Eq. 5.43 must be altered slightly to produce answers in m^3/tonne. Equation 5.43 ignores the area (A) in Fig. 5.37a. To correct the volume of ore removed, that area must be calculated trigonometrically and added to W. The configuration of surface mines used here is simplified to illustrate the stripping ratio principles. In practice, the cross sections will not normally consist of simple geometric figures. Therefore, most of the cross sections must be analyzed using area determinations that are more sophisticated than those demonstrated here. For many years, engineers performed their volume calculations using polar planimeter (area measurement devices). Today they have software programs and digitizing equipment that perform the same function mathematically in less time and with more reliable results. These calculations are now routinely performed by computer systems that determine pit limits, calculate reserves, perform pit planning, and estimate grades and costs.

5.6 Special Topics

5.6.1 Calculation of Stripping Ratios and Pit Limits

Using the basic equations developed in this chapter, we can now apply the principles to calculate the maximum and overall stripping ratios and the pit limit for ore, coal, and stone deposits. The following example outlines the procedure for a dipping mineral deposit of modest value.

5.6.1.1 Example 5.1

The following data are given for a mineral deposit occurring under conditions similar to those in Fig. 5.37:

- Value of ore = \$4.80/ton (\$5.29/tonne).
- Costs (excluding stripping) \$3.30/ton (\$3.64/tonne).
- Stripping cost (for overburden of e = 1) = \$0.20/ton (\$0.26/tonne),
- Berm dimension a = 0 ft. (0 m),
- Dip of deposit $\alpha = 20°$,
- Pit slope $\beta = 60°$,
- Deposit thickness t = 50 ft. (15.2 m),
- Equivalent yardage e = 2.0, and
- Tonnage factor TF 15 $ft.^3$/ton (0.47 m^3/tonne).
- Calculate SRmax;
- Locate the pit limit for the deposit in terms of h; and

Calculate SR0

Solution

- Using Eqs. 5.31, 5.35, 5.36, and 5.37 to find SR_{max}. and the pit limit:
- Stripping allowance (for breakeven or zero profit) = value − cost = 4.80 − 3.30 = \$1.50/ton.
- $b = \cos(90° - 20° - 60°) = \cos 10° = 0.9848$.
- $l = \left(\frac{27 \times 50 \times 7.5}{2 \times 0.9848 \times 15}\right) = 343$ ft..
- $h = 343 \sin 60° = 343(0.8660) = 297$ ft..
- Using Eqs. 5.38, 5.39 and 5.43 to find SR_0:
- $d = a + \frac{h}{\tan \alpha} + \frac{h}{\tan \beta} = 0 + \left(\frac{297}{\tan 20°}\right) + \left(\frac{297}{\tan 60°}\right) = 816 + 171 = 987$ ft.
- $m = \frac{h}{\sin \alpha} = \left(\frac{297}{\sin 20°}\right) = 868$ ft..
- $n = \left(\frac{50}{\tan 20°}\right) = 137$ ft..
- $A = 0.5(50)(137) = 3427$ $ft.^2$.

$$SR_0 = \frac{\text{Volume of overburden}}{\text{Weight of ore for entire ore body or cross} - \text{section}} = \frac{V}{W} = \left(\frac{\frac{0.5h\,d}{27}}{\frac{mt}{TF}}\right)$$

- $= \left(\frac{\frac{0.5\times297\times987}{27}}{3427 + \frac{868\times50}{15}}\right) = 1.736 \cong 1.7\ \text{yd}^3/\text{ton} = 1.5\ \text{m}^3/\text{tonne}$

If a berm had been used in this problem, there would be no change in SRmax or h, but SR0 would increase.

5.6.1.2 Example 5.2 (solution Left for Students)

Calculate the maximum allowable stripping ratio, locate the pit limit (specify h = vertical depth to the deposit and d = horizontal distance along the surface from the outcrop), and calculate the overall stripping ratio (actual yd^3/ton or m^3/tonne) in mining the ore body shown in Fig. 5.38 by surface methods. Allow a 30 ft. (9.1 m) berm and a pit slope of 45°. Cost of excavating an equivalent yard of overburden is \$0.35/yd^3 (\$0.46/m^3), the stripping allowance is \$2.10/ton (\$2.31/tonne) of ore uncovered, and the ore tonnage factor is 16 $ft.^3$/ton (0.50 m^3/tonne). Work the problem first in English units, then in SI units, and check the results to be certain they are compatible.

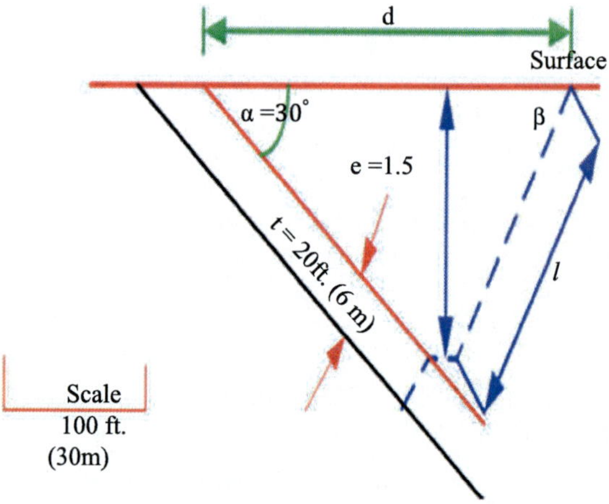

Fig. 5.38 Cross-section of surface mine-problem #1

5.6.1.3 Example 5.3 (Solution Left for Students)

Calculate the maximum stripping ratio and the overall stripping ratio for the open pit shown in Fig. 5.39, given the following conditions.

- Locate and draw the economic pit limit to scale for the cross section shown in Fig. 5.39.

Stripping allowance	\$1.50/ton (\$1.65/tonne) of ore
Stripping cost (e = 1)	\$0.25/yd^3 (\$0.33/m^3) of overburden
Pit slope	65°
Berm, at top of ore	30 ft. (9.1 m)
Surface inclination	25°
Dip of bed	0°
Ore thickness	40 ft. (12.2 m)
Sandstone thickness	100 ft.(30.5 m)
Tonnage factor	12 ft.^{3}/ton (0.37 m^3/tonne)
Equivalent yards, overburden:	
Shale	1.5
Sandstone	2.5

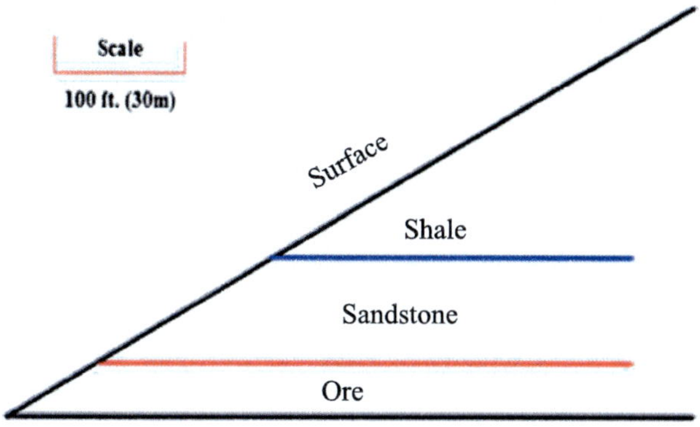

Fig. 5.39 Cross-section of surface mine-problem #2

5.6.1.4 Example 5.4 (Solution Left for Students)

Locate and sketch the pit limit (expressed as the vertical depth of overburden at the maximum stripping ratio) for the coal deposit shown in Fig. 5.40, assuming that a berm is not used. Calculate the maximum and overall stripping ratios. The following conditions are given:

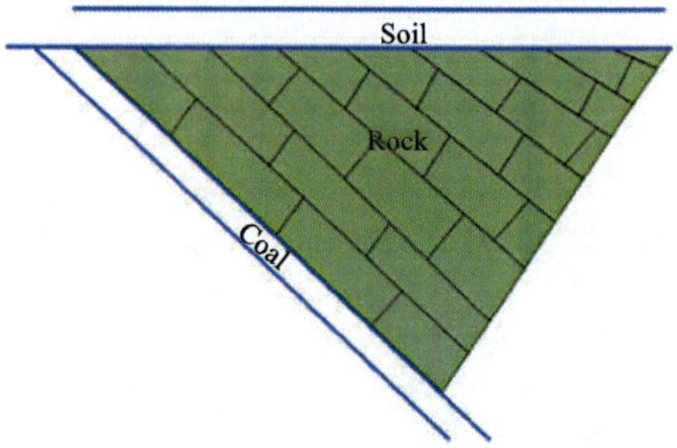

Fig. 5.40 Cross-section of surface mine-problem #3

Value of coal	\$24/ton (\$26/tonne)
Production cost	\$12/ton (\$13/tonne)
Stripping cost	\$0.60/yd^3 (\$0.78/m^3)
Soil thickness	80 ft. (24.4 m)
Pit slope	56°
Dip of the seam	33° 70′
Coal thickness	40 ft. (12.2 m)
Equivalent yards, overburden:	
Soil	0.5
Rock	2.5
Tonnage factor	24.5 ft.^{3}/ton (0.765 m^3/tonne)

5.6.1.5 Example 5.5 (Solution Left for Students)

Locate the pit limit (t, 1, h, and d) by trial and error for the ore body shown in Fig. 5.41 under the following conditions:

Maximum stripping ratio	6.0 yd^3/ton (5.1 m^3/tonne)
Pit slope	35°
Tonnage factor	13.5 ft.^{3}/ton (0.421 m^3/tonne

- Give answers to the nearest 5 ft. (1.5 m), and sketch the pit limit to scale.

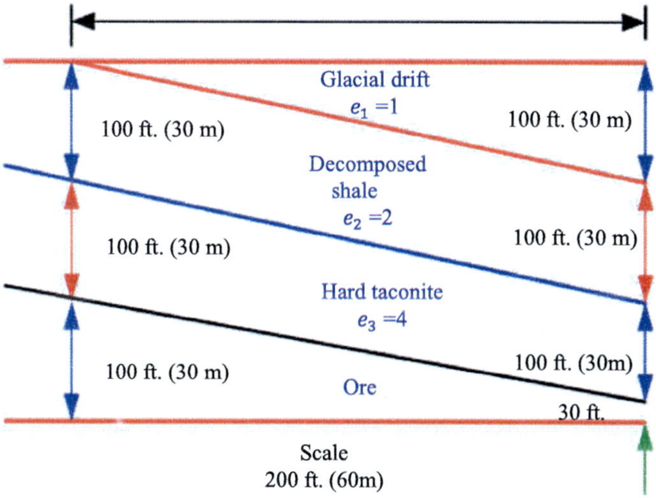

Fig. 5.41 Cross-section of surface mine-problem #4

5.6.1.6 Example 5.6 (Solution Left for Students)

A vertical Kimberlite pipe, cylindrical in shape with a diameter of 300 ft. (91.4 m), is to be exploited by open pit mining. The pit slopes in the Kimberlite and the surrounding overburden will be 45°. If the Kimberlite is worth $20/ton ($22.05/tonne) and the rock costs 84/ton ($4.4 l/tonne) to remove, what is the depth of the pit at the maximum stripping ratio? Assume that the pit limit is an inverted cone and that no hems are left in the ore or the rock. Does the cross-sectional approach work in this case? If not, why?

References

1. Dutta PK, Barman BK, Moitra BC (1973) Some problems of opencast drilling and blasting. Seminar on Limestone and Dolomite Industry in India, Roukema, pp 150
2. Jeong KY, Phillips DT (2001) Operational efficiency and effectiveness measurement. Int J Oper Prod Manag 21:1404–1416. https://doi.org/10.1108/EUM0000000006223
3. Moris R (1983) Design of an opencast mine over previously mined areas. In: Proceedings, surface mining and quarrying, institution of mining and metallurgy, London, pp 288–310
4. Rzhevsky VV (1977) Open mining processes. Nedra, Moscow [in Russian]
5. Lesin YV, Lukyanova SY, Tyulenev MA (2015) Formation of the composition and properties of dumps on the open-pit mines of Kuzbass. IOP Conf Ser Mater Sci Eng 91(1):012093
6. Siskind DE (1973) Ground and air vibrations from blasting. In: SME mining engineering handbook, vol I, New York, USA, pp 11–99
7. Agrawal A, Minis I, Nagi R (2000) Cycle time reduction by improved MRP-based production planning. Int J Prod Res 38:4823–4841. https://doi.org/10.1080/00207540050205659

8. Beyglou A (2016) On the operational efficiency in open pit mines. Licentiate Thesis. Division of Mining and Rock Engineering Department of Civil, Environmental and Natural Resources Engineering Luleå University of Technology Luleå, Sweden
9. Tzu-Hsien L (2009) Optimal production run length and maintenance schedule for a deteriorating production system. Int J Adv Manuf Technol 43:959–963. https://doi.org/10.1007/s00170-008-1773-1
10. Gartner EHE (1969) Garsdorf lignite strip mine—operations to unusual depths. In: Hartman HL (ed) Case studies of surface mining. SME-AIME, New York, pp 12–35
11. Phlevani D, Osanloo M (2015) Resumption of deep open-pit mining as a future challenge may. In: Bandopadhyay S (ed) Application of computers and operations research in the mineral industry. Society for Mining Metallurgy and Exploration Inc.: Englewood, CO, USA, pp 249–256. ISBN 0-8733-5417-6
12. Vasil'ev EI, Kolesnikov VF (1969) Technical-economic bases of schemes for opening up open-pit fields during the working of inclined and steeply-dipping deposits. Soviet Mining 5(5):532–541
13. Valery W, Thornton D, Jankovic A, Kristiansen J (2002) Mine to mill scoping study at Aitik Mine Boliden. Technical Report, No. TM_REP 2002/006, Boliden Mineral AB, Sweden
14. Giri BC, Yun WY (2005) Optimal lot sizing for an unreliable production system under partial backlogging and at most two failures in a production cycle. Int J Prod Econ 65:229–243. https://doi.org/10.1016/j.ijpe.2004.01.004
15. Patti AL, Watson KJ (2010) Downtime variability: the impact of duration-frequency on the performance of serial production systems. Int J Prod Res 48:5831–5841. https://doi.org/10.1080/00207540903280572
16. Rappold J, Yoho KD (2008) A model for level-loading production in the process industries when demand is stochastic. Prod Plann Control 19:686–701. https://doi.org/10.1080/09537280802573726
17. Samanta B, Sarkar B, Mukherjee SK (2013) Selection of opencast mining equipment by a multi-criteria decision-making process. Min Technol 111:136–142
18. Chen S, Jianhong Wu, Zon Y (2020) The impact of the freight transport modal shift policy on China's carbon emissions reduction. MDPI. Sustainability 12:583
19. Li L, Chang Q, Ni J, Biller S (2009) Real time production improvement through bottleneck control. Int J Prod Res 47:6145–6158. https://doi.org/10.1080/00207540802244240
20. Manyele SV (2017) Analysis of waste-rock transportation process performance in an open-pit mine based on statistical analysis of cycle times data. Engineering 9:649–679
21. Schultz CR (2004) Spare parts inventory and cycle time reduction. Int J Prod Res 42:759–776. https://doi.org/10.1080/00207540310001626210
22. Kolesnikov VF (2016) Methods and schemes of opening-up the quarry fields at various bedding conditions of deposits. In: The 8th Russian-Chinese symposium. Coal in the 21st Century: mining, processing and safety
23. Ta CH, Kresta JV, Forbes JF, Marquez HJ (2005) A stochastic optimization approach to mine truck allocation. Int J Min Reclam Environ 19:162–175. https://doi.org/10.1080/13895260500128914
24. Ji P (2018) Forbidding road transport of coal accelerates the "Highway to Railway" but harms the semi-trailer production enterprise in the Beijing-Tianjin-Hebei region. Spec Purp Veh 7:28–32
25. Krzyzanowska J (2007) The impact of mixed fleet hauling on mining operations at Venetia mine. J South Afr Inst Min Metall 107:215–224
26. Kusar J, Berlec T, Zefran F, Starbek M (2010) Reduction of machine setup time. Strojniškivestnik—J Mech Eng 56:833–845
27. Roberson AM (1982) Site selection and design options for uranium mine waste and plant tailings. In: Proceedings of the 12th CMMI Congress, Johannesburg
28. Chanda EK, Gardner S (2010) A Comparative study of truck cycle time prediction methods in open-pit mining. Eng Constr Archit Manag 17:446–460. https://doi.org/10.1108/09699981011074556

Chapter 6
Surface Mining Equipments

6.1 Introduction

The nature and characteristics of rock and rock derivatives forming the crust of the earth vary so greatly from place to place that it is impossible to use the same numerical factors to show the productive ability of excavating equipment on different jobs or projects. Factors for each project must be developed individually and applied as the conditions and requirements dictate. For further reading: Chugh [10] and Weimer [24].

6.2 Types of Draglines

Types of draglines are differentiated by the propelling method used and are generally divided into four classes, as follows:

- Wagon mounted
- Crawler mounted
- Truck mounted
- Walking.

Figure 6.1 shows a truck-mounted dragline. This type is considered to be the most mobile unit. In order to increase stability over the side, outriggers are installed as part of the truck chassis. This feature is not shown in this figure. Figure 6.2 shows a crawler-mounted dragline. The long, wide crawlers are used to increase the stability and reduce the bearing pressure of this unit. These features allow for use of these units on surfaces too soft for the wagon- or truck-mounted types, Fig. 6.3. The large diameter tub and the well-supported boom through the gantry give high stability for rugged works over a long life. For further reading: Chironis [9]; Hird [13] and Mohammadi et al. [17].

M. M. Ali Elbeblawi et al., *Surface Mining Technology*, Topics in Mining, Metallurgy and Materials Engineering, https://doi.org/10.1007/978-981-16-3568-7_6

Fig. 6.1 Truck mounted dragline

Fig. 6.2 Crawler-mounted dragline

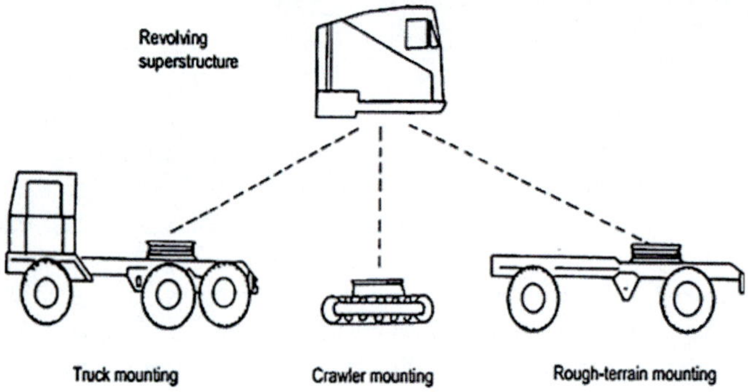

Fig. 6.3 Wagon- or truck-mounted types

As the bucket clears the ground, the swing pedal, on the side corresponding to the direction in which it is desired to swing, is slowly depressed. While swinging, the hoist is raising the bucket higher and the swing and hoist should be coordinated so dumping height and position over dump are reached together.

If the swing is long and the dumping height is low, the loaded bucket is carried at the desired height by selecting the proper point of power on the hoist controller. Heavy loads should not be held longer than is necessary to complete the regular digging cycle. As the desired position for dump is approached, the swing pedals are slowly reversed, bringing the swing motion to a smooth stop. The tension on the drag cable is released allowing the bucket to dump. As the bucket dumps, less power is applied on the hoist. By maintaining the reversal of the swing, the machine will start to return the bucket to the pit where it is lowered and positioned to the point where it is to be filled again. Care should be exercised in paying out and reeving in the drag cable to keep it free and unsnarled on the drag drum. Whenever the operator leaves the operator's seat, he should make all controls inoperative and set all brakes.

6.2.1 *Size of Dragline (Range and Capacity)*

Figure 6.4 is a range diagram of a walking dragline designating the dimensions by letter identification. Table 6.1 is a condensed specification showing the working ranges identified by corresponding letter to those used in Fig. 6.4. Table 6.2 shows additional specification details. For further reading: Erdem and Kormaz [12].

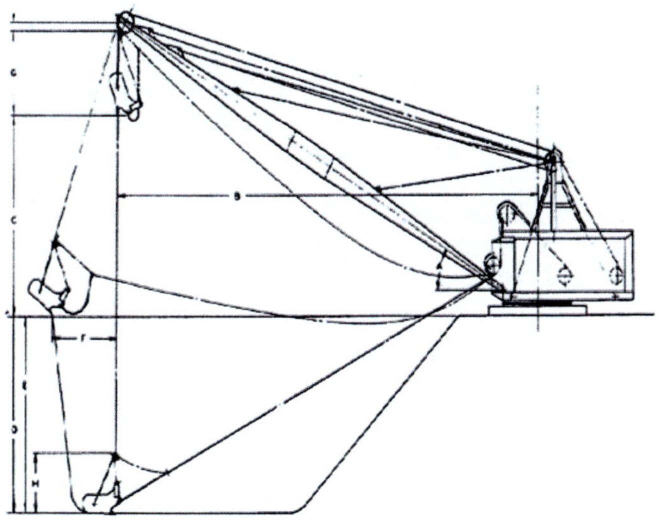

Fig. 6.4 Range diagram of a walking dragline

Table 6.1 Working ranges of 35-yd^3 walking dragline

	Boom length	220′
	Bucket size, yd^3	35
A	Boom angle (approx.)	30°
B	Dumping radius	211′-0″
C	Dumping height	81′-0″
D	Depth (std. cables)	110′-0″
E[a]	Bucket throw	35-55′
	Maximum allowable load, lb	177,500
	Drag pull, single line, lb	325,000

[a]Depends on the skill of operator

Bucket sizes are based on material weighing 100 lb/ft^3

The capacity of draglines is obtained from the allowable loads calculated for each machine. Stability of the machine against tipping under the conditions for its use is considered when determining the allowable load. From Table 6.1 the allowable load is given as 177,500 lb. From Table 6.3 the approximate loaded weight of a dragline bucket of 176,000 lb. corresponds to a rated 35-yd^3 size.

6.2.2 *The Output of Draglines*

A full cycle of operation for a dragline bucket starts with the function of dragging and filling of the bucket; followed by the function of hoist, swing out, and dump; then followed by lower, swing back, and position. Delays inherent on a particular job and those usual to a dragline operation are added to the theoretical time for accomplishment of the above functions in order to achieve a cycle time of the dragline. Different cycle times may be expressed depending upon the portion of the total delays accounted for on the job. When cycle times are used, it is very important to thoroughly understand just what delays are included.

It should be kept in mind that, at this point, the figures are based on perfect conditions. This is true whether the conversion factors for differences in degrees of swing and depths of cut are used or not. It is known that such perfect conditions seldom, if ever exist. Therefore, these figures must be rationalized by some other factors to compensate for the fact that actual job conditions differ widely from the perfect conditions assumed so far. Table 6.4 lists the management factors with respect to the job factors. Figure 6.5 illustrates the additional power required over that from directly under the boom point when the bucket is lifted from the various positions shown. For further reading: Rai [19] and Rai et al. [20].

Table 6.2 35-yd^3 walking dragline condensed specification

Base	
Outside diameter, nominal	50′-0″
Bearing area (effective), ft^2	1965
Walking traction	
Width of shoe	90′-0″
Length of shoe	48′-0″
Width over both shoes	71′-0″
Bearing area of both shoes, ft^2	864
Length of step (approx.)	6′-2″
Walking speed (approx.) mph	0.15
Rotating frame	
Clearance radius, rear end	48′-0″
Clearance under frame	5′-4″
Ground to boom foot	10′-4″
Electrical equipment	
Hoist motors, two-500 hp, blown	1000
Drag motors, two-500 hp, blown	1000
Swing motors, four-137.5 hp, blown	550
Boom hoist motors, Two-20 hp	40
AC driving motors, total hp	2500
Generator capacity to suit motors	
Swing speed, rpm	1.65
Weights	
[a]Domestic shipping weight, less ballast, incl. bucket, lb	3,220,000
Export shipping weight, less ballast, incl. bucket, lb	3,370,000
Working weight, incl. ballast and bucket, lb	3,520,000
Ballast (furnished by purchaser), lb	300,000

[a]Domestic shipping weight subject to ±5% variation

6.2.3 Mining Method

This article will be devoted to the development and application of factors concerning a specific project encompassed by particular conditions and requirements. Each phase of estimated production and costs will be evolved separately. The mining method envisaged to be used here is strip mining uncovering a single coal seam by side casting overburden. It is assumed that the sequence of blasting will be such that excellent fragmentation will result. Figure 6.6 shows a plan and cross-sectional view of the project at the working face of the dragline. From Position No. 1, the dragline

Table 6.3 Dragline buckets

Rated				
Bucket size, yd^3	Actual, ft^3	Height dumping carrying	Approx. weight	
			Empty	Loaded
7	210	20′-6″ 15′	3″ 14,300	35,300
9	270	21′-3″ 15′	3″ 17,600	44,600
11	330	22′-O″ 15′	3″ 21,250	54,250
12	360	22′-3″ 15′	6″ 24,150	60,150
13	390	22′-3″ 15′	6″ 26,100	65,100
14	420	22′-6″ 15′	6″ 28,400	70,400
20	600	28′-O″ 20′	O″ 42,000	102,000
25	750	31′-O″ 23′	O″ 51,000	126,000
30	900	34′-O″ 25′	O″ 62,000	152,000
35	1050	35′-O″ 27′	O″ 71,000	176,000

Table 6.4 Combination job and management factors

Job factors	Management factors			
	Excellent	Good	Fair	Poor
Excellent	0.84	0.81	0.76	0.70
Good	0.78	0.75	0.71	0.65
Fair	0.72	0.69	0.65	0.60
Poor	0.63	0.61	0.57	0.52

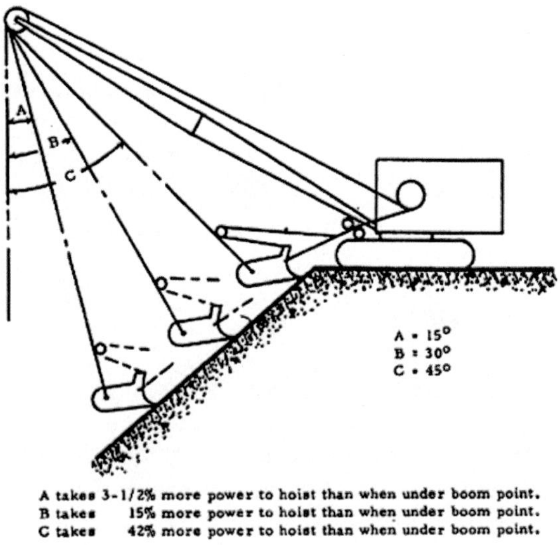

Fig. 6.5 Effect of hoisting angle on power requirements. Where: A: takes 3–1/2% more power to hoist than when under boom point. B: takes 11% more power to hoist than when under boom point. C: takes 42% more power to hoist than when under boom point

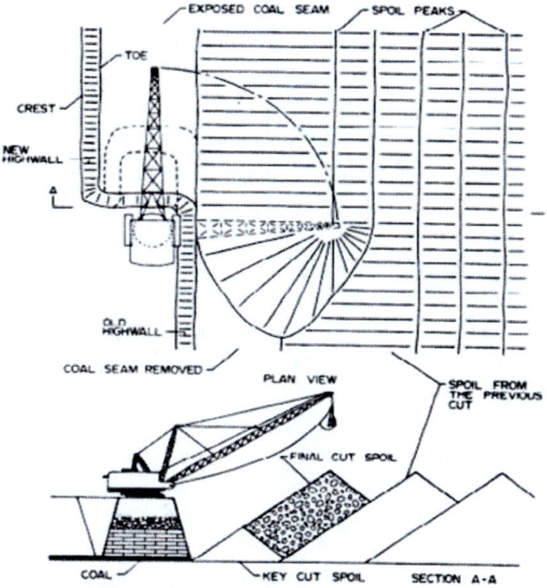

Fig. 6.6 Plan and cross-sectional view at the working face of a dragline

makes the key cut in 80-ft depth of overburden and casts its loads into the adjacent area of the pit from which the coal has been removed. The dumping radius of 211 ft allows the loads front the key cut to build up near the base of the spoil pile figure shown. No rehandling is anticipated under the conditions shown. The dragline moves to Position No. 2 with the edge of its tub at 10 ft from the brow of the high wall. From this position, the machine finishes making the cut and side casts, its loads into the adjacent area of the pit on top of the spoil from the key cut. For further reading: Erdem and Kormaz [12].

Under the conditions shown, a dumping radius of 211 ft and a dumping height of 39 ft are required of the dragline to accomplish this work. The dragline illustrated has a 215-ft dumping radius and an 83-ft dumping height. The following data and conditions are used as a basis for the particular project involved:

1. Slope High wall: 3-ft vertical and 1-ft horizontal.
2. Slope Spoil: 1-ft vertical and 1.25-ft horizontal
3. Swell: 20%.
4. Overburden Material:

   - Solid weight/yd^3 (3240 lb).
   - Loose weight/yd^3 (2700 lb).
   - Swell (20%).

6.2.4 *Average Mining Load per Cycle*

In order to estimate the average mining load in the bucket that will be an average over the economic life of the machine, two factors must be considered. Swell. The first of these factors accounts for the swell of the material to be handled when it is loosened from its in-place position. A cubic yard of material in place is considered to be a solid without air pockets. The weight of this material will be expressed as so many pounds per solid cubic yard. Swell is used as a percentage figure and shows the amount of increase in volume the solid cubic yard of material has expanded when loosened.

If a material has 20% swell, then the solid cubic yard has been increased in volume 1.2 times. A cubic yard taken out of the larger volume containing the loosened material will contain solids and air space. When weighed, this cubic yard will be expressed as so many pounds per loose cubic yard. The weight of air is a negligible, thus, under these circumstances, the weight for the loose cubic yard accounts for the solid matter present. The swell of the material is being loosened from its solid, in-place position is expressed in the ratio, W_S/W_L; where W equals the weight per cubic yard in the solid and WL equals the weight per cubic yard in the loose. The reciprocal W_L/W_S gives the factor accounting for the swell in the bucket factor.

6.2.5 *Fillability*

The second factor accounts for the fillability of the bucket in the loosened material as an average over the economic life of the machine. This factor is obtained from observations and is predicated upon experience relating to the material being handled. The volume of the bucket is considered with the struck capacity reduced approximately 10% for a slope line off the front; this has been accepted as a basis in the trade to determine the rated size of buckets. However, various materials will have various slope lines off the front. In estimating the fillability of the bucket in the material being handled, care should be exercised. The fillability factor is expressed as a percentage figure. When $W_L = 2700$ lb/yd^3 and $W_S = 3240$ lb/yd^3, and if B1 equals the factor accounting for the swell, expressed in percent, then W_L/W_S 2700/3240 = B1 = 83%. If B2 equals the factor accounting for fillability, expressed in percent, and if we assume B2 95%, then the bucket factor becomes B1 × B2 = B.F. or 0.83 × 0.95 = 0.79 (say 0.80 for estimating purposes).

To determine the average mining load per cycle in solid cubic yards, it is only necessary to multiply the bucket factor by the bucket size.

6.2.6 Cycle Times

Initial determination of cycle times begins with theoretical values that must be increased as conditions present on the job consume time away from the actual digging and spoiling of the overburden material.

6.2.7 Theoretical Swing Time

Theoretical swing time for a walking dragline is calculated as part of the theoretical cycle time which is used as a guide in designing a particular size machine. Only the swing motion is considered. The swinging time is calculated using the average inertia of the machine and the available swinging torque from the speed-torque curve, taking the swing gear ratio into consideration. It includes accelerating and decelerating times plus the running time at full speed to obtain the desired total angle of swing. From such calculations, the following tabulation was made for the theoretical swing time for a walking dragline.

Swing time	
Angle	(one way)
72.50	16.0 s
90.0	18.0
120.0	21.0
150.0	25.0
180.0	27.0

The progress of the cycle through the three major stages is shown:

- Drag and fill: 21.5 s.
- Hoist, swing right, and dump: 16 s.
- Lower, swing left and position: 20.5 s.
- Total time elapsed: 58.0 s.

The machine shown has now completed one cycle without delays.

6.2.8 Mining Cycle Time

The recorded cycle time or, alternately, the watch cycle time does not include delays. Delays covering short of time, those usually not recorded on production reports, are in mining cycle time. A partial list of these delays follows:

- Trimming and dressing the digging face.

Table 6.5 A partial list of the work delays

1. Mechanical availability of machine	4. Waiting on other equipment
2. Electrical availability of machine	5. Building roads job
3. Deadheading job	

- Short moves for positioning machine.
- Waiting on other equipment.
- Operator-personal delays.
- Adjustments to machine.
- Occasional bad pass.

An estimated mining cycle time for a certain job is based upon experience, records, and observations.

6.2.9 Percent Operating Time

The major delays resulting in work stoppage of the machine are those tabulated on Table 6.5 and should be taken into consideration in the percent operating time.

Production summary yearly production is estimated by combining the various factors given above as follows, Eq. 6.1:

$$P = L \times C \times \% O.T. \times S \tag{6.1}$$

where:

P	average yearly production.
L	average bucket load-rated bucket capacity × bucket factor.
C	cycle per hour (mining cycle time).
% O.T.	is percent operating time.
S	is scheduled operating hours per year.

6.2.10 Costs

The developments of costs for the operation of the dragline are shown in Table 6.6. They consist of two major types of costs and subdivisions under each.

6.2.10.1 Indirect Costs

The approximate installed cost of the dragline is determined as follows:

Table 6.6 Development of costs for the operation of the dragline

Indirect costs	Direct costs
Amortization	Repair, maintenance and Supplies
Interest, taxes, insurance	Labor
	Power

- Price F.O.B. factory: $2,500,000
- Freight to destination: 75,000
- Ballast F.O.B. jobsite: 5000
- Erection: 250,000
- *Total investment: $2,830,000
- Period of amortization: 30 years
- Scheduled operating time for dragline: 3 shifts a day: 7 days/week: 52 weeks/year
- Hours per year: 8616 h
- Scheduled operating hours covering amortization period of 30 years: 258,480 h

Amortization 30-years period)	per year	per hour
• Percent of total investment:	3.33%	0.000387
Amount:	$94, 300	$10.94

The total investment figures used here do not include the cost of equipment and labor to bring power to the machine. The average investment formula is given in Eq. 6.2:

$$1/2\left(\frac{n+1}{n}\right)\times 100 \tag{6.2}$$

where:

n number of years economic life (30 years for this example)

$$1/2\left(\frac{n+1}{n}\right)\times 100\% = 1/2\left(\frac{30+1}{30}\right)\times 100\% = 51.7\%.$$

Thus, the average investment per year is 51.7% and the number of years of economic life is 30 years. Therefore; number of hours/year is 258,480 h. Using 10% average investment per year = 5.17%:

	per year	per hour
• Interest, taxes, and insurance:	5.17%	0.0006%

6.2.10.2 Direct Costs

- Repairs, Maintenance, Supplies (including Labor).
- Total investment = 100%.

Table 6.7 Breakdown of component items

Electrical maintenance	18.0
Bucket maintenance	27.0
Wire-rope maintenance	13.0
Mechanical maintenance	28.0
Grease and oil	14.0
Sum	100.0

- R.M. and S. = Total of 265%.
- Number years of economic life = 30 years.
- Number hours/year = 8616.
- R.M. and S. = $\frac{265\%}{30}$ = 8.83% per year (% of total investment).
- R.M. and S. = $\frac{8.83\%}{8616}$ = 0.001025% per hour (% of total invest).
- Repairs, maintenance and supplies $\begin{matrix}\text{per year} & \text{per hour}\\ 8.83\% & 0.001025\%.\end{matrix}$

The cost of repairs, maintenance, and supplies is one of the most difficult to estimate. The average figures used presume proper maintenance. Without this, the cost of repairs may greatly exceed the estimates given here. The assumption is also made that the dragline will be used within its specified rating and that the severity of the work is kept in prudent limits. The uniformity of these expenses changes from year to year and it should be expected that the amount of repairs required and the losses caused by down time will increase as the dragline becomes older. Table 6.7 shows a breakdown of component items constituting R.M. and S. by percentage amounts.

Operating charges	% Total investment		
	Amount		
	Per hour	Per year	Per hour
Labor[a] (direct labor/h)	$11.60	3.53	0.000410
Power (kWh/h)	9.40	2.86	0.000332
Operating charges	$21.00	6.39%	0.000742%

[a]The labor rates shown here are not considered to include such items as social security, fringe benefits, overtime, supervisor, administration and insurance

6.2.10.3 Summary of Direct and Indirect Costs

Economic life = 30 years or 258,480 h. Total cost of investment = 100%. The summary for direct and indirect costs is listed in Table 6.8:

Table 6.8 Summary for direct and indirect costs

	% Total investment	
	Per year	Per hour
Amortization	3.33	0.000387
Interest, taxes, insurance (10% of average investment)	5.17	0.000600
Repairs, maintenance, and supplies (265% of total investment)	8.83	0.001025
Sum	17.33%	0.002012%
Labor	3.53	0.000410
Power costs	2.86	0.000332
Sum	23.72%	0.002754%

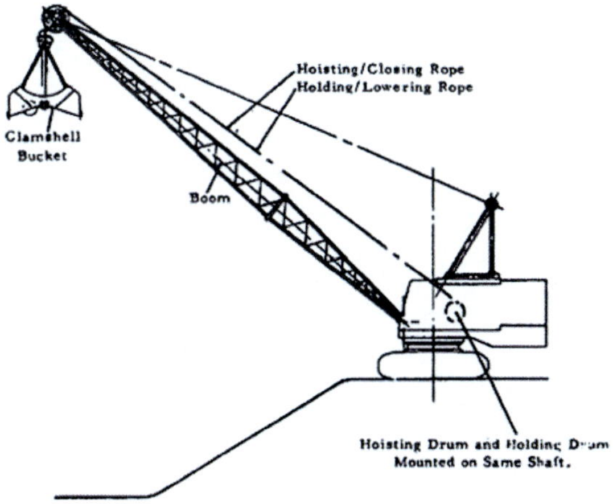

Fig. 6.7 Components of base crane with clamshell bucket

Figure 6.7 shows the various components comprising the base crane and clamshell bucket used in clamshell work. A tag line is not shown. It is used to position the bucket as it is lowered to its digging position, and to keep the bucket from spinning. Clamshells are especially suited to vertical lifting of loose materials from one elevation to another. Various job functions may call for removal of material from below the level upon which the machine rests, or they may require carrying material above the machine. In operation, the clamshell bucket is dumped by releasing the closing line and holding it with the lowering rope (Fig. 6.8).

6.2.11 Outputs of Clamshells

The weight of the bucket and the weight of the maximum payload combined constitute the allowable working loads under various radii. These working loads also vary with the designated percent of tipping. Table 6.9 shows an allowable working load

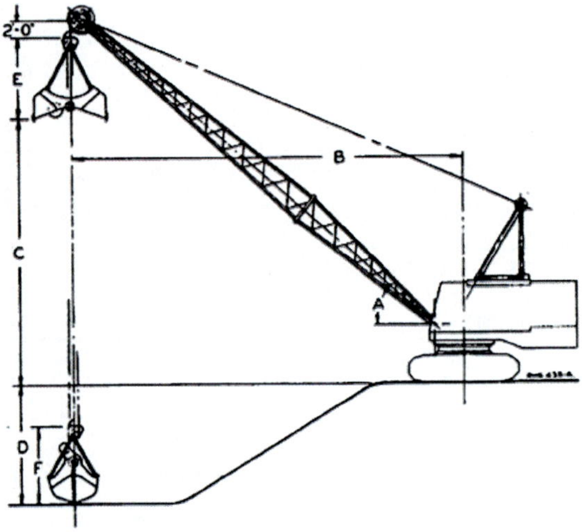

Fig. 6.8 Working ranges of a clamshell unit

Table 6.9 Allowable working loads, lb and 4.5 yd^3 clamshell unit

Boom length radius, ft	80-ft	90-ft	100-ft	110-ft
45	30,000[a]	–		
50	30,000[a]	30,000[a]		
55	30,000[a]	30,000[a]	30,000[a]	
60	29,600	29,225	28,825	28,500
65	26,675	26,300	25,900	25,550
70	24,250	23,875	23,450	23,100
75	–	21,775	21,375	21,000
80	–	–	19,550	19,200
85	–	–	18,000	17,625
90	–	–	–	16,250

specification for a certain machine for the various radii of operation. The approximate weights and dimensions of clamshell buckets are shown in Table 6.10. Combined weight of bucket and contents should not exceed values in table below. Values are based on 66.7% of tipping except those starred (*) which are based on maximum recommended hoist line pull and represent greater stability than 66.7% of tipping.

6.2.12 Working Ranges of Clamshells

Figure 6.9 designates the lettered dimensions of the working ranges, which are provided by the manufacturers for the various clamshell units. Figure 6.10 shows

Table 6.10 Approximate weights and dimensions of clamshell buckets

Size, yd^3	E Height open	F Height closed	Weight of empty bucket, lb
2	10′-9”	8′-6″	7200
2.5	11′-6”	9′-0″	8815
3	11′-9”	9′-6″	9075
3.5	12′-3”	10′-0″	11,450
4	12′-9”	10′-6″	12,850

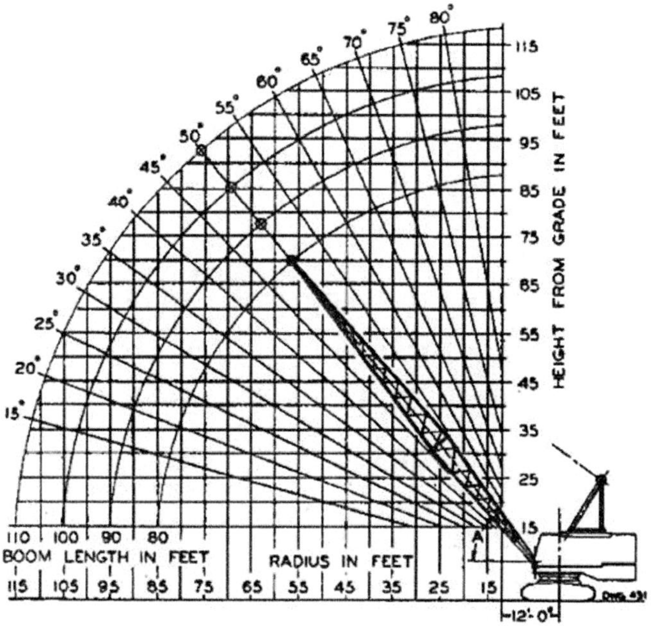

Fig. 6.9 Dumping radii versus dumping height for various lengths of clamshell booms

dumping radii versus the dumping height of various length booms under varying angles of operation.

6.2.13 Production Rate

Example

Assume that a 4.5-yd^3 standard rehandling clamshell bucket will be used to transfer dry sand from barges into a stock pile:

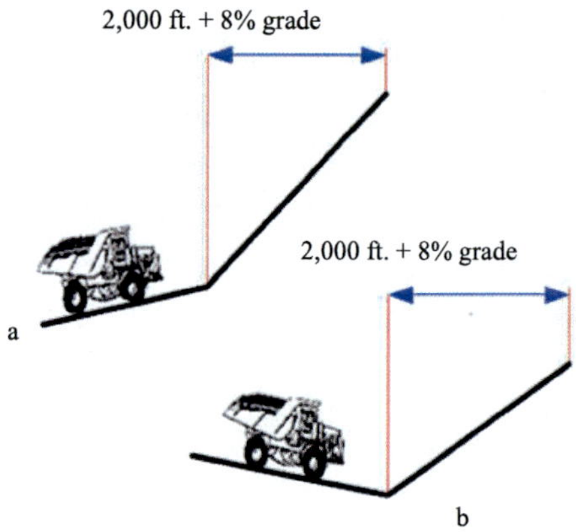

Fig. 6.10 Example of momentum on 8% grade preceded by a stretch of **a** lesser adverse grade, **b** favorable grade

Known factors for this calculation	
Boom length	80 ft
Working radius	55 ft
Hoisting distance	35 ft
Dumping height	56 ft 9 in.
Digging depth	26 ft 3 in.
Allowable load, maximum	30,000 lb
Angle swing	180°
Hoist speed	300 ft/mm
Swing speed	3.2 rpm
Bucket size	4.5 yd^3
Bucket weight, empty	12,800 lb
Bucket weight, heaped	29,400 lb

Cycle time	s
Closing time to load	6
Hoisting time (60 × 35) 300	7
Swing time, excess of hoist for 180° Swing $\left(\frac{60\times0.5\times0.8}{3.2}\right) = 12$	
$12 - 7 = 5$	5
Dumping time	5
Return swing and lower bucket	12
Position bucket	5
Total cycle time	40

Production per digging hour:

Dry sand weighing 100 lb/ft^3	
Average load for bucket will be based on line of plate capacity which is 122.7 ft^3	
Heaped load for bucket will be based on 166 ft^3	
Size of bucket	4.5 yd^3
Average load	6.135
Number cycles	90 cycles/h
Number tons per digging hour	552.15 tons/h
Ratio: scheduled hours to digging hours	0.83
Number tons per scheduled hour	458.28 tons/h

Some examples of different clamshell operations include loading into a hopper, digging a canal under water and loading crushed stone from a stockpile into a truck.

6.3 Continuous Excavators (Bucket Wheel and Chain Diggers)

6.3.1 Introduction

Surface excavating is done on a continuous basis with a variety of machines: (1) trenchers and ditchers, (2) conveyor loaders, and (3) bucket-chain and bucket-wheel excavators (BWE). Surface mining is largely concerned with only the last two machines, although some of the conveyor-loader types are often used by contractors for large-scale earth moving, and the engineer should be aware of these applications. This chapter is restricted to discussing only the bucket-chain and bucket-wheel excavators, with special emphasis on the latter.

Continuous digging with these machines is done with evenly spaced buckets on either an endless chain or on the periphery of a rotating wheel. The bucket-chain digger excavates and takes the material away from the face with the same machinery. In the case of the bucket-wheel machine, these two functions are separate. Material is excavated by the buckets and then flows in a continuous stream via a transfer to the conveyor system, which in turn transports it to the discharge point. The wheels are adjustable in rpm and mount 6–12 buckets of equal size. These are classed as celled, cell-less or semi-celled. The digging ladders for both types can move simultaneously in the horizontal and vertical planes. Ladder control is similar to the operation of a bucket-line dredge. Some ladder-chain and all wheel excavators have the machinery deck and superstructure mounted on a 3600 swing circle. Other chain diggers have a fixed machinery deck. Deep and or high cutting models are available in either the chain or wheel excavator. For further reading: Aikens and George [1]; Aiken and Wohlbier [2]; Benecke [6] and Bond et al. [7].

Machines can be either rail or crawler mounted, but the trend is to crawlers. Many of the early ladder-chain, rail-mounted excavators has been converted to crawlers. These may be either of the single or bogie type, depending on the conditions. Large machines usually use the superior three-point crawler suspension. Capacity and flexibility restrictions caused the bucket-chain machine to be supplanted in most cases by the wheel-type excavator. But undoubtedly occasions will arise in the future where a bucket-chain machine will be installed For further reading: Vogt [23].

6.3.1.1 Choosing a BWE

All the operating elements and conditions must be carefully studied before choosing a wheel of specific size and design. For further reading: Kay and Bartach [15]; Linden [16]; Schellhorn [21] and Tilmann [22]. Some of these requiring evaluations for a specific job are:

The material handled

- The mining plan.
- The output capacity of the BWE.

6.3.2 Material Transport

If direct spoiling is not done, some form of transport must be used to take the material away from the EWE. One of three methods can be used: (1) conveyors, (2) railroad, and (3) trucks.

6.3.2.1 Conveyors

A EWE and conveyor system is most adaptable to flat-bedded, homogeneous materials which can be transported to a fixed point. Some loss of operating flexibility must be accepted, since the EWE is tied to fixed-capacity conveyor loading sections. This closely defines the area of EWE movement and depth of cut. Shiftable conveyor sections parallel the direction of EWE movement and are normal to the direction of advance in parallel mining.

Belt shifting is effected by a bulldozer with a special arm and head attachment. Shifting time at the mining face will require about one hour for each 21,000 ft^2 covered in the conveyor move. Large moves require proportionally less shifting time than small moves. Belt alignment will require at least an additional hour per each 800 lineal feet. To achieve the best operating costs, the conveyors should be closely matched to the EWE output. In practice, this is seldom achieved without expensive oversizing of the conveyor system. Most commonly a design factor of 1.3–1.5 of the true average hourly EWE output is used for proper sizing of the belt. Therefore, when the EWE is in easy digging, the production must be cut back to the load limit of the conveyor system. Nevertheless, where applicable, conveyors provide the lowest transport cost for large long-term operations. Their major disadvantage is that if they fail, all production stops. This is not true, for instance, if a single truck in a fleet fails. Rarely is it practical to have a duplicate conveyor system that can take over in such an emergency. Operations indicate that conveyors have a lower overall availability than rail or truck. Even though standard conveyors show an availability of 90–95%, it is doubtful if down time could be kept this low with the high-capacity types currently associated with wheel excavators.

6.3.2.2 Trucks

Truck transport with the EWE is attractive for irregular excavations, selective mining, and short hauls. In fact, operating costs for trucks can be shown to be a standoff with conveyors when the production is large and the hauls short. However, as operations are extended and the haul lengths increase, truck costs rapidly exceed conveyor costs. A collector conveyor and dump station can shorten the haul and improve costs. To maintain a continuous EWE production, there should be no delay between trucks. This requires a special discharge boom loading device, which switches material flow from the full to the empty truck with the trucks spotted for side-by-side loading. Another problem with truck haulage comes in providing enough units, under all conditions, to keep the EWE working continuously. Variable excavating rates and haul distances can he handled easily by providing sufficient spares in the truck fleet. For further reading: Alarie and Gamache [3]; Barbaro and Rosenshine [5] and Ercelebi and Bascetin [11].

6.3.2.3 Railroads

A strong trend has developed in Germany away from rail transport and towards conveyors. The equipment obsolescence created by World War II accelerated this move. Large amounts of mine trackage are still in use in Germany, but usually on very long hauls or where capital investments are so great as to make the switch to conveyors prohibitive in cost. Much of the rolling stock in the German coal mines is antiquated and inefficient. However, rail operating costs can be shown to be competitive with conveyors when the automated features of modern railroading are installed but the hauls must be long, and with no excessive grades. Some advantages of rail transport that can offset its high initial cost are: (1) good availability, (2) low labor cost (automated), (3) better flexibility, and (4) surge capacity in cars. Low-cost surge capacity in cars can be of importance when a plant or process depends on the production from a EWE. German brown coal operations always have ample surge facilities between the mines and the generating plants to ensure an uninterrupted flow of the feed.

6.3.3 *Sizing and Operating a BWE*

6.3.3.1 EWE Sizes

Selecting a EWE which is sized to meet a production requirement can be complex. While shovel output standards are available for a wide range of materials and conditions, the EWE buyer is forced to develop its own operating factors and standards. Like bucket-line dredges, EWE's are classed by bucket capacity, e.g., 0.92 yd^3 (700 L), 1.3 yd^3 (1000 L) etc. Within the basic size range, machines of equal capacity may vary in profile dimensions, weight, horsepower, etc. The large wheel excavators are not produced on a standard model basis, as are power shovels. Therefore, no digging rates, bucket factors, etc. have been established. The information here is not presented for the purpose of setting EWE capacity or digging standards. These have to be based on a specific installation. However, the planning engineer concerned with a EWE application may be guided in his selection by the sample problem and supplementary information presented below. For further reading: Schellhorn [21] and Tilmann [22].

6.3.3.2 Output Ratings

When calculating the output capacity of a EWE, it should be considered as one member of a transportation haulage chain. This may include extensive belt conveyor lengths, auxiliary machines, and stockpiling equipment. All units in the chain should be matched in capacity; otherwise the availability of the weakest link sets the true output.

Theoretical Output

The theoretical output (Q_{th}) can be calculated by the following methods, Eq. 6.3:

$$Q_{th} = \frac{I \times S \times 60 \times f}{27}, \text{ min /h} \tag{6.3}$$

where:

- Q_{th} = theoretical capacity per hour (yd^3/h solid).
- I = single bucket capacity (cu ft).
- S = discharge rate per minute.
- f = swell factor; ratio of solid or bank yards to loose or bucket yards.

By inclusion of the cutting speed at the knife edge of the bucket the equation becomes, Eq. 6.4:

$$Q_{th} = \frac{I \times z \times V_c}{27\pi D} \times 3600 \times f, (s/h) \tag{6.4}$$

where:

Z no. of buckets on wheel.
D wheel diameter at bucket edges (ft).
Vc 60.D.π/60 Vcutting speed (ft/s).

When considering slewing speed gives, Eq. 6.5:

$$Q_{th} = h \times d \times V_s \times 60\ (min/h) \tag{6.5}$$

where:

h height of cut slice (ft)
d depth of cut (ft)
Vs slewing speed (ft/mm).

Equation 6.5 indicates that the theoretical capacity in bank yards solid is a function only of height and depth of cut and slewing speed. The number of discharges per minute does not appear in the formula. Increase the discharge rate results in output increases only when the slewing speed also increases. But this can be increased only as long as the buckets are free cutting horizontally. The height of a cutting slice depends on the bucket-wheel diameter; i.e., output, Q_{th}, is increased as wheel diameter increases. This is limited since peripheral speed increases with wheel diameter and bucket discharge is impeded at high speeds. Height of a cutting slice is usually assumed at 0.5–0.7 of the diameter, D. The cutting slice depth, d, is limited by the hardness of the material and the dimensions of the bucket. Cutting slice width depends on swing speed (slewing), but in theory cannot exceed bucket width. Generally it is more convenient to derive theoretical capacity in cu yd/h solid from data supplied by manufacturers in the form of three factors, Eq. 6.6:

$$Q_{th} = \frac{W \times Z \times I \times 60 \times f}{27}, (min/h) \tag{6.6}$$

where:

W wheel rpm at mid-speed point.
Z no. of buckets on wheel, and W.Z = S.
I single bucket capacity (ft^3).

In operations, the wheel output is controlled by:

Rpm of the wheel (depends on material and bucket characteristics), W.
Depth of cut, d.
Swinging (slew) speed, Vs.

Theoretically, a fully automatic EWE is possible. Australian experiments have shown that, in some cases, automation results in greater continuous output.

True Output

True EWE design capacity is the theoretical output dc-rated for the specific work conditions. These reflect such items as maintenance outages, type of material, excavating plan, and job management efficiency. The large investment in a high capacity EWE requires a three-shift operation. This work schedule and the planned maintenance then establish the effective working hours per week. Normally, maintenance requires 4 h per day during 6 days of the week and 12 h on the 7th day, leaving 132 h for effective digging per week. According to the operating plan adopted, the EWE will suffer varying periods of no production. A poor operator can seriously affect machine production if he is inefficient in the reverse slewing operation. Outages are mainly due to digging and maneuvering of the wheel (see the subarticle on Crowd-Type Wheels). From time to time the machine has to cut ramps, make box cuts, and be down for conveyor or rail shifting. Eased on German experience, the effective yearly working time is about 5000 h for well-matched BWE applications. The West German lignite mines have their BWE's operating 19.2 h per day on a 6-day week. On Sundays and holidays (about 60 days per year) the usual 4 h per day are available for maintenance. In regions combining severe weather, difficult digging, and a 7-day work week, effective working time will decrease accordingly. It is not uncommon in perfect conditions to achieve 6000 h per year working time, but for planning, a suggested value is, Eq. 6.7:

$$Q_{eff.} = 0.60 \times Q_{th} \times 5000\,\text{h}, \text{ in bank yd}^3/\text{year} \tag{6.7}$$

There are differences among the wheel designers as to how true output capacity should be calculated. The annular ring of the cell-less wheel is one of the main reasons for difference. Ring filling, at times, can give the EWE up to 50% increase in capacity over rated output. A guaranteed output may fully include, partially include, or completely exclude the material in the annular ring. A manufacturer should be asked to state the approach used. The older cell-type buckets did not present this problem of ring filling. In practice, output capacity is determined more by the conveyor belting sizes available than the bucket wheel size. For some machines a performance guarantee can usually be negotiated with the manufacturer. After commissioning the EWE, it is not uncommon to carry out performance tests which have been previously agreed upon between the buyer and the seller. Outputs achieved during these tests are always higher than the average yearly output. With no digging interruptions, EWE output has frequently exceeded its theoretical rating.

6.3.4 Example of BWE Selection

Difficulties in selecting a BWE often are the result of the lack of established design controls. Manufacturers and BWE consultants are the best sources of guidance, but in comparing schemes the user often finds it necessary to make BWE selections for estimating purposes. In doing this he can follow steps and reasoning similar to the exercise problem below. However, in reality, job complexity most often requires additional refinements beyond the scope of this chapter. Because of this and the high cost of special design wheels, a consultant is recommended before making any major BWE selection. The following exercise problem is only a guide and should be modified per the requirements of other conditions. But the, maintenance schedule used should adhere to manufacturers recommendations.

Operating Conditions

- Overburden with medium digging characteristics and a 0.75 swell factor, to be stripped at the rate of 8×10^6 yd^3/year.
- Average management.
- A temperate climate.
- Working schedule—7 days per week, 24 h per day.
- Mine Layout.
- Mining faces 2000 ft long.
- Height of faces 50 ft.
- Pit slopes 550.
- A parallel mining plan.
- Digging width of block—90 ft.
- Belt conveyor transport.
- Belt wagon to extend the time between conveyor shifting.

- Availability of BWE Hours
- Total hours per year—$365 \times 24 =$ 8760
- Recommended maintenance
- 4 h/day/6 days = 1248
- 12 h on 7th day = 624 −2712
- Annual overhaul, 35 days = 840
- Available work hours, incl. scheduled maintenance 6048 unscheduled outages
- Belt shifting during periods of scheduled BWE maintenance = 0
- BWE maneuvering at the face, 17% of work hours = 1030
- Unscheduled mechanical repairs avg. @ 5% of work hours = 302 −1634
- Unscheduled electrical repairs avg. @ 2% of work hours = 121
- Cleaning and miscellaneous avg. at 3.0% of work hours = 181
- Total effective working hours = 4414
  Availability of BWE, $\frac{4414}{6048} =$ 73%
- Avg. conveyor and belt wagon availability = Avg. BWE Availability, 0.73/0.85 = 85%
- Digging Requirements $= \frac{2000\,\text{ft}/90\,\text{ft}/50\,\text{ft}}{27\,\text{ft}^3} = 334{,}000\ \text{yd}^3$
- $\frac{\text{Block volume}}{\text{No. of blocks/year}} = \frac{8\times10^6\,\text{yd}^3/\text{year}}{334\times10^3\,\text{yd}^3/\text{block}} = 24$ blocks
- No. of moves/year—blocks $= \frac{24}{2} = 12$ (using belt wagon)
- No. of box cuts/year = 24
- Belt shifting time Allow 21,000 ft^2/h $= \frac{2000\,\text{ft}/90\,\text{ft}/12\,\text{moves}}{103\,\text{h}} = 21{,}000\ \text{ft}^2/\text{h}$
- Allow 8 h to make final alignment per each move, 8/12 = 96 h

- Included in BWE maintenance hours = 199 h
- Volume of box cuts/block (this will vary with BWE dimensions) = $\frac{(77\,\text{ft}\times 68\,\text{ft}\times 50\,\text{ft})+(110\,\text{ft}\times 90\,\text{ft}\times 50\,\text{ft})}{27} = 28{,}000\ \text{yd}^3$
- Volume of box cuts/year, 24/28,000 = 670,000 yd^3
- Digging Hours
- Tentatively select a BWE with an effective output of 1200 yd^3
- (Loose) per hour:
- Box-cutting bra/year at y4 normal production rate:
- $\frac{670{,}000\,\text{yd}^3}{600\,\text{yd}^3/\text{h}\times 0.75\ \text{swell factor}} = 149\ \text{h}$
- Digging hours/year:
- $\frac{\times 10^6\,\text{yd}^3/\text{year}\times 670{,}000}{1200\,\text{yd}^3/\text{h}\times 0.75} = 8140\ \text{h}$
- Total hours required per year = 8289.

From the above, one BWE works 4414 effective hours/year. Hence two BWE's of the selected size would give approximately the desired output by working 8828 unit-hours versus the 8289 indicated as being required. However, if two BWE's are selected, then conveyor shifting time must be added to the total outage, and the box cutting hours are doubled. As an alternative for this case example, a single machine with a theoretical rating of 4000 cu yd/h could be selected, giving:

$4000\,\text{yd}^3 \times 7\,\text{h} \times 0.60(availability) = 2400\,\text{yd}^3/\text{h}$ (loose) as the effective rating, and:

$$\frac{670,000}{1200\,\text{yd}^3/\text{h} \times 0.75} = 75\,\text{h for box cutting}$$

$$\frac{8 \times 10^6 - 670,000}{2400\,\text{yd}^3/\text{h} \times 0.75} = \frac{4075\,\text{h}\ for\ digging}{4150\,\text{h}\ total}$$

Experience indicates that the average BWE will work between 4000 and 5000 h per year, on a round-the-clock basis. Therefore, this larger BWE has some spare hours available for unexpected conditions, rearrangement of electrical equipment, and selective digging. If single machine reliability was not objectionable, this is the preferred alternate in the exercise example. When two BWEs are used, maintenance is normally carried out on only one machine at a time. Scheduling belt movements becomes more complex and is not always possible or feasible during BWE maintenance periods. Also the choice between one machine or two frequently involves the question of multiple benches. With one machine per bench, belt shifting is simplified, but experience indicates that generally the minimum number of benches will be the most efficient and cheapest operation. However, if production requires relatively low output from the BWE, a single high face for one machine is not practical. High diggers require long and costly booms. Thus each BWE has an optimum output capacity for its digging height. If bank height is not a determining factor, it may be cheaper to use multiple low-profile machines combined with one or more crawler-mounted belt wagons.

6.3.5 *Estimating BWE Costs*

6.3.5.1 Capital Costs

A fairly long write-off period and a high production rate are necessary to achieve reasonable ownership costs. It is only on this basis that they can provide the most economical digging. Large BWE's are considered to have an economic life of 20 years or more. One wheel excavator at a dam site, theoretically rated at 9000 yd^3/h, is reported to have cost about $1.8 million; and another wheel, theoretically rated at 4500 yd^3/h, about $1.5 million. The largest contractor-type made in the U.S., rated at 3000 yd^3 and pre-engineered, costs about $800,000 in the ready-to-work condition. These prices include the cost of erection. One manufacturer estimates that 20–40 h of erection time are required for each ton of delivered weight. Small BWE's require proportionally more erection hours than large ones. About 20–30% of the erection hours require skilled labor. Depending on wheel size and conditions, an operating crew is two to six men.

6.3.5.2 Operating Costs

It is not uncommon to find direct operating costs of 2ç to 3ç yd^3 for BWE's in service. Midwestern open east coal mines using BWE's credit them in part with the high strip ratios possible today. They have established new low costs for earth moving and reclaiming operations. Direct operating costs are computed by wheel users on various bases, making any comparisons between operations or wheels meaningless. Cost figures only reflect what the originators incorporated for their own purposes. However, there are some generally accepted cost rules on BWE excavators:

- Maintenance charges on the digging wheel represent 80–95% of the total excavator maintenance.
- Total maintenance and supply charges per year amount to 5–8% of the excavator cost.
- Spare parts inventory will be about 6–8% of machine cost.

Tooth wear is frequently found to be the greatest single variant in operating costs, and the time between changes under severe conditions may be less than 24 h. Power costs also vary from area to area. German lignite operations report operating costs of 1.5ç to 7.5ç per cu yd, not including ownership (11). Based on experience, the wide spread in cost reflects the amount of auxiliary equipment included with the wheel costs. This may include operating expenses for a tractor, belt wagon, belt shifter, cable reel car, etc. Costs on the digging wheel itself are mainly for teeth, lips, and liner plate. Belt expenses are the big item for the remainder of the excavator. Repair and maintenance costs on the rest of the machine are comparatively small. This is true since BWE's do not experience the severe impact loadings associated with shovels on the boom and crawler assemblies.

Ownership costs	Operating costs
Assumed 20 year economic life:	Cost/year Cost/Cu I'd
Cost/year Cost/Cu Yd	Maintenance, incl
\$2,750,000 = \$137,500 1.7ç	Cost/year Cost/Cu I'd
20 years	Maintenance, mc. supplies and parts@ 5% of BWE
Average annual investment:	
20 + 1 × 2,750,000	
40 × \$	
= \$1,440,000	
Insurance, interest, and taxes @ 12%	
per year:	
0.12 × \$1,440,000 = \$173,000 2.2 c	
Ownership costs \$310,000 3.9 c	

Total system costs, excavating and transporting, range from 6ç to 15ç per cu yd in Germany (11). Operating costs reported for BWE's in the United States are between 1.5ç and 9ç per cu yd (11): with no ownership costs included. Within this range, direct operating costs fall most frequently between 3ç and 4ç per cu yd. These costs may or may not include charges for BWE associated equipment. The 9ç cost mentioned above is not representative, since it occurred during a start-up and experimental period. A single machine as selected in the example presented under Article 8.4.7 might be expected to cost about \$2,750,000 and weigh 3,000,000 lb. Ownership and operating costs for this machine can be estimated as follows:

Cost/year Cost/Cu I'd
Maintenance, mc.supplies and parts@ 5% of BWE cost \$137,500
Labor, 4 men = \$163,000.
Power @ 1.0/kWh or 0.25 kWh/yd^3.
8 × 10^6 yd^3/year (0.25 × 1.0)
Operating costs
Total Costs: \$ 20,000
\$320,500
\$630,500.

Table 6.11 gives the guidelines of materials handling in surface mining.

6.3.6 Selection of Type of Hauling Equipment

The following information is intended as a guide for the engineer in the selection of the most suitable type or types of haulage equipment for the proposed application. This selection could be based upon the following qualifications:

Table 6.11 Guidelines for materials handling in surface mining

	Dozer-front-end loader	Dozer-scraper	Dragline-(direct casting)	Excavator-truck	Excavator-hopper-crusher-conveyor	Wheel excavator-conveyor
Maximum production	Medium	Medium	High	High	High	High
Production rate	Medium	Medium	High	Medium	Medium	High
Pit life	Short	Short	Long	Medium	Long	Long
Pit depth	Medium	Flat and Shallow	Medium	Deep	Deep	Medium
Deposit	Unconsolidated	Unconsolidated	Consolidated	Consolidated	Consolidated	Uniform, no large boulders
Preparation (if required)	Ripping	Ripping	Drill and Blast	Drill and Blast	Drill and Blast	Drill and Blast
System complexity	Low	Medium	Low	Medium	High	High
Operational flexibility	High	Medium	Low	High	Low	Low
Blending capability	High	High	Low	Medium	Low	Low
Selective placement (disposal)	Good	Excellent	Poor	Good	Medium	Medium
Wet weather impact	High	High	Low	Medium	Low	Low
Scheduling requirements	Low	High	Low	High	Medium	Medium

(continued)

Table 6.11 (continued)

	Dozer-front-end loader	Dozer-scraper	Dragline-(direct casting)	Excavator-truck	Excavator-hopper-crusher-conveyor	Wheel excavator-conveyor
System availability	Medium	Medium	High	Medium	Low	Low
Support equipment	Low	Low	Medium	Medium	High	High
Ease of start up	Simple	Simple	Moderate	Simple	Complex	Complex
Investment	Low	Low	Medium	Medium	High	High

General Criteria:

- The ability to make maximum use of the equipment on whatever variety of applications may be required.
- The degree of compatibility between existing hauling or loading equipment and the recommended new units.
- Possible restrictions on maneuvering space or side and overhead clearances.
- Possible restrictions on bridge or haul road widths.
- The effect of extreme grades, either favorable or adverse, particularly on the loaded haul.

Use Conventional Type Rear Dumps when:

- The material hauled is large rock, ore, shale, etc., or a combination of free-flowing and bulky material.
- Dumping into restricted hoppers or over edges of a waste bank or fill.
- The hauling unit is subject to severe loading impact when under a large shovel, dragline, or loading hopper.
- Maximum gradeability and rapid spotting in restricted area is required.
- Maximum flexibility is required for hauling a variety of materials under variable job conditions.

Use Rocker-Type Rear Dumps when:

- Extreme maneuverability is required in locations such as tunnel construction and on narrow benches.
- Traction presents no problem as on level or low grade hauls.
- Satisfactory performance can be achieved with relatively poor horsepower-to-weight ratio.
- Same prime mover could be interchanged with an existing scraper in a contractor's fleet.

Use Bottom Dumps when:

- The material hauled is free flowing.
- The haul is relatively level, allowing high travel speeds.
- Dumping is unrestricted into a drive-over hopper, or the load is to be spread in windrows.
- Long grades do not exceed 6%. (This is recommended as a general rule for optimum performance but is by no means a measure of the actual grade-ability of bottom dump semitrailers.)
- Maximum flotation of the large, single, low-pressure tires is required. This also applies when large capacity dual tires are needed to handle imposed loads.

Normally, when a haulage study is to be made, several different sizes and types of units are considered in order to determine comparative value of productivity per hour, cost per ton or yard of material hauled and possibly capital investment. This, then, becomes the basis for final selection of equipment.

6.3.6.1 Haulage Cycle Time

The information in this section is intended as a guide in the determination of the time required by the vehicle for its complete haulage cycle, including loading, hauling, dumping, returning, and spotting times. The following means of vehicle loading are commonly used in mining, quarry, and general construction operation:

- Shovels and draglines

The rating of a shovel or dragline bucket is generally given as the struck volume in cubic yard capacity. When digging, a load of more or less than the struck measure might be obtained, depending upon the type and state of the material. There are always some voids in the loaded bucket, whether the material is dug from a natural or a pre-blasted bank. Two variables must be considered when determining the quantity of loose cubic yards of material passing through the bucket during any one pass: (a) the swell factor of the material; this has been previously discussed, and (b) the fill factor of the bucket. The latter factor is an expression of the available bucket capacity utilized. Both of these factors must be applied in the following formula to determine loose cubic yards per pass for loading purposes (i.e. Cu yds. per pass (loose) = bucket capacity × swell factor × fill factor).

- Front-end loader

The front-end loader has become a very popular tool for handling easy flowing materials. However, with the advent of increased capacity, better material blasting techniques, and improved designs, the loader has progressed extensively into shot rock and similar types of material handling. When determining the load capabilities of the loader, the same bucket fill factor should be considered as for the shovels and draglines. It is estimated that the production capabilities of the loader in easy-to-medium digging would approach that of the same size shovel or dragline. However, in the hard digging category, a front-end loader would achieve only about 60% of a corresponding size shovel capacity. In the very hard digging category, the front-end loader does not perform satisfactorily and thus, is not recommended. As a guide, it is believed that the following approximate loading cycle times could be used for the different class of units working at minimum haul distances:

Class 1—0.40 to 0.60 mill—up to 4 yd^3.
Class 2—0.50 to 1.00 mm—S to 10 yd^3.
Class 3—0.80 to 1.50 mm—i 1 to 20 yd^3

- Hopper loading

On occasion, it is possible to consider the possibility of top loading a haulage unit by means of an overhead hopper or storage bin having controlled gates for discharge purposes. This application is limited to the handling of a material which is not bulky and flows freely, such as crushed stone or ore, sand and gravel, coal, etc. Normally, the haulage units used for this purpose would be bottom-dump, tractor-trailer, or rear-dump units. Due to the wide variety or hopper sizes (capacity and

Table 6.12 Suggested fill factors

	Shovel	Loader	Dragline
Easy digging	0.95–1.00	0.95–1.00	0.95–1.00
Medium digging	0.90–0.95	0.90–0.95	0.90–0.95
Hard digging	0.80–0.90	0.50–0.55	0.75–0.90
Very hard digging	0.70–0.80	n.a	0.60–0. 70

gate openings), it is not feasible to establish any guideline on possible material flow rates for loading time determinations. It is suggested that the hopper manufacturers be contacted for this specific information. Bucket Fill Factors. The following "fill factors" have been established for the following four different classes of material. These data are required by the engineer in the determination of ultimate vehicle loading times, Table 6.12.

6.3.6.2 Haulage Unit—Excavator Match

As a rule of thumb, relative excavator and haulage unit capacities should be established so that on short- or medium-length hauls under one mile (one way), not less than three, and preferably not more than five, passes would be required to load the vehicles. On hauls over one mile (one way), the loading time becomes much less significant and, as a result, a greater number of passes could be permitted with a minimum effect on the total haul cycle time and resulting productivity.

6.3.6.3 Excavator Swing Cycle Time

Data shown in Table 6.13 may be used as a guide for the determination of total load time. These swing times are quite general and any change of a few seconds, greater or less, will not materially change the loading or ultimate total cycle times. They are all based on a 900 swing from loading to dumping position. For swing angles greater than 900, add about three seconds to the indicated cycle time for each 300 increase in swing angle.

Values for sizes not listed may be determined by approximation using swing times of closest unit.

6.3.6.4 Excavator Passes per Load

There are several methods for determining the number of shovel, dragline, or loader passes per load and the resulting load time. One of the simpler and reasonably accurate methods is as follows:

Table 6.13 Swing cycle time (s)

Bucket size	Easy digging		Med. digging		Hard digging		V. hard digging	
	Shov	Drag	Shov	Drag	Shov	Drag	Shov	Drag
1.5	15	22	17	25	21	36	25	–
2	16	23	18	26	22	37	26	–
2.5	17	25	19	28	23	37	26	–
3.5	18	27	20	30	24	39	27	–
4.5	20	29	23	32	26	41	30	–
5	20	30	23	33	27	41	31	–
6	21	32	24	35	27	42	33	–
8	22	34	24	37	28	43	33	–
10	24	36	26	39	30	45	35	–
12	26	40	28	44	32	50	37	–
15	30	50	32	54	36	60	42	–

- Weight per cu yd of material (loose) weight in bank × swell factor no. tons per pass = bucket capacity (cu yd) × fill factor × loose weight per cu yd (tons).
- No. passes per load = tonnage rating of haulage unit − tons per pass Load time (min) = no. passes × excavator cycle time (min).

The answer to the above number of passes is rounded off to the next largest whole number and generally the potential excess load is ignored in the production study. Care should be taken to insure that the same short, long, or metric ton, as the case may be, is used for both vehicle rating and bucket loading purposes in the above computation.

6.3.6.5 Average Travel Speeds

The following data are intended for use with performance charts, similar to those shown in Fig. 6.10, in order to determine more realistic speed (mph) values on the various sections of the haul road. Having selected a transmission gear or range from the performance charts, it is necessary to modify the indicated road speeds through the use of a speed factor to reflect an average rather than a maximum mph. Table 6.14 shows the factors required to establish these average travel speeds over various lengths of haul.

A unit having the "stick shift" transmission and clutch cannot accelerate as rapidly as one with the torque convertor and full power shift transmission. As a result, the lower speed factors, at least up to 1500 ft distances, should be considered. To determine the average travel speed, multiply the maximum attainable speed as indicated in the performance chart by the factor shown, unless safety or other considerations impose lower limits. Knowing the average travel speed from the above, the travel time can be calculated by means of the following formula:

Table 6.14 Speed factors

Length of haul road section (ft)	Short, level hauls 500–1000 ft total length	Unit starting from stop	Unit in motion when entering haul road section
0–350	0.20	0. 25–0. 50	0.50–2.00
350–750	0.30	0.35–0.60	0.60–0.75
750–1500	0.40	0.50–0.65	0. 70–0. 80
1500–2500		0. 60–0. 70	0.75–0.80
2500–3500		0.65–0.75	0.80–0.85
over 3500		0.70–0.85	0. 80–0.90

- Travel time in minutes = distance in feet/(mph × 88)

6.3.6.6 Proper Speed Factor Selection

Table 6.15 gives a sizeable spread between the lowest and highest speed factors. An error in the total cycle time estimate is most likely to be due to an error in estimating hauling and returning. Consequently, care should be exercised in the selection of the proper factor used to obtain the average speed. Always consider the following:

1. Starting speeds

A unit starting on a 500 ft section of minus 2% grade on a good road will quickly get into high gear and has a chance to reach a high percentage of the maximum speed in high gear-therefore, use high factors in the above table. On the other hand, the same units starting on the same type of haul road section but only 250 ft long and level will hardly have time to get into high gear. Although theoretically it could travel in high gear, the low speed factor should be used.

2. Momentum on shorter sections

In calculating average speeds of units entering short haul road sections while already in motion, the momentum has to be taken into consideration. This accounts for the wide difference in the above factors. For example: A unit enters a +8% grade haul road section 150 ft long while traveling 20 mph. Theoretically, its maximum speed on such grade would be 7 mph. The momentum is likely to carry the unit over

Table 6.15 Recommended downgrade speeds

Grade (%)	Speed, mph
0–6	30–35
7–8	21–25
9–10	17–20
11–12	13–16
over 12	<13 mph

at 12 mph. Momentum on longer sections. Figure 6.10a shows a possible application of a rear dump operation over a 2000-ft haul road section having an 8% adverse grade. A 35-ton unit would negotiate this grade in 2nd gear at a maximum attainable speed of 8 mph. As indicated, the unit will be slow entering this section because of the preceding climb. Figure 6.10b shows the same rear dump entering the adverse 2000 ft of 8% grade section following a downhill stretch of haul road. Obviously, in the latter case, the unit has a better chance of averaging closer to 100% of its maximum speed, and the high limit of the given speed factor should be used. These illustrations are intended only as a guide to the majority of job conditions.

6.3.6.7 Power, Weight, and Speed Ratio

Over a given section of haul road, a unit that is geared slowly has a better chance of attaining its top speed than a unit of equal horsepower that is geared for fast travel speeds. Similarly, a unit with a ratio of 300 lb per hp has a better chance of attaining its top speed than a unit with a ratio of 500 lb per hp, provided both units are geared to travel at the same speed. This is best illustrated by the "better performance" of an empty unit against a loaded unit. The shorter the haul road section, the more noticeable will be the difference; hence, the wider margin in the speed ratio tables for the shorter haul road sections.

6.3.6.8 Delay Factors

There are often hazards or obstructions in the haul road that slow down vehicle speed. Time allowances must be made when these conditions exist.

Intermittent factors (Consider delay time or slow down on each item)	Continuous factors (Consider delay time or slow down over entire haul)
One-way haul roads	Extremely variable and high rolling resistance
Delay at passing points	
Sharp curves	Wet or slippery haul roads
Multiple curves or switchbacks	
Blind corners	Unskilled operators
Bridges, Underpasses	Long down-grade hauls
Railroad crossings, Cross traffic	

Return Time. Return time of the hauling unit is often governed by job conditions and safety precautions, rather than by the performance ability of the unit. If no grade conditions or operating hazards are present, these factors apply to the top speed empty:

Table 6.16 Turning-spotting-dumping time (mm)

Operating conditions	Bottom-dump		Side-dump
	Tractor-trailer Rear-dump	Semitrailer	
Favorable	0.3	1.0	0.7
Average	0.6	1.3	1.0
Unfavorable	1.5	1.5–2.0	1.50

Table 6.17 Spot at loading machine (mm)

Operating conditions	Bottom-dump-tractor-trailer rear-dump		Side-dump-semitrailer
Favorable	0.15	0.15	0.15
Average	0.50	0.30	0.5
Unfavorable	1.00	0.50	1.00

	<500 ft	>500 ft
Favorable	0.65	0.85
Average	0.60	0.80
Unfavorable	0.55	0.75

If complicated grade conditions are present, use the high factors in Table 6.16. Travel Speeds in Loading is usually at limited speeds other reasons. Average speeds Areas. Entry into the loading or pit areas due to possible rough roads, congestion, or for the unit can be estimated as follows:

Favorable	10 mph
Average	7 mph
Unfavorable	4 mph

Downgrade Speed Limits. Downgrade speeds, particularly when loaded, can be critical, dependent upon operating and safety conditions. It is recommended that under normal conditions speeds be held to those values shown in Table 6.17. These values are arbitrary and should be reviewed for each specific application.

Turn, Spot, and Dump Time. Turning, spotting, and dumping time depends upon the type of unit and the specific operating conditions. As a guide, average values for the different types of haulage units under the indicated conditions are shown in Table 6.19.

6.3.6.9 Spot Time at Loading Position

The spotting time at the loading position also depends upon the type of haulage unit and the operating conditions. The average values are given in Table 6.20.

Haphazard spotting at the shovel is one of the most common and wasteful malpractices of the shovel cycle. Cutting a 300 arc from the swing out and return (only 1/12 of the full 360° circle) will increase shovel production by approximately 15%. If the haulage unit is spotted for loading on an 1800 swing when the job could be planned for 900, nearly one-third of the possible production is lost. In addition, accurate spotting of trucks at a correct angle and in the same relative position considerably improves the operator's timing and speeds up shovel operation.

Trucks always should spot on the same radius to save racking the dipper in and out. This should line up as closely as possible with radius of dipper as it leaves the bank. The following suggestions relate to establishment of most suitable spotting procedures:

1. Train crew for close-in spot for shortest possible swing cycle.
2. For back-in spotting, use log or tie for bumper to (a) get trucks in close, (b) provide uniform position, and (c) permit fast truck spotting.
3. Whenever possible, use two-truck spotting so the second truck is in and ready to load when first pulls out. Continuous in-line spotting is better, if job layout permits.
4. A spotter at the excavator frequently will pay off in extra yardage. The operation should be studied and analyzed for extra yardage to be gained by efficient and uniform spotting.

6.3.6.10 Operational Delay Factor

An estimate must indicate sustained or average hauling production over a long period of time. An estimate which is too optimistic on the hourly production ability of each hauling unit will result in failure to maintain required production and an insufficient number of haulers assigned to the job. It is necessary to allow for the unavoidable delays encountered on all operations. To provide for these delays, which does not include downtime for breakdown or maintenance, consideration is given to application of an operational or efficiency factor. This efficiency factor is reflected ultimately in the number of operating minutes per 60-min hour. A recommended range of values dependent upon operational conditions are as follows:

Job efficiency (%)	Working mm/h
Excellent (92)	55
Average (83)	50
Unfavorable (67)	40

On some applications, this efficiency factor should be reduced in percentage, depending on the amount of unusual delays for shovel moving, blasting, traffic, management and supervision efficiency, operator experience, and similar items.

6.3.6.11 Productivity per Hour

The item entitled Operational Delay Factor covered the development of the job efficiency or the number of working minutes per hour. Using this value and dividing it by the total haul cycle time results in the possible number of trips per hour. The haulage unit's productivity per hour is the product of the number of trips per hour and the tons or cubic yards payload per trip.

6.3.6.12 Number Haul Units Required

The number of units necessary to handle the required hourly production (tons or cubic yards) is determined as follows:

$$\text{No.Units} = \frac{\text{required production/h}}{\text{possible production per hour per unit}}$$

Normally, when the answer to the above comes out in a fraction of a unit, such as 5.6 for example, the number of units required would then be the next largest whole number or 6. However, should the answer turn out to be 5.1 or even 5.2, the engineer might want to review some of his factors to determine if it would be possible to stretch the units ability to absorb the extra 0.1 or 0.2 of the truck required so that 5 units could do the work. It is possible, however, that on a certain job such as a stripping project, the excess capacity of the additional unit could be utilized, and this would then result in a still cheaper cost per ton or yard of material hauled. These various factors must be weighed before making recommendations.

6.3.6.13 Spare Units

Spare or stand-by units are necessary if a sustained production schedule is to be maintained. The normal amount of downtime due to maintenance is not considered to be a part of the job delay factor as provided for by the 50-min hour. The engineer must consider the reliability of the machines to be used, and on a "tough job" he will allow for extra hauling units. Typical machine availability figures are as follows:

Rating	Percent
Favorable	90
Average	85
Unfavorable	75

Regular lubrication and inspection schedules are extremely important in maintaining high machine availability. Proper driver training to prevent misuse of the equipment and good haul road maintenance are of great importance. Again, the estimator must know the job and the quality and experience of the supervision and the operator.

6.3.7 Definition of Payloads

Payloads are given in the various manufacturers' specifications, literature, and advertisements in terms of tons, cubic yards, and pounds, or in combinations of the three. In the case of rear and bottom dumps, all units are rated first on the basis of tons payload and secondly on the basis of cubic yards capacity. All manufacturers' specifications show in addition to the tonnage rating, the SAE struck and heaped cubic yard capacity. In addition, some manufacturers show heaped capacity on the basis of 3:1 and/or 1:1 material slope. However, in their advertisements and other literature, some manufacturers will feature the cubic yards capacity on the basis of a heaped load, whereas others will feature struck capacity. When making comparisons, the engineer must compare like values. It is important that a common baseline be used, either tons, struck yards, heaped yards, or pounds when comparing the rating and capacity of one similar haulage unit to another.

6.3.8 Cost Estimating

The major portion of the foregoing sections relates to the development of data required for the selection of the most suitable type of equipment and the determination of its productivity per hour for the proposed haulage study. Equally important is the development of the necessary data to permit the engineer to determine, as accurately as possible, the estimated ownership and operating cost per hour of the selected equipment. This value, coupled with the above haulage unit productivity in tons or yards per hour, and the number of haulage units required, provides the ultimate cost per unit (tons or yards) of material hauled. This cost per unit then becomes the primary measure of comparison to be used in the selection of one type vehicle over another and/or one size vehicle over another.

The following method of calculating haulage costs has been found most satisfactory over a period of many years. It covers the complete cost of equipment ownership and operation for the haulage only on a given operation. It should be understood that the cost per unit values determined are for comparative purposes only, to be used as a measure of comparison between equipment. They are only an indication, as reasonably correct as possible, of what the haulage costs might be. There is no intent of implied guarantee that they would be the actual cost. These cost values include only those directly connected with the haulage units and do not include loading, road construction, crushing, supervision, or other related charges.

6.3.9 Ownership Cost Items

Specific information must be developed to cover the following:

- List price of haulage equipment FOB. Factor: if actual quoted prices are available, they can be substituted.
- List price of optional equipment.
- The prices of all optional equipment deemed necessary for suitable operation of the vehicle must be developed.
- Transportation cost: These charges from F.O.B. factory to the operation site must be determined.
- Erection cost: These are the charges when it is necessary to reassemble the vehicle at the job site. In their ultimate use they would be included with transportation charges.
- Depreciation period: This period is initially established in years and ultimately in terms of hours of operation.
- Interest, taxes, etc.: Values will have to be assigned to interest, taxes, storage, and insurance.

6.3.10 Operating Cost Items

As in the case above, specific data must be obtained for the following:

- Tire replacement cost:

The replacement cost including excise tax for all tires for vehicles under consideration is required. Cost of recapping of tires is required in the event recapping procedures are considered. Note: Only tire replacement costs are considered in the ownership cost development. It is recommended that these cost values be obtained from the local tire dealers.

- Fuel cost per gallon: Determined locally.
- Operator wages: This should include social security and all other fringe benefits.

6.3.11 Development Data for Above

6.3.11.1 Depreciation

There are three official tax methods of computing depreciation for periods of three years or longer. These are known as: (1) straight line, (2) sum-of-the-years-digits, and (3) declining balance. The straight line method, which will be the only one discussed here, is the simplest and provides a uniform basis for figuring machine costs and avoids complications in reserve for depreciation.

The only thing it lacks is a fast write off. In the write-off, no resale or salvage value of the unit is used. This is believed to be a sound and conservative practice since used equipment values cannot be considered stable. Actual useful life of the various types of equipment is difficult to establish, since proper preventative maintenance could extend the service life beyond the depreciation period. The following data may be used for cost estimating purposes, but they are not meant to designate any legal depreciation periods for tax purposes. Whenever possible, the engineer for his study should determine the values his company, client, or customer use in their accounting procedures.

Depreciation period (h)	Operating hours per year	Years write-off	Type of operation
10,000	2000	5	Construction service: roads, dams
15,000	3000	5	Industrial service: mining, quarrying, plant operation

Interest, Taxes, Insurance and Storage

These items are calculated as follows: Hourly charge for interest, taxes, insurance and storage $= \frac{13\% \times average\ yearly\ investment}{Operating\ hours\ per\ year}$.

The above 13% interest value is composed of the following: Interest @ 9%; taxes @ 2%; insurance and storage @ 2%. The 9% interest rate is the simple interest equivalent of the 6% compound interest rate commonly used in equipment financing.

The average yearly investment is shown as follows:

Depreciation period average (years)	Yearly investment[a] as % of purchase price plus freight
1	100.0
2	75.0
5	60.0
8	56.2
10	55.0

[a]Equals $\left(\frac{n+1}{2n}\right) \times 100$

Tire Replacement and Repair. While tires represent a substantial portion of the initial vehicle cost, their life is considerably less than the life of the vehicle and, therefore, becomes an operating rather than an ownership cost item. Therefore, when determining the vehicle depreciation, the replacement value of a set of tires is deducted from the total cost of the unit.

$$\text{Hourly tire cost} = \frac{\text{tire replacement cost(set)}}{\text{estimated tire life(h)}}$$

- Tire repair cost = 15% × hourly tire cost

For example:
Considering the following conditions:

A Average—1.0
B 30 mph—0.8
C No curves—1.1
D Gravel road—0.9
E 10% overload—1 .0
F Driving—0.8
G 10% max.—0.8
H None—1.0

$$\text{Tire life} = 1.0 \times 0.8 \times 1.1 \times 0.9 \times 1.0 \times 0.8 \times 0.8 \times 1.0 \times 6000\,\text{h} = 3041\,\text{h}$$

Recapping of tires is a practice used by some and tried and rejected by others in an effort to reduce tire costs. When considering recapping, the tire must be pulled with 15–20% of tread life remaining to insure a suitable recap carcass. In the absence of specific recapping costs, it may be assumed that the price will be about 50% of a new replacement tire. It is suggested that local prices be obtained whenever possible. Recap life may be assumed to be between 50 and 100% of a new tire life. As previously discussed, tires having added tread depth are available and should be considered for all installations where applicable. In round figures, it can be assumed that this tire design will permit up to 50% longer life than a tire with the standard tread design, dependent upon suitable operating conditions.

6.3.11.2 Repairs—Parts and Labor

An hourly operating charge can be established for estimating purposes by using the purchase price of the hauling equipment as being indicative of the repair costs. Experience over a number of years, shows that the average hourly repair cost can be estimated very simply as a percentage of the initial purchase price of the vehicle divided by 15,000. This number is a constant derived from field experience and is used without regard to depreciation (Table 6.18).

$$\text{Hourly cost of parts and labor} = \frac{\text{Purchase price of machine} \times \%(\text{from above table})}{15,000}$$

Table 6.18 Percent of purchase price to be applied for maintenance estimates

Job conditions	Tractor-trailer units (%)	Conventional rear dumps (%)
Favorable	40	50
Average	50	60
Unfavorable	60	80
Extremely unfavorable	70	95

6.3.11.3 Engine Fuel Consumption

Due to widely varying operating conditions, it is virtually impossible to pinpoint specific fuel consumption. The best that can be done is to arrive at an estimated value based upon the application of a load factor in the following formula:

$$\text{Fuel consumption (gph)} = 0.06\ \text{gal/hp/h} \times \text{gross hp} \times \text{load factor}$$

The following load factors may be used in the determination of fuel consumption for the various rear and bottom dump units.

Service	Load factor
Light duty	0.35
Average duty	0.40
Severe duty	0.45

$$\text{Fuel cost per hour} = \text{fuel consumption (gph)} \times \text{cost per gallon.}$$

6.3.11.4 Oil, Grease and Labor

In view of all the variables in operating conditions, etc., it is recommended that the engineer use the rule of thumb to the effect that lubrication costs equals one-third the hourly cost of the diesel fuel.

Example:

A 35-ton rear dump consumes 10 gph @ 14c per gal for a total fuel cost of $1.40. Lubrication cost per h = ~ /s × $1.40 = $0.47.

6.3.11.5 Driver Wage

The driver wage should include in addition to his base amount all fringe benefits such as social security and unemployment compensation, paid holidays, sick leave, etc. These extras could increase the base pay by as much as 10–30%.

6.4 Loading and Excavation

6.4.1 Materials Handling

All unit operations involved in excavating or moving bulk minerals during mining are termed materials handling. In cyclic operations, the two Principal operations are loading and haulage, with hoisting an optional third when essentially vertical transport is accomplished. In continuous operations, where machines combine the breakage and handling functions, cutting, drilling, and blasting are eliminated, and extraction and loading are performed in a single function (excavation). In combined loading-haulage machines, materials handling is performed in a single operation.

Materials handling in modern mechanized mining are equipment-centered. Unit operations are characterized by and sometimes identified with the equipment which performs them. Thus field terminology refers to a mining shovel, stripping dragline, or coal loader. The scale of materials-handling equipment in surface mining grows ever larger. Upper size limits have risen to 300 tons (270 tonnes) for trucks, 220 yd^3 (170 $in.^3$) for draglines, 180 yd^3 (140 $in.^3$) for shovels, and 11,000 yd^3/h (8400 m^3/h) for bucket wheel excavators.

The latter three, boom-type machines are employed for stripping overburden in coal mining and are believed to be the largest man-made, dry-land, mobile structures in existence. Reasons for the growing, gigantic scale of surface mining equipment are found in its high productivity and low unit operating cost. In part, this is due to its worldwide commercial availability from multiple, competing sources, standardized in product lines and graduated sizes. As we have done throughout this chapter, our attention here is on principles, reserving applications for later discussion.

6.4.2 Principles of Loading

6.4.2.1 Excavation

The extraction and elevating of minerals, either broken or in place, is termed loading or excavating. If the material is soil or very weak rock, it can probably be "dug" in place. If it is rock or unusually well-consolidated soil, it likely will have to be blasted or mechanically broken prior to excavation. In non-coal mining, 20% of the material has to be blasted. These two terms are not synonymous, but we use them almost interchangeably. Excavation implies some action of extracting from the solid; loading suggests elevating material. In practice, the term loading is used more commonly in underground mining and excavation in surface mining. A classification of mine loading-excavating equipment is presented in Table 6.19. Bases for the classification are the locale of mining (surface or underground) and continuity of operation (cyclic or continuous). Categories and common examples of individual machines are given for each. For the sake of completeness, continuous mining and boring machines appear, even though they were mentioned with penetration devices.

Table 6.19 Classifications of loading-excavating methods and equipment

Operation	Method	Haul distance	Gradeability, degrees	
			Average	Maximum
Surface cyclic	Rail (train)	Unlimited	2	3
	Truck, trailer	0.2–5 mile (0.3–8 km)	8	12
	Scraper (rubber-tired)	500–5000 ft (150–1500 m)	12	15
	Front-end loader	<1000 ft (300 m)	8	12
	Dozer	<500 ft (150 m)	15	20
	Skip	<8000 ft vertical (2400 m)	Unlimited	
	Aerial tramway	0.5–5 mile (0.8–8 km)	5	20
Continuous	Belt-conveyor	0.2–10 mile (0.3–16 km)	17	20
	High-angle conveyor	<1 mile (1.6 km)	40	60
	Hydraulic conveyor (pipeline)	Unlimited	Unlimited	
Underground cyclic	Rail (train)	Unlimited	2	3
	Truck, shuttle car	500–5000 ft (150–1500 m)	8	12
	Slusher (Scraper)	100–300 ft (30–90 m)	25	30
	LHD	300–2000 ft (90–600 m)	8	12
	Skip, cage	<8000 ft vertical	Unlimited	
Continuous	Conveyor (belt, chain and flight, monorail)	0.2–5 mile (0.3–8 km)	17	20
	Hydraulic conveyor	Unlimited	Unlimited	
	Pneumatic conveyor	Unlimited	Unlimited	

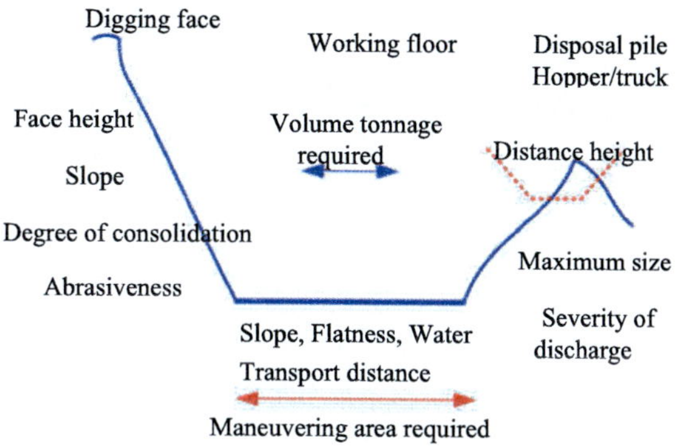

Fig. 6.11 Working zones and constraints in loading or excavating in a surface mine

The variety of equipment may overwhelm the student. Certain machines dominate the scene, however, and they are easily learned. For surface mining, shovels, loaders, draglines, and scrapers are in most common use. For underground mining, loaders load-haul-dump units (LHD5), and continuous miners are prevalent. In underground mines with large openings and sufficient headroom, surface equipment: dozers, shovels, trucks, loaders, and drills, modified to meet safety and operating conditions has increasing application. Each category and individual machine has operating characteristics that distinguish it and help to earmark it for selection. Several have a unique feature: They perform joint functions of loading and haulage (in addition to boom-type machines that cast the material being excavated). Examples of loading equipment that perform substantial haulage are dozers, rubber-tired scrapers, rope-drawn scrapers, and LHDs.

Most loaders-excavators are required to operate in three working zones, with a variety of operating constraints. These zones-digging, maneuvering-transport, and dumping-and some of their constraints are illustrated for surface mining in Fig. 6.11. The situation pictured occurs in stripping overburden or loading ore with a boom-type excavator (power shovel, dragline, or bucket wheel). Other conditions prevail where scrapers, dozers, front-end loaders, or dredges find unique application. The major features of large shovel, dragline, and wheel excavators are summarized as advantages and disadvantages in Table 6.20. Competitive and commonly used to strip overburden in surface coal mining, these three machines warrant careful comparison, although the trend in recent years has been strongly to draglines because of their greater range, both digging and for casting.

6.4.3 Selection of Equipment

Four groups of factors largely determine the selection of excavating equipment (the discussion applies specifically to surface equipment, but the factors are applicable to underground as well).

Table 6.20 Comparison of features of shovels, dragline and bucket wheel excavators

Machine	Advantages	Disadvantages
Shovel	– Lower capital cost/$yd^3(m^3)$ of bucket capacity, although when boom length or machine weight is considered, the capital costs are roughly equivalent – Digs poor blasts and tougher materials better – Can handle partings well	– More coal damage can result in lower coal recovery – Susceptible to spoil slides and pit flooding – Cannot easily handle spoil having poor stability – Cannot dig deep box cut easily – Reduce cover depth capability compared with a dragline of comparable cost – Difficult to move
Dragline	– Flexible operation; easy to move – Large digging depth capability – Can handle and stake overburden having poor stability – Completely safe from spoil pile slides or pit flooding during normal operation – High percentage of coal recovery, less coal damage – Will dig deeper box cut – Low maintenance cost – Can handle partings well – Is not affected by an uneven or rolling coal seam top surface – Can move in any direction	– Requires bench preparation – Does not dig poor blasts well – Higher capital cost/$yd^3(m^3)$ of bucket capacity, although when boom length or machine weight is considered, capital costs are roughly equivalent
Bucket Wheel	– Continuous operation; no swinging necessary – Long discharge range – Can be operated on a highwall bench or on the coal seam – Can easily handle spoil with poor stacking characteristics and poor stability – Can extend range of shovel or dragline when operated in tandem – Can facilitate land reclamation as it dumps surface material back on top of the spoil pile	– Will not dig hard material – Some surface preparation required – Lower availability – Large maintenance crew required – High capital cost compared with output – Can be susceptible to spoil slides and flooding – Can cause coal damage with resulting lower coal recovery – Poor mobility

6.4.3.1 Performance Factors

These relate directly to machine productivity and include cycle speed, available force (power), digging range, bucket capacity, travel speed, and reliability.

- **Conditions**:

  Shovel, dragline, or wheel excavator.

- **Zones**:

  Digging, maneuvering-transport, and dumping.

- **Design factors**:

  The design factors provide insight into the quality effectiveness of detail design, including the sophistication of human–machine interface for operators and maintenance personnel, the level of technology employed, and the kinds of control and power available.

- **Support factors**:

  Sometimes overlooked in machine evaluation, support factors are reflected in servicing and maintenance. Ease of servicing, special skills involved, parts availability, and manufacturer's support are important considerations.

- **Cost factors**:

  Probably the most quantitative (and ultimate) factor, costs are determined by standard estimating procedures for large mining and construction equipment. If reasonable assumptions as to life, interest rates, inflation, fuel, and maintenance are made, then results should be meaningfu1 The customary basis is to use unit costs, estimating overall costs as the sum of ownership and operating costs, all computed on a \$/h basis and converted to \$/ton (\$/tonne) or \$/yd^3 (\$/m^3). For estimation purposes involving excavation, data such as those in Table 6.21 may be used if actual specifications are lacking.

Numerical selection and cost estimation procedures will be demonstrated later. Operating characteristics of large surface mining excavators (power shovel, dragline, bucket wheel, front-end loader, and hydraulic excavator) are available in several

Table 6.21 Estimating parameters for surface excavators

	Capacity, yd^3 (m^3)	Est. weight, lb/yd^3 (kg/m^3)	Est. power hp/yd^3 (kW/m^3)	Est. life, h	Est. price, \$/yd^3 (\$/m^3)
Rubber-tired scraper	25–52 (19–40)	2900 (1700)	14 (14)	12,000	9500 (12,400)
Front-end loader	3.5–30 (2.7–23)	12,000 (7100)	52 (51)	12,000	36,000 (47,100)
Hydraulic excavator	4–30 (3–23)	30,000 (17,800)	72 (70)	30,000	85,000 (111,000)
Electric power shovel	6–75 (4.6–57)	54,000 (32,000)	41 (40)	75,000	110,000 (144,000)
Walking dragline	9–180 (6.9–138)	114,000 (68,000)	102 (99)	100,000	210,000 (275,000)
Bucket wheel excavator	0.1–5.2 (0.1–4)	–	–	30,000	–

references. Ratings for underground equipment are more difficult to obtain and the choices are more restricted. For the latest information on surface or underground equipment, consult the manufacturers or appropriate trade associations.

6.5 Haulage and Hoisting

6.5.1 Principles of Haulage and Hoisting

Bulk materials in mining are transported by haulage (primarily horizontal movement) and hoisting (primarily vertical). Machines to accomplish these functions are identified in Table 6.22, classified on the same basis as excavating equipment. Information is also provided as to (1) normal range of haul distance and (2) gradeability, average and maximum, important performance parameters that distinguish machine application.

The primary function of all equipment in Table 6.22 is haulage, although some (front-end loader, dozer, rubber-tired scraper, rope-drawn Slusher, LHD) are self-loading. The only true hoisting machines are the skip and cage, but as indicated under gradeability all haulage devices can elevate the material they transport to some extent (hydraulic and pneumatic conveyors are completely versatile and virtually unlimited as to distance and grade). Elevating material is generally crucial in any mining operation, because most mines, surface or underground, expand vertically to depth. Again the variety of equipment seems overwhelming, but again certain machines predominate. In surface mines with multiple benches, as in open pits, Rakhmangulov et al. [4] trucks team with shovels in most applications, with lesser use of belt conveyors and rubber-tired scrapers. In underground mines, rail, trucks (shuttle cars), LHDs, and belt conveyors are all popular in various segments of the overall haulage system. Just as excavators perform in prescribed working zones, thus the haulage units do. For the most widely used haulage machine, the truck, as well as other machines, there are four zones: loading, traveling loaded, dumping, and traveling empty, Fig. 6.12. Important design and operating decisions are required in specifying conditions for all these zones, but loading and dumping are especially critical.

Alternative arrangements are sketched for each; the use of double-truck back-up or spotting is preferred for quick loading if a drive-through is not possible, and a drive-by grizzly (coarse screen made of rails) affords the fastest dumping. Different working zones characterize other haulage machines, and they all require careful analysis in selecting equipment and methods. Comparable to Table 6.20 for excavators, Table 6.23 summarizes some of the main features of haulage equipment. Most of these machines are used both on the surface and underground.

Table 6.22 Classification of haulage and hoisting methods and equipment

Operation	Method	Haul distance	Gradeability (degrees)	
			Avg	Max
Surface cyclic	Rail (train) Truck, trailer	Unlimited 0.2–5 mile (0.3–8 km)	2 8	3 12
	Scraper (rubber-tired)	500–5000 ft (150–1500 m)	12	15
	Front-end loader	<1000 ft (300 m)	8	12
	Dozer	<500 ft (150 m)	15	20
	Skip	<8000 ft vertical (2400 m)	Unlimited	
	Aerial tramway	0.5–5 mile (0.8–8 km)	5	20
Continuous	Belt conveyor	0.2–10 mile (0.3–16 km)	17	20
	High-angle conveyor	<1 mile (1.6 km)	40	60
	Hydraulic conveyor (pipeline)	Unlimited	Unlimited	
Underground cyclic	Rail (train) Truck, shuttle car	Unlimited 500–5000 ft (150–1500 m)	2 8	3 12
	Slusher (scraper)	100–300 ft (30–90 m)	25	30
	LHD	300–2000 ft (90–600 m)	8	12
	Skip, cage	<8000 ft vertical	Unlimited	
Continuous	Conveyor (belt, chain and flight, monorail)	0.2–5 mile (0.3–8 km)	17	20
	Hydraulic conveyor	Unlimited	Unlimited	
	Pneumatic conveyor	Unlimited	Unlimited	

- **Conditions**

Truck haulage, shovel loading, grizzly or bank dumping. Zones: loading, traveling loaded, dumping, and traveling empty. For further reading: Chironis [8]; Ko et al. [14] and Osanloo and Hekmat [18].

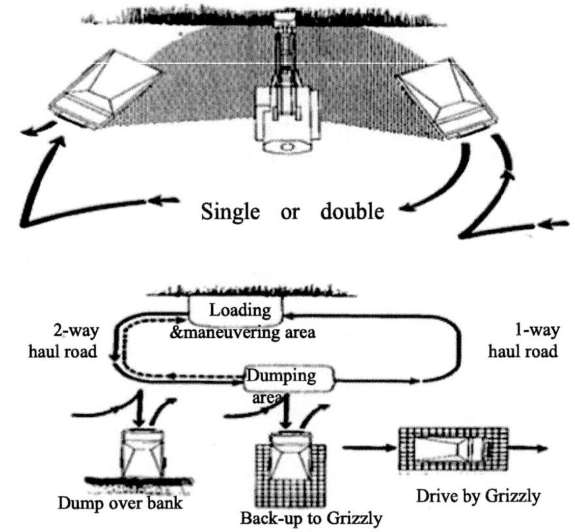

Fig. 6.12 Working zones in haulage in a surface mine

Table 6.23 Comparison of features of principal Haulage units

Machine	Advantages	Disadvantages
Dozer	– Flexible – Good gradeability – Negotiates rough terrain	– Limited to short haul – Discontinuous – Low output, slow
Truck, trailer	– Flexible and maneuverable – Handles coarse, blocky rock – Moderate gradeability	– Requires good haul roads – Slowed by bad weather – High operating cost
Scraper (rubber-tired)	– Flexible and maneuverable – Good gradeability	– May require push loading – Limited to soil, small fragments – High operating cost
Rail	– High output, low cost – Unlimited haul distance – Handles coarse, blocky rock	– Track maintenance costly – Poor gradeability – High investment cost
Belt conveyor	– High output, low cost – Very good gradeability – Low operating cost	– Inflexible – Limited to small – Or crushed rock – High investment cost

References

1. Aikens GF (1966) Bucket wheel excavators: how to choose the right one for the job. Min Eng 18(1):76–81
2. Aiken GE, Wohlbier RH (1968) Bucket wheel and chain diggers. In: Pfleider EP (ed) Surface mining. AIME, New York, pp 478–502
3. Alarie S, Gamache M (2002) Overview of solution strategies used in truck dispatching systems for open pit mines. Int J Surf Min 16:59–76
4. Rakhmangulov A, Burmistrov K, Osintse N (2021) Sustainable open pit mining and technical systems: concept, principles, and indicators. Sustainability 13:1101
5. Barbaro R, Rosenshine M (1987) Evaluating the productivity of a shovel truck materials haulage system using a cyclic queuing model. AIME Trans 282:1824–1827
6. Benecke KJ (1979) Bucket wheel excavator technology for mining lignite in Texas. Min Eng 31(8):1209–1213
7. Bond C, Spartin P (1966) Wheel excavators: an aid to strip coal mining. In: 95th annual meeting, Feb 28–Mar 3 1966
8. Chironis NP (ed) (1979) Scrapers tackle deep overburden. Coal Age 84(10):156–164
9. Chironis NP (ed) (1986) Small Draglines in tandem, reach deep seams efficiently. Coal Age 91(12):48–51
10. Chugh YP (1980) Opencast mining of brown coal in eastern Europe. Min Eng 32(11):1587–1590
11. Ercelebi SG, Bascetin A (2009) Optimization of shovel truck system for surface mining. J South Afr Inst Min Metall 109:433–439
12. Erdem B, Kormaz F (2012) analysis of dragline cycle time components. J Min Sci 48:545–558. https://doi.org/10.1134/S1062739148030185
13. Hird JM (1980) Overburden stripping—combination use of dredge and dragline. Min Eng 32(3):311–314
14. Ko SS, Serfozo R, Sivakumar AI (2004) Reducing cycle times in manufacturing and supply chains by input and service rate smoothing. IIE Trans 36:145–153. https://doi.org/10.1080/07408170490245441
15. Kay FJ, Bartach D (1984) Spreader and bucket wheels pre-strip mine. World Min Equip 52–53
16. Linden G (1957) The bucket wheel and its versatility in practice. deutsche hebe-und fordertechntk
17. Mohammadi M, Rai P, Oraee K, Kumar M (2013) Analysis of availability and utilization of dragline for enhancement of productivity in surface mines—a case study. In: Proceeding of 23rd world mining congress, Montreal, Paper No. 598
18. Osanloo M, Hekmat A (2005) Prediction of shovel productivity in the Gole Gohar iron mine. J Min Sci 41:177–184
19. Rai P (2004) Performance assessment of draglines in open cast mines. Indian J Eng Mater Sci 11:493–498
20. Rai P, Yadav U, Kumar A (2011) Productivity analysis of draglines operating in horizontal tandem mode of operation in coal mine: a case study. Geotech Geol Eng 29:493–504. https://doi.org/10.1007/s10706-011-9398-9
21. Schellhorn HW (1963) Some aspects of high capacity production with bucket wheel excavators in open pits. In: The k. a. Clark volume, a collection of papers on the Athabasca oil sands. Research Council of Alberta
22. Tilmann W (1965) Bucket-wheel excavations. Taschenbuch für Braünkohle, erz, Salz, etc. (1964): 39–84; (1965): 25–70
23. Vogt W (1983) Planning and operation of opencast working with bucket wheel excavators and conveyor systems. In: Proceedings, surface mining and quarrying, institution of mining and metallurgy, London, pp 383–400.
24. Weimer WH (1968) Production engineering in surface coal mines. In: Pfleider EP (ed) Surface mining. AIME, New York, pp 224–246

Chapter 7
Rock Extraction with Scrapers, Bulldozers and Loaders

7.1 Technological Parameters of Wheeled Scrapers

Wheeled scrapers are used to extract rock, shift it and dispose of the mined mass at a waste dump or load it into transport facilities; in the latter case special hoppers are required. Most scrapers are of the hoe-type, Fig. 7.1a, with respect to loading. High-capacity scrapers are as a rule loaded by pushing them with a bulldozer. For further reading: Alkass and Harris [3]; Anon [4]; Burmistrov and Osintsev [6] and Caterpillar performance handbook [8].

Also available are scrapers in which the rock separated from the solid is delivered into the scraper bowl by an elevator that serves as the front wall of the bowl, Fig. 7.1b; these scrapers are efficiently used in working loose rocks. High-capacity scrapers are of the semi-trailed type operated in set with a wheel-mounted truck. Their maximum travel speed is up to 40–60 km/h during light running and 20–30 km/h when loaded. Trailed scrapers operated in set with crawler-mounted trucks have a low travel speed (7–10 km/h) and are used at small haulage distances (up to 400–500 m) at poor road conditions and intricate topography.

7.2 Mining Rock with Scrapers

As the scraper approaches the face the bowl is lowered, and its front gate is lifted. As the scraper moves over the face the blade cuts off a slice (cut) of rock of thickness t. When soft kinds of rock are worked t = 20–30 cm, when compact and broken rocks t = 10–15 cm. The magnitude of t predetermines the duration of bowl. Loading and the length of scraper travel for complete bowl filling. The bowl fill factor depends on rock moisture content. When dry rocks are extracted the bowl is not filled completely. The bowl is filled with a heap when working sandy barns and loamy rocks of natural humidity (respectively of 4–12 and 8–18%). A higher moisture content of argillaceous rock materially deteriorates their carrying capacity, Fig. 7.2. For further reading:

M. M. Ali Elbeblawi et al., *Surface Mining Technology*, Topics in Mining, Metallurgy and Materials Engineering, https://doi.org/10.1007/978-981-16-3568-7_7

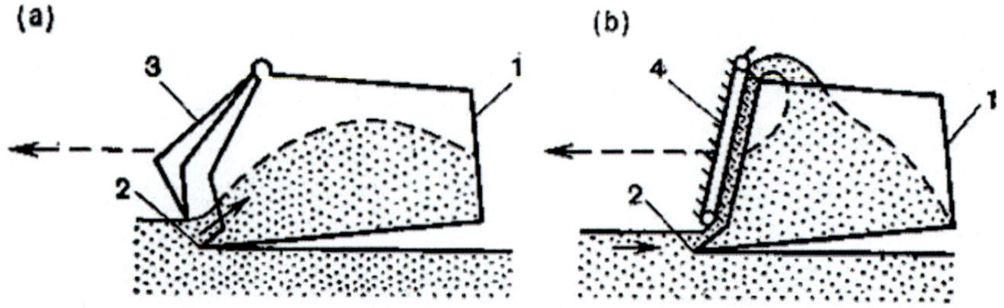

Fig. 7.1 Filling of hoe-type (**a**) and elevating (**b**) scraper bowl, where: 1. Bowl. 2. Bowl lip. 3. Rear wall. 4. Elevator

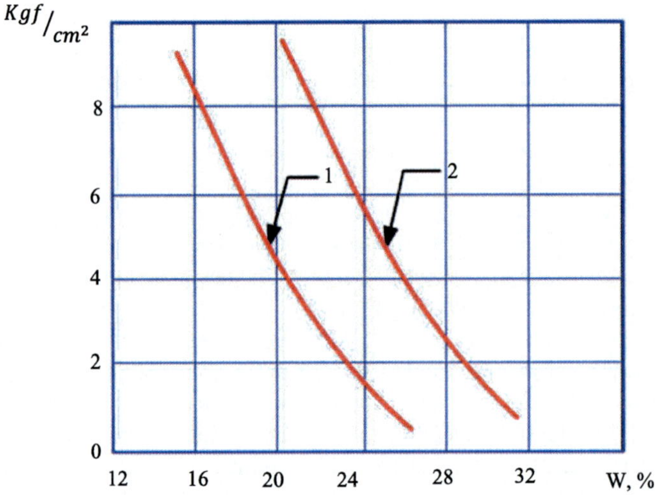

Fig. 7.2 Dependence of rock-carrying capacity on moisture content: 1—clay; 2—heavy loam

Alshibani and Moselhi [1]; Montaser and Moselhi [2]; Eldrandaly and Eldin [15]; Ingersoll-Rand [18] and Mayfield [24].

Since the pressure exerted by the scraper is equal to (1.15–1.2) P, where P is the inflation pressure of pneumatic tyres, the cross-country capability of the scraper does not permit its use in working clay rocks of moisture content greater than 24–27% without preliminary dewatering. When a pusher-tractor is used in excavating compact and broken rocks by heavy-duty hoe-type scrapers, the bowl fill factor increases by 10–12% (sometimes by 20–30%), bowl filling time reduces by 30–50% and the wear of the scraper diminishes. One pusher-tractor serves two-four scrapers. Bowl filling is greatly affected by the conditions under which its gate is opened and closed. The gate lifting height is decreased as Πex increases. Closing of the gate is initiated when 75–85% of bowl volume is filled. A berm, longitudinal slope or side of a bench serves as faces for a scraper, as illustrated in Fig. 7.3. If the berm on a bench

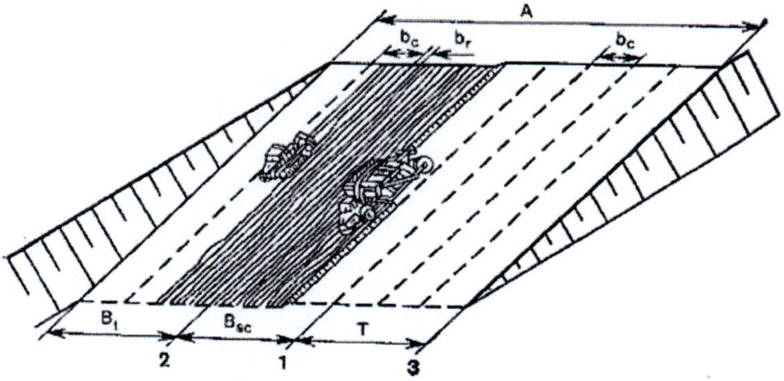

Fig. 7.3 End face for scraper operation, where 1, 2, 3—scraping, loosening and transport strips

serves as a face, rock is extracted in horizontal slices, and in the other two cases—in inclined slices. A berm-face favorably affects natural thawing of frozen rock during the spring–summer season, however downgrade extraction from longitudinal and end faces reduces the bowl filling time by 20–30% since the bowl is quicker filled due to making a thicker cut and entrainment of a part of the dragging prism. The length of an inclined face is often assumed multiple of the bowl filling path and depends on the height of the worked bench Hb and the face slope angle αf. This angle must not exceed 6–8° for sandy-clay and 10–12° for compact rocks. The optimal slope suiting bowl filling is within 5–6° (8–10%). Sometimes up to 5% upgrade extraction is practiced to increase the bowl fill factor when mining loose or free running rock. For further reading: Morris [26] and Rupprecht [30].

At an end face, the width A of a cut carried by a scraper is equal to its scraping width B5~ or it may also include the transport strip, T wide, for the travel of filled and empty scrapers, the drainage strip Z, and when broken rock is extracted-the strip for loosening compact or soft weathered kinds of rock B1, Fig. 7.3. Thus, in the general case, Eq. 7.1:

$$\boldsymbol{A} = \boldsymbol{B}_{sc.} + \boldsymbol{B}_1 + \boldsymbol{T} + \boldsymbol{Z},\ \mathrm{m} \tag{7.1}$$

The width of the strip scraping, Eq. 7.2:

$$\boldsymbol{B}_{sc.} = \mathbf{n}(\mathbf{b_c} + \mathbf{b_r}),\ \mathrm{m} \tag{7.2}$$

where:

n is the number of adjacent scraper passes (usually n = 2–4).

bc is the scraper rock cutting width, m (usually equal to the width of the scraper bowl bb).

b_r is the width of the ridge of height t between adjacent scraper passes, m (b_r = 0.5–1.5 m).

If the carried cut includes scraping, loosening and transport strips, having extracted the inclined natural slice in the scraping strip to the loosening depth (0.5–1.5 m) the scraper extracts the broken rock from the loosening strip. The ripper prepares rock for excavation within the limits of the former transport strip. Scrapers and other machines travel along the former scraping strip (after it has been treated with a grader). That is why the width of all indicated strips is assumed equal: $B_{sc} = \mathrm{B1} = \mathrm{T}$. The width of the drainage strip $Z = 2\ h_{dr}$ cot αd, where h_{dr} is the depth of drainage ditch and αd the ditch slope angle. At $h_{dr} = 2–3$ m, Z = 5–7 m. When compact and weathered rocks are worked from end faces with the aid of heavy-duty scrapers the width of a carried cut is up to 50 m and more. Wide cuts are also carried with face-berms. Placers are mined carrying cuts of a width reaching the width or length of a polygon. With longitudinal faces the width of carried cuts A = t or A = nt, where "n" is the number of slices, "t" is the slice thickness. The main percentage, of the scraping cycle time (up to 70–80%) falls on scraper travel both with a filled and empty bowl. This time depends on the dimensions of the quarry and waste dump, the place of excavation and stockpiling of the material, haulage roads grades, gearing diagram and speed of scraper travel. The maximum grades of scraper communications are given in Table 7.1.

To ensure an upgrade travel speed not less than 15 km/h for a scraper with a filled bowl, the road grade should not exceed 10–12%. The radius of scraper curves $R \geq (2.5–3.0)\ R_t$, where R_t is the minimum radius of the scraper turn. The magnitudes of the side slopes on curves must be decreased two times compared with the values given in Table 7.1. Scraper communications must systematically be levelled, compacted, cleaned up and watered. Scrapers construct waste dumps in horizontal or inclined slices, from 0.2–0.3 to 2–3 m thick. Dumping of rock on an inclined straight section (10–20% grade) at a maximum slice thickness and low travel speed permits a shorter bowl emptying time, makes it possible to prevent pushing the barren rock downgrade by a bulldozer and ensure stability of the waste dump. For further reading: Rupprecht [30–32].

The selection of a scraper for a particular job and the determination of its capacity require an assessment of the total resistance force F_{sc} to be overcome by the scraper tractor at the end of bowl filling. It is a sum of the resistance forces opposing the scraper travel, rock cutting, displacement of the dragging prism and bowl filling. The force of resistance to scraper travel, Eq. 7.3:

Table 7.1 Maximum scraper road grades

Scraper	Upgrade	Downgrade	Side slope
Tractor-drawn trailed scraper with filled bow	$\frac{14–18^{a}}{8–10}$	$\frac{\leq 27}{15}$	$\frac{\leq 12}{7}$
Empty	$\frac{20–23}{11–13}$	$\frac{\leq 40}{22}$	$\frac{\leq 12}{7}$
Semi-trailed with filled bowl and empty	$\frac{12–15}{7–9}$	$\frac{\leq 27}{15}$	$\frac{\leq 10}{5}$

[a]Figures in numerator are in percent, in denominator-in degrees

$$\boldsymbol{F_1 = M_{sc}(\omega_0 \pm 1)}, \text{ kgf} \tag{7.3}$$

where:

M_{sc} is the mass of scraper and rock filling the bowl, kg.
ω_0 is the main resistance to scraper travel, kgf/kg ($\omega_0 = 0.1$–0.15 kgf/kg).
$i = \omega_i$ is the resistance offered by a grade, kgf/kg.

The force of resistance to rock cutting, Eq. 7.4:

$$\boldsymbol{F_2 = kf_c b_c t}, \text{ kgf} \tag{7.4}$$

where:

kf_c is the rock resistance to cutting, kgf/cm^2.
bc and t are respectively the width and thickness of cut, cm (usually $b_c = \text{bb}$ with bb the width of the scraper bowl).

The forces of resistance to bowl filling are due to the weight of the upward moving rock F_3, the friction between rock particles and the friction between the particles and walls of the bowl F_4, Eq. 7.5:

$$\boldsymbol{F_3 + F_4 \approx 0.001(1 + f)b_c t h \gamma_l}, \text{ kgf} \tag{7.5}$$

where:

f is the coefficient of dynamic friction between rock particles and bowl walls, Table 7.2.
h is the bowl filling height, cm (usually $h \geq 120$–190 cm).
γ_l is the bulk density of loosened rock in the bowl, tf/m^3.

When sand rocks and a rock waste-gravel mass are extracted, the cutting force amounts to 20–25% of the total resistance, and with clay rock, up to 60–75%. The traction force of the scraper P_t must overcome the total force of the excavated rock resistance to digging, Eq. 7.6:

$$\boldsymbol{P_t \geq F_{sc} = F_1 + F_2 + F_3 + F_4 = KFbbt}, \text{ kgf} \tag{7.6}$$

Table 7.2 Values of coefficient of rock friction (f)

Rock	Value of rock friction coefficient, f	
	Between rock particles	Between rock and steel
Sands, sandy loams	0.8	0.35
Loam	1.0	0.50
Clays	1.2	0.80
Chernozem	–	0.90
Damp peat	–	0.65

where:

KF is the rock resistance to digging during scraper extraction, kgf/cm^2.

7.2.1 Scraper Capacity

The certified capacity, specified by Eq. 7.7, is true for excavating from a face-berm of a certified rock whose excavation ability index Πexc is equal to Πexa, Table 7.3, hauled along a horizontal road through a distance of 100 in and unrestricted dumping of the hauled rock in horizontal slices. For each standard-size scraper these conditions are met when making a cut of definite, design (certified) thickness and cross-sectional area, travel speed during bowl filling, length and duration of haulage, Table 7.3. Equation 7.7 gives the effective capacity ($Q_{eff.}$):

$$Q_{eff.} = \frac{3600EK_{bf}}{twcK_{bl}}, \ m^3/h \tag{7.7}$$

The coefficients K_{bf} and K_{bl} depend primarily on the excavation ability of the extracted rock and rock moisture content, Eq. 7.10 and Table 7.4. The duration of the scraper working cycle is given in Eq. 7.8:

Table 7.3 Indices characterizing certified scraper capacity

Index	Scraper bowl volume, m^3			
	3–5	8–12	15–20	Over 20
Cross-sectional area of cut (S_c), m^2	0.3–0.4	0.5–0.6	0.65–0.7	0.75–0.85
Bowl filling travel length, l_f, m	10–12	16–20	23–30	30–35
Scraper travel speed during bowl filling (v_t), m/s	0.2–0.3	0.3–0.35	0.35–0.45	0.4–0.5
Bowl filling duration, t_f	40–50	55–60	65–70	70–75
Certified rock excavation ability index, $I I_{exc}$	0.6–1.0	1.2–1.6	1.8–2.6	2.8–3.6
Factor allowing for effect of standard size of scraper, K_{st}	1.0	0.97–0.93	0.9–0.85	0.8–0.7

Table 7.4 Average values of factors K_{bl} and K_{bf} for scrapers used to excavate rocks of natural humidity

$\prod exa$	K_{bl}	K_{bf}
≤2	1.15–1.2	0.95–1.1
2.5–3.0	1.25–1.3	0.85–0.95
3.5–4.0	1.3–1.4	0.75–0.85

$$t_{wc} = t_f + t_l + t_e + t_{dl} + n_1 t_{sh} + n_2 t_t \quad (7.8)$$

where:

t_f	is the bowl filling time, s.
t_l and t_e	are respectively duration of scraper travel with a filled and empty bowl, s.
t_{dl}	is the duration of scraper dumping-levelling operations, s.
t_{sh}	is the time of one gear shift, s ($t_{sh} \approx 3$ s).
n_1	is the number of gear shifts ($n_1 = 2$–6).
t	is the duration of one scraper turn, s.
n_2	is the number of scraper turns per working cycle.
l_f, l_1, l_e, l_{un}	are respectively length of bowl filling travel, length of travel with filled and empty bowl and length of travel for bowl unloading, m.
V_f, V_l, V_e, V_e	are respectively travel speed during bowl filling, travel with loaded and empty bowl and during unloading, m/s.
K_{lv}	is the coefficient allowing for the relative duration of rock levelling after dumping.

The length of bowl filling travel is expressed as per Eq. 7.9:

$$l_f = \frac{E \prod exaK_{bf}K_{ls}}{S_c \prod excK_b K_{th}}, \text{ m} \quad (7.9)$$

where:

Пexa	is the true index of rock excavation ability when rock extraction is done with a scraper of a definite standard size;
Пexc	is the certified rock excavation ability when extraction is accomplished with a scraper of definite standard size from a face-berm, Table 7.3.
Sc	is the certified (design) cross-sectional area of cut, m^2, Table 7.3.
k_{1s}	is the factor allowing for losses due to the formation of a ridge ($k_{1s} \approx 1.2$); k_{th} is the factor allowing for non-uniform cut thickness (usually $k_{th} \approx 0.7$).

The bowl-filling travel speed of a scraper may be specified as given in Eq. 7.10:

$$V_f = V_{fc} K_{tr} K_{sp}, \text{ m/s} \quad (7.10)$$

where:

V_{fc}	is the certified scraper travel speed during bowl filling, Table 7.3.
K_{tr}	is the factor allowing for type of traction: for crawler mounted tractors K_{tr}: ≈ 1.0, and for wheeled tractors $K_{tr} \approx 1.25$.
K_{sp}	is the factor allowing for a change in travel speed depending on the excavation ability of rock, Table 7.5.

The haulage distance, number of scraper turns and gear shifting are determined by the scraper operation technology. When haulage distance is up to 0.5 km along

Table 7.5 Average values of factor K_{sp}

$\prod$ exa	Factor K_{sp} at a scraper bowl volume E, m^3			
	3–5	8–12	15–20	Over 20
≤2	1.0	1.0	1.10	1.15
2.5–3.0	0.9	0.93	1.0	1.05
3.5–4.0	0.75	0.80	0.9	0.93

temporary roads, the travel speed of a loaded semi-trailed scraper is assumed equal to 2.5 m/s, and of a trailed scraper 1.3 m/s. The average travel speed of empty scrapers of this kind is equal to 4.5 and 2.0 m/s, respectively. With a longer rock haulage distance over the roads with different pavements and haul profiles the travelling speed of semi-trailed scrapers must be determined along road sections with different road conditions, or, approximately, assumed equal to the travelling speed of truck tractors with semi-trailers. The scraper emptying path is 10–20 m long, and the traveling speed $V_{un} = 0.8–1.2$ m/s. The factor $K_{1v} \approx 2.0$ when the scraper is engaged in levelling a horizontal slice; when this is done with a bulldozer and the rock is dumped in inclined slices $K_{1v} \approx 1.0$. It takes from 15 to 30 s to turn a scraper with a bowl capacity change from 2 to 15 m^3. For further reading: Mayfield [14] and Farid and Husaini [16].

7.3 Rock Extraction with Bulldozers

A bulldozer is a crawler-mounted or wheeled tractor fitted with a blade. Similar to a wheeled scraper it is capable of extraction, moving and stockpiling rock. The capacity of the tractor classifies bulldozers into extra heavy-duty (over 300 hp), heavy-duty (150–300 hp), medium-duty (100–150 hp) and light-duty (20–80 hp). The bulldozer, just as a wheeled scraper, can be used to excavate rock from a berm, longitudinal and end-slopes of a bench or from a muck pile, Fig. 7.1. It is expedient to excavate from inclined faces, cutting off and moving the rock downgrade. For further reading: Bulldozer blades capacities [7] and Demidenko [12].

The operating cycle of a bulldozer consists in cutting off a horizontal or inclined slice from the ground, formation of the dragging prism, moving the latter and dumping. The traction force P_t, of a straight-blade bulldozer in motion with the blade dug into the ground overcomes the following resistances, Fig. 7.4 to the motion of the bulldozer proper F_b, to rock cutting by the blade (horizontal component P_1 of the summary reaction of the ground P), to moving the dragging prism F_{dp}, to lifting the rock along the blade F1f, due to the slope grade F_1. It follows that the following condition must be valid, Eq. 7.11:

$$P_t \geq F_b + F_{dp} + F_{1f} + F_1 = \omega_0 M_b + K_{fc} bt + (1.2 - 1.3) V_{dp} \gamma_1 f \pm \left(M_b + V_{dp} \gamma_c\right) \omega_i, \text{ kgf} \tag{7.11}$$

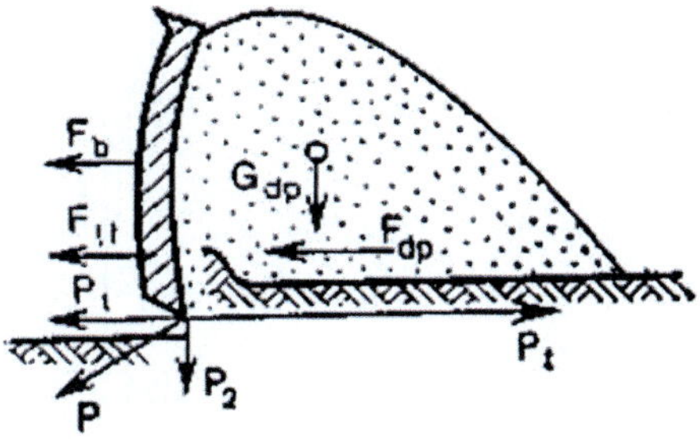

Fig. 7.4 Forces acting on bulldozer blade

where:

ω_0	is the specific resistance to tractor motion, kgf/kg ($\omega_0 = 0.08$–0.12 kgf/kg).
M_b	is the mass of bulldozer, kg.
K_{fc}	is the resistance of rock to cutting, kgf/cm^2, Table 7.6.
b, t	are respectively the cut width and thickness, cm.
V_{dp}	is the volume of dragging prism, m^3.
γ_1 and γ_c	are respectively bulk weight of loosened rock, kgf/m^3, and its density, kg/m^3, Table 7.6.
f	is the rock friction coefficient, Table 7.2.

When a slice to be extracted is being cut off the ground the resistances F_{dp} and F_{lf}, increase from zero to maximum as the cut-off rock accumulates in the dragging prism. To utilize to the full extent the traction force of the tractor while a slice is cut off the ground, it is advisable to gradually diminish the force of resistance P_1, by cutting off a wedge-shaped slice, Fig. 7.5a; then the thickness of the cut decreases as the rock accumulates in front of the bulldozer blade. When working compact rocks with a bulldozer, it is advisable to make comb-shaped cuts, introducing many times the blade into the ground, Fig. 7.5b, otherwise $P_1 > P_t$. For further reading: Cerny [9]; Construction planning, equipment and method [10] and Han et al. [17]. The kind of cut made in light coherent grounds is illustrated in Fig. 7.5c; then the thickness of the cut $t = t_{max.}$ const, and the cut length and the time of prism accumulation in front of the blade are at a minimum. The cut width b is assumed equal to the blade length 1b when extracting free running and non-compacted soft rocks. With rising excavation difficulty (K_{fc}) the width of the cut is decreased and the (a) length of the extraction section is increased. In practice, the parameters b and t are consistently adjusted so as to ensure the admissible values of P_1. When material is moved $P_1 = 0$, the free

Table 7.6 Approximate values of K_{fc} and γ_c assumed in the design of bulldozer traction forces

Rock	K_{fc}, kgf/cm^2	γ_c, kg/cm^3
Clays	1.2–2.0	1200–1800
Sandy loam and loam	0.7–1.0	1600–1800
Sand	0.5–0.6	1500–1700

Fig. 7.5 Methods of shaping the slice extracted by bulldozer

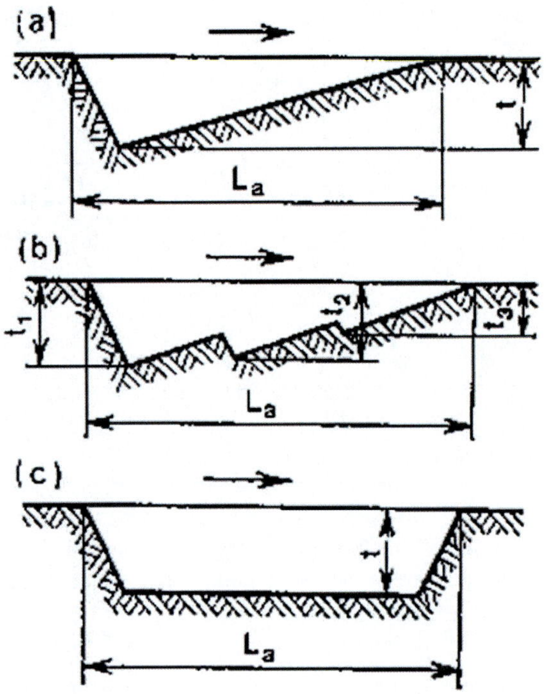

capacity of the tractor can be utilized both for increasing the travelling speed and for increasing the volume of the dragging (c) prism by 12–15%; to this end the blade is fitted with wings-wideners and with a spill guard. Compact, jointed weathered rocks of the first group and frozen rocks must be prepared for bull, Fig. 7.5.

Methods of shaping the dozing by mechanical loosening slice extracted by a bulldozer when extracting blasted rock the lump size must not exceed 100–120 cm. It is expedient to move the bulldozer over the bench toe along the muck pile, while the blade penetrates partly into it. At a rock accumulation distance La and a slope angle of the muck pile β, the actual volume of the dragging prism. For further reading: Karshenas and Xin [19] and Komatsu [20, 21].

Equation 7.12 gives the design volume of prism:

$$V_{dp} = \frac{\left[(ml_b)^2 L_a \tan\beta\right]}{2},\ \mathrm{m}^2 \tag{7.12}$$

where:

$m = (b{:}\ l_b)$ is the ratio of the part of the blade penetrated into the muck pile to its full length, lb.

At a bulldozer blade height, h_b, the design volume of the dragging prism is given in Eq. 7.13. At $l_a = 8$–10 m, $m = 0.2$–0.4, and the width of the carried cut, $A_b = ml_b$

= 0.7–1.2 m.

$$V_{dpd} = \left[\frac{h^2 l_b \cot\beta}{2}\right],\ m^3 \qquad (7.13)$$

Safety regulations prescribe that the muck pile should be not higher than 5–7 m. Since extraction of blasted rock requires more time and bulldozing is accomplished with a smaller volume of the dragging prism (due to increase of the looseness factor and the angle of repose of rock), the output of the bulldozer drops 1.5–2.0 times as compared to the output achieved in working soft rocks. To decrease slippage of the hauled rock, the bulldozer makes many passes along one strip 3–3.5 m wide, forming at the edges of the strip ridges 0.7–1.0 in. wide, Fig. 7.6a, b. The formed trenches are 0.6–0.7 m deep. Dealing with soft rocks bulldozers are sometimes operated in pairs, as illustrated in Fig. 7.6c, then the bulldozer output increases by 30–50%.

At the haulage distance increasing up to 50–100 m, the section through which the excavated rock is moved is divided into two–three lengths. At the end of each length, the rock is piled in the shape of an intermediate bank that is moved in succession

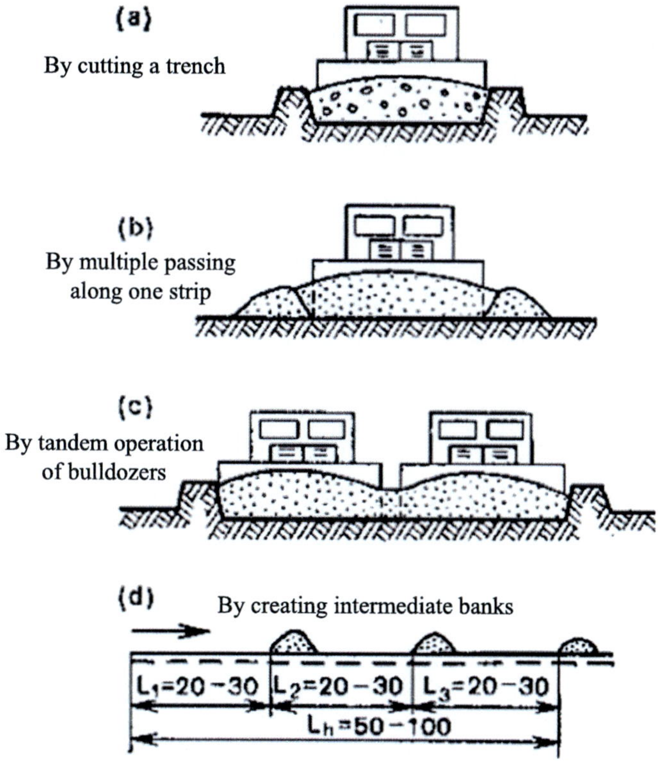

Fig. 7.6 Methods of reducing rock spillage by a bulldozer

to the dumping place. The operating cycle then takes more time, but the output of the bulldozer increases by 30–40% due to the lesser spill age and better rock accumulation.

The bulldozer can be operated in shuttle-like and circular motion. It is expedient to apply shuttle-like motion at a haulage distance longer than 50 m. The maximum grades, % for loaded and empty travel by bulldozers are given below:

	Upgrade	Downgrade	Transverse grade
Loaded travel	15–18	45	30
Empty travel	35–40	45	30

The bulldozer dumps the excavated material in slices down the bench slope, then levels the dumped material (by lifting the blade through the height of the slice) or stockpiles the excavated mass depending on the kind of the loading equipment used.

7.3.1 Bulldozer Capacity

The certified capacity of a bulldozer is based on extracting from a face-berm rock whose Пexa = Пexc at design parameters of the excavated slice, Table 7.7, movement of the design dragging prism through 10 m, unimpeded (free) dumping of rock and idle running of the bulldozer through the same distance. The effective capacity of a bulldozer is presented in Eq. 7.14:

Table 7.7 Parameters characterize the bulldozer certified capacity

Parameter	Bulldozer horse-power			
	≤100	140–180	200–300	≥400–500
Cross-sectional area of cut, S_c, m^2	0.2–0.3	0.4–0.5	0.6–0.7	0.8–1.0
Design volume of dragging prism, V_{dpd}, m^3	2–2.5	4–5.5	7–10	12–15
Prism accumulation length, L_a, m	8–10	10–12	12–14	15–16
Bulldozer excavation travelling speed, v_a, m/s	0.3–0.4	0.4–0.5	0.5–0.6	0.6–0.7
Rock accumulation duration, t_a, s	20–25	20–25	20–25	20–25
Certified rock excavation ability, $\prod exc$	1.2–1.8	2.0–2.6	2.8–3.8	4–5
Factor allowing for standard size of bulldozer, K_{st}	1.08–1.03	1.0	0.97-0.92	0.9–0.8

$$Q_{eff} = \frac{3600 V_{dp} K_d}{t_{wc} K_{lp}}, \ m^3/h \tag{7.14}$$

where:

V_{dp} is the actual volume of dragging prism, m^3.
K_d is the factor allowing for the change of the grade and rock haulage distance, Table 7.8.
t_{wc} is the duration of bulldozer operating cycle; s; and
K_{lp} is the factor allowing for rock loosening in the dragging prism.

In working placers, the volume of compact rock in the dragging prism (i.e. V_{dp}: K1~) may be assumed as indicated in Table 7.9. For further reading: Klanfar et al. [23]; Rashidi et al. [29] and Tsuji et al. [40].

Equation 7.15 gives the general case:

$$V_{dp} = V_{dpd} K_{ea}, \ m^3 \tag{7.15}$$

where:

V_{dpd} is the design volume of dragging prism, m^3.

Table 7.8 Values of grade change and rock haulage distance factor

Rock haulage distance, m	K_d			
	Horizontal section	10% slope	20% slope	10% upgrade
15	1.00	1.80	2.50	0.60
30	0.60	1.10	1.60	0.37
65	0.30	0.60	0.90	0.18
100	0.20	0.36	0.55	0.12

Table 7.9 Bulldozer dragging prism volume in working placers

Work	Bulldozer horse-power						
	87	108[a]	108[b]	140	180	300	340
Peat deposit stripping	1.9	2.0	2.2	2.5	3.8	5.3	6
Working and moving of sands to washing units	1.7	1.8	1.95	2.2	3.4	5.1	
Stripping of frozen rocks	1.17	1.22	1.35	1.5	2.4	3.20	3
Removal of vegetative layer and arrangement of cone-shaped waste dumps in banks	2.0	2.1	2.30	2.6	3.85	5.40	
Stocking of pit-run sand	1.30	1.40	1.55	1.72	2.85	3.75	

[a]with a blade 3030 mm long
[b]with a blade 3200 and 3940 mm long

K_{ea} is the factor allowing for the excavation ability of rock:

Пexb	≤2	2.5–3	3.5–4
Kea	1.0	0.87	0.79

The duration of the operating cycle of a bulldozer may approximately be specified from Eq. 7.16:

$$t_{wc} = t_a + t_m + t_r + t_{aux} = \frac{L_a}{V_a} + L_1 + \frac{L_a + L_m}{V_m} + \frac{t_{aux}}{V_r}, \ s \quad (7.16)$$

where:

t_a is the rock accumulation (extraction) time, s.
t_m and t_r are respectively the times for rock moving and bulldozer return, respectively, s.
$t_{aux} = 7\text{–}10$ s is the time of auxiliary operations during the bulldozer operating cycle, s.
L_a and L_m are respectively the rock accumulation and moving distances, m.
V_a, V_m and V_r are respectively the average velocities of rock extraction, moving and bulldozer return travel, m/s.

The rock accumulation time may approximately be specified as given in Eq. 7.17:

$$t_a = \frac{t_{ac} \prod exa}{\prod exc}, \ s \quad (7.17)$$

where:

t_{ac} is the certified duration of rock accumulation, s (Table 7.7).
Пexa and Пexc are respectively actual and certified indices of rock excavation ability for a bulldozer of a definite standard size, Table 7.7.

The rock accumulation distance is expressed as follows in Eq. 7.18:

$$L_a = \frac{V_{dp} \prod exaK_{sp}}{S_c \prod excK_{th}}, \ m \quad (7.18)$$

where:

V_{dp} Volume of compact rock in dragging prism, m,
S_c Design (certified) cross-sectional area of cut, m^2 (Table 7.7).

Where:

K_{sp} is the factor allowing for rock spilled during extraction (on the average $K_{sp} \approx 1.2$).

Table 7.10 Average design bulldozer travelling speeds

Rock mined	Bulldozer horse-power	Bulldozer travelling speed, m/s		
		Rock accumulation	Rock moving	Return run[a]
Arenaceous and soft rocks	100	0.40	0.60	1.0
	140	0.45	0.65	1.05
	180	0.50	0.70	1.1
	300	0.55	0.90	1.25
	≥300	0.60	1.0	1.45
Compact, rock debris-gravel and coherent consolidated rocks	100	0.30	0.45	0.80
	140	0.35	0.50	0.90
	180	0.40	0.60	1.0
	300	0.50	0.80	1.15
	≥300	0.55	0.90	1.35
Fine-broken by blasting	100	0.20	0.30	0.65
	140	0.25	0.35	0.70
	180	0.30	0.40	0.75
	300	0.35	0.60	0.90
	≥300	0.40	0.70	1.10

[a]return run distance up to 30 m

K_{th} is the factor allowing for non-uniform thickness of cut ($K_{th} \approx 0.7$).

The average speeds of rock accumulation, loaded and return run of the bulldozer are given in Table 7.10. With a return distance over 30 m, the return travel speed of the bulldozer, Eq. 7.19:

$$V'_r = V_r K_{1n}, \ m/s \tag{7.19}$$

where:

V_r is the return travel speed of bulldozer through a distance up to 30 m, Table 7.10.

K_{1n} is the factor allowing for the increase in speed depending on rock moving distance.

Bulldozer return run distance, m	31–40	41–50	51–60	61–70	71–80	81–90	91–100
K_{1n}	1.17	1.28	1.37	1.44	1.50	1.55	1.59
Bulldozer return run distance, m	101–110	111–120	121–130	131–140	141–150		
K_{1n}	1.62	1.64	1.65	1.66	1.67		

The effective bulldozer output in levelling work is given in Eq. 7.20:

$$Q_{1V} = \frac{3600L_{1V}(b - a)}{n\left(\frac{L_{1V}}{V_{1V}+t_t}\right)}, \ m^2/h \tag{7.20}$$

where:

L_{1V} is the length of levelled area, m.
b is the width of strip per bulldozer pass, m ($b_{max} = l$).
a is the strip overlapping width, in (a = 0.3–0.5 m).
n is the number of bulldozer passes over one strip.
v_1 is the average bulldozer speed in levelling (usually in first gear), m/s; and
t is the time spent in turning the bulldozer during each pass (for one turn t_t = 8–12 s).

7.4 Technological Fundamentals of Mining Automation

In the course of extraction, moving and dumping rock the resistance offered to the motion of the extraction-transport machine varies within a wide range. To ensure better filling of the scraper bowl or bulldozer blade, at the end of the cycle, the operator often has to apply special techniques (bites, jerks by means of the clutch, etc.). Due to the fact that up-to-date bulldozers and scrapers travel at comparatively high speeds, the operator of a bulldozer shifts gears up to 1200 times per hour, and of a scraper—400–500 times. For further reading: Montaser et al. [27]; Nielson [28]; Seung and Sinha [33] and Smith [35] (Fig. 7.7).

The greater part of these gear shiftings concern the drive of the excavating tool and each of them requires an effort of 15–20 kgf. This fatigues the operator quickly, resulting in that at the end of the shift the output of the machine, bulldozer or scraper, drops, and sufficient supervision of quality extraction is not ensured. Complete automation of the performance of bulldozers and scrapers, primarily of the digging process, is a very intricate and not yet solved engineering problem. This permits to decrease materially the labor input involved in controlling the machine due to a 2–3 times lesser number of hand engagements of the excavating tool of the machine, to

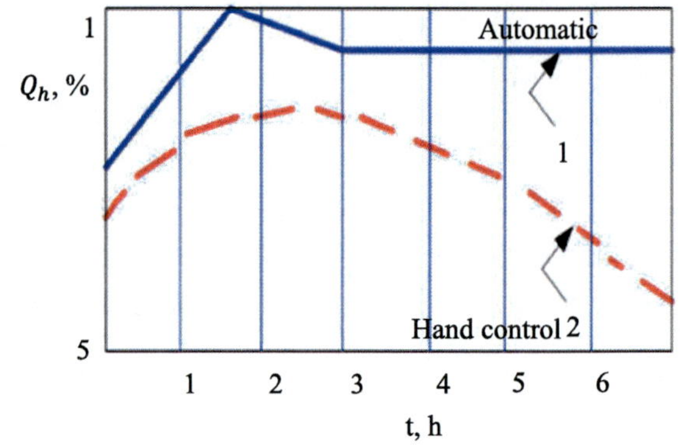

Fig. 7.7 Variation of scraper hourly output during a shift

utilize most of the traction force, to increase the speed of accomplishing individual operations 1.5–2.0 times, and to raise the output of the machine. An automatic-control system of the regulator type (ACS) must ensure control of the excavating tool, and the controlled variable is the cut thickness. The ACS may first be acted upon by the operator by presetting the initial cut thickness. Engine power conditions are determined by setting up and they are ensured by varying the torques tractive force, engine speed (rpm) or machine travel speed. It is simpler to automate the performance of bulldozers and scrapers operated in set with diesel-electric tractors. Then two modes of motion are preset: the operating (excavation) and haulage (transport). Travelling speed is automatically controlled by the tractor transmission in compliance with the external resistance. For further reading: Shmulevich and Asaf [37] and Tam et al. [38].

The already developed ACSs for the loads on the engines of bulldozers and scrapers are based on the principle of automatic stabilization (equalization) of the reduced load depending on the forces offering resistance to digging. Stabilization is reached by changing the thickness of the material cut off the ground as the bowl is getting filled with it. Service tests of machines fitted with automatic-control systems of this kind have shown that automation of the rock digging raises the output of scrapers by 12–17% and of bulldozers by 14–22 00. These automatic-control systems are most effective in the applications where the transport operations occupy a comparatively small part of the technological cycle of a scraper or bulldozer, in particular, when making embankments (fills), digging pits, performing stripping work in quarries (pits) located near roads, etc.

Digging by means of extraction-transport machines is also automated with the aid of positive-displacement hydraulic drives. The principle of action of the already developed hydraulic controllers is based on equalizing the delivery of two positive-displacement pumps. They ensure a constant resistance to digging by lifting and lowering the blade of a bulldozer, when slippage is greater or less than a definite value. For further reading: Son et al. [36].

7.5 Technological Characteristics of Loaders

Single-bucket loaders are designed to perform in the capacity of extraction-loading, extraction-haulage or auxiliary equipment. Loaders are usually operated in set with motor transport. The maximum crowding (penetration) force developed by a load is given in Eq. 7.21:

$$P_{Cr} \leq P_t = \psi G t_f \tag{7.21}$$

where:

P_t is the maximum tractive force,

t_f G is the weight of loader, tf;

ψ is the factor allowing for ground cohesion (between loader crawler tracks or wheels and ground); for wheeled loaders $\psi = 0.5$–0.6 and for crawler-mounted $\psi \approx 0.65$–0.7.

The loaders made in the USSR have buckets with a capacity of 2–5 t at a machine power rating within 75–240 hp, mass of machines 9–26 t, maximum tractive force within 6–17 tf and a dumping height of 2.7–3.3 m. Such loaders are used for loading-dumping operations and earth-moving work of a volume amounting up to (150–250) × 103 m^3/year. Provisions have been made for the manufacture of heavy-duty rubber-tired (wheeled) loaders of 10, 15, 25 and 40 t capacity rated at 360–1500 hp. and a tractive force within 15–50 tf and greater. The buckets of the loaders have a solid lip or teeth (for excavating blasted rock). The loaders are fitted with up to 30 change attachments such as buckets of various capacities, a shovel or back-hoe, a grader, a bucket for handling stones, a bulldozer blader, ripper, clamshell and fork grabs, a crane boom, a wire brush for cleaning motor roads, etc. The versatility of the mounted equipment (attachments) and its rapid replacement (40–60 mm) raise the utilization factor of the tow tractors. For further reading: Brinkschulte and Geimer [5]; Edwards and Holt [13] and Brinkschulte and Hafner [22].

7.5.1 Rock Extraction with Loaders

Loaders are suited for excavating soft rocks directly from the massif, mechanically broken rock (loosened by tractor rippers or by loader teeth) or blasted rock from a muck pile. When excavating rock characterized by Пex > Пexc it is expedient to use special reinforced buckets of a smaller capacity and extraction pick-ups. Rock is extracted due to the crowding force, turning of the bucket and lifting of the loader boom. Various methods of rock extraction are distinguished: the selective, combined, excavation and method of extraction in slices, Fig. 7.8. When the selective method is applied to extract low-density loose rocks from a muck pile, the extraction is accomplished in three successive operations. First the bucket is introduced along the toe of the face into the rock till the rear wall of the bucket comes to bear against the rock, Fig. 7.8a, 1, with the loader in translational motion at a speed of 0.6–1.1 m/s (in the first or second gear). The bucket is then turned through the entire tipping angle, as illustrated in Fig. 7.8a, 2, with the loader at rest. The third operation consists in lifting the filled bucket Fig. 7.8a, 3, into the transport position (to a height of 0.3–0.4 in.), then the loader backs up from the face. For further reading: McGill and Freadrich [25].

If rock is mined applying the combined method of extraction, the crowding force of the loader introduces the bucket into the rock at the face toe to a depth of (0.2–0.5) 1b (lb is the length of the bucket), then simultaneously the boom is lifted and the bucket is turned with the loader in continuous translational motion, Fig. 7.8b. The travel speed of the loader (in first–second gears) and the speed with which the bucket is lifted must be equal. The combined method is usually applied when

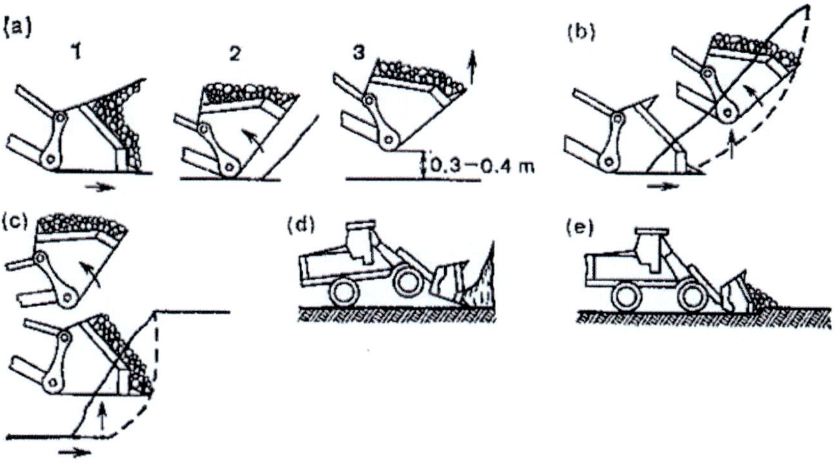

Fig. 7.8 Excavation with loaders

excavating soft and fine-blasted rocks, ensuring a high bucket fill factor and reducing the digging time, for this method permits the horizontal force of bucket penetration to be decreased two–three times, compared with selective extraction.

Rocks of a lower excavation ability (of a greater strength) are extracted practicing the excavation technique in the course of which the bucket is introduced into the ground (to a depth of 0.3 1b) and lifted simultaneously, Fig. 7.8c, by simultaneous lifting the boom and moving the loader. After the lip of the bucket clears the face, the bucket is turned and the loader backs up from the face. When coherent rocks and rocks blasted to medium lump size are extracted the bucket is introduced into the material stopping it and lifting the front part of the loader, as shown in Fig. 7.8d, simultaneously swinging the bucket in the vertical plane and moving the loader gradually toward the face. This done, the penetration force is maximum and the resistance offered by the rock to digging decreases. The method of rock extraction in slices, Fig. 7.8e, is a variety of the selective method when dealing with thin horizontal or inclined extraction slices, when the bucket is gradually filled during the continuous motion of the loader in first or second gear, which is similar to scraper extraction. For further reading: Singh and Narendrula [34].

A single-bucket loader is capable of excavation from an end and a longitudinal face and from a face-berm. When the loader is used for extraction and loading, after the bucket is filled the loader operator backs up from the face, moves it toward a dump truck, lifting at the same time the bucket to the dumping height required. Loaders are operated in a face in different ways as illustrated in Fig. 7.9. The way of operation shown in Fig. 7.9a consisting in turning the rubber-tired loader through 30–45° and approaching a dump truck, is most widely applied in longitudinal faces and when wide cuts are carried in end faces, ensuring a minimum loader travel. The way of operation illustrated in Fig. 7.9b differs from the preceding one in that the material

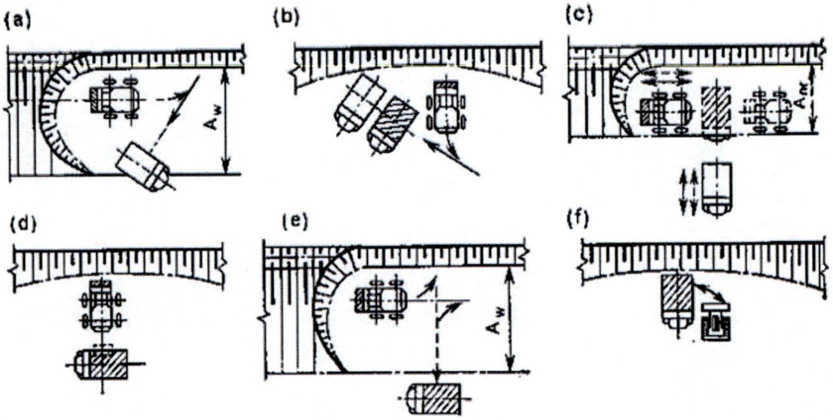

Fig. 7.9 Performance schematics for loaders

is dumped into two dump trucks; this permits a wider operating berm at any type of face and width of the cut carried in an end face; the loader travels through longer distances. Figure 7.9c illustrates the mode of operation during which the loader is in shuttle-type motion through 8–12 m at right angles to the face with the dump truck placed between the face and loader. This mode of operation is applied when rock is extracted with a wheeled loader carrying narrow cuts in an end face and with a crawler-mounted front-end dumping loader both in a longitudinal and an end face, and also with through running of the dump trucks on a longitudinal bench. The mode of mining with loaders illustrated in Fig. 7.9d also requires shuttle-type 5–10-m motion of rear-ends dumping loader. The dump trucks are placed at a distance of 8–13 m from the face.

The mode of operation illustrated in Fig. 7.9e requires two turns of the loader and travel through a longer distance. It is followed only when rock is extracted from an end face, carrying a wide cut with the dump trucks in through travel on the bench. The shuttle-type mode of operation shown in Fig. 7.9f can be accomplished with a side-dumping wheeled loader or with a crawler-mounted loader turning through 900 due to its tracks moving in opposite directions. The loader is backed up from the face through a distance of 2–3 m. Low (up to 2 m), normal (2–5 m) and high (over 5 m) faces are distinguished. If rock is extracted with a loader fitted with a bucket of over 5 m^3 capacity, an increase in the height of the bench over 43–8 m, no matter the kind of rock, does not affect the loader output. The benches are usually 8–15 m high. The minimum width of cut carried with a loader, Eq. 7.22:

$$A_{min} = B_b + C, \mathrm{m} \tag{7.22}$$

where:

B_b is the width of loader bucket, in.;

C is the minimum distance between the loader and the lower crest of a muck pile or bench, equal to 0.4–0.6 m.

The bucket of heavy-duty loaders is about 4 m wide, then $A_{min} = 4.5–5.0$ m. Heavy-duty loaders ensure the greatest output when the carried cut is 12–15 m wide, permitting rational turning of the loader and dump truck in the cut.

7.5.2 *Loader Capacity*

The certified (design) capacity of loaders, specified from Eq. 7.7, is based on extracting a certified rock, Table 7.11, practicing combined digging, shuttle-type motion of the loader through minimum distances and unimpeded bucket emptying. The effective output is given in Eq. 7.23:

$$Q_{eff} = \frac{3600E_0}{t_{wc}}, \ m^3/h \tag{7.23}$$

where:

E_0 is the operating bucket capacity, m^3.

E_0 is given in Eq. 7.24:

$$E_0 = \frac{K_{bf}}{\gamma_m K_{bl}}, \ m^3 \tag{7.24}$$

where:

P	is the rated capacity of loader's bucket.
t; K_{bf} and K_{bl}	are respectively the bucket fill factor and factor allowing for rock loosening in the bucket.
γ_m	is the rock density in the massif, t/m^3.
t_{wc}	is the duration of loader's operating cycle, s.

The values of factors K_{bf} and K_{bl} under favourable operating conditions for the loaders are given in Table 7.12. An increase in the lump size of excavated rock causes a decrease in K_{bf} to 0.9–1.0 at $d_{av} = 25–30$ cm, and to 0.6–0.7 at $d_{av} = 40–45$ cm. At the same time K_{bl} increases close to the data given in Table 7.12 for power shovels. For further reading: Tosuna and Konak [39].

The auxiliary operations performed by up-to-date rubber-tired loaders, such as lifting and lowering of the bucket, gear-shifting, etc., are accomplished simultaneously with the main operations of the operating cycle the duration of which is given in Eq. 7.25:

$$t_{wc} = t_{dg} + t_l + t_e + t_{un} = \frac{t_{dg}21}{V_{av} + t_{un}}, \ s \tag{7.25}$$

Table 7.11 Parameters characterizing the certified capacity of single-bucket loaders

Parameter	Bucket capacity, m^3			
	2–3	4–6	7.5–12.5	15–20
Digging time, t_{dg}, s	10–12	10–12	10–12	10–12
Duration of loader travel between face and dumping point and return at a minimum travel distance, t_{tr}, s	26–36	30–42	36–48	40–56
Duration of operating cycle, t_{wc}, s	50–52	54–56	57–62	66–70
Certified index of rock excavation ability, $\prod exc$	4.9–5.1	5.2–5.4	5.5–5.7	5.8–6.0
Factor allowing for the standard size of loader, K_{st}	1.1–1.05	1.0	0.95–0.90	0.9–0.85

Table 7.12 Average values of K_{bf} and K_{bl} for single-bucket loaders[a]

Actual excavation ability of rock mined with loader, $\prod$ exl	K_{bl}	K_{bf}
4–4.5	1.30–1.35	1.2–1.25
5–5.5	1.25–1.30	1.15–1.2
6–6.5	1.20–1.25	1.1–1.15

[a]At a maximum lump size $d_{max} < 35$ cm

where:

t_{dg} is the digging time, s.
t_l and t_e are respectively times of loaded and empty travel of loader, s.
t_{un} is the bucket emptying time, s.
l is the rock moving distance, in.
V_{av} is the average loader travelling speed, m/s.

When combined excavation is applied, the approximate digging time may be calculated from Eq. 7.26:

$$t_{dgc} = \frac{t_{dg} \prod exl}{\prod exo}; \text{ s} \tag{7.26}$$

where:

t_{dgc} is the certified rock digging time (Table 7.11);
Пexl and Пexo are respectively the actual and certified indices of the excavation ability of rock excavated with a loader of a definite standard size, Table 7.11.

The digging time for loaders usually falls within the limits indicated in Table 7.11. When wheeled loaders are used as extraction loading equipment and the rock haulage distance falls within 20–25 in the weighted average travelling speed of the loaders ranges within 1.0–1.8 m/s. The output of loaders operated in these conditions can be calculated using the values of their travelling speed given in Table 7.13. When wheeled loaders are used in the capacity of extraction-haulage equipment their travelling speed over quarry (pit) roads may be assumed equal to the standard speed of dump trucks in similar conditions. Loader speed may also be determined by traction design for individual road sections. The bucket dumping, or emptying, time is assumed 3–4 s.

7.6 Rock Extraction with Single-Bucket Excavators

7.6.1 *Technological Parameters of Power Shovels*

Power shovels of three types are distinguished: construction shovels, quarry-mine shovels, and stripping shovels. General-purpose construction-type excavators fitted with a bucket of 0.5–2.0-m^3 capacity, a diesel- or diesel-electric drive and crawler-mounted are used for mining sandy, soft and fine-broken rocks in quarries or pits of an annual production of (0.5–2) × 106 m^3 of mined rock, and also in quarries of a higher production for selective mining of thin deposits. These excavators feature deep dumping. The height of a bench is up to 6–8 m. Quarry-mine crawler-mounted power shovels with a multi-motor drive are suited for extracting soft and broken rock of any lump size. Bucket capacity is 2–25 m^3 bench height 6–20 m. They find application in quarries of any production capacity. Deep dumping is most common. Soviet industry makes available hydraulic excavators. For further reading: Construction equipment management for engineers [11].

Quarry-mine power shovels fitted with a longer working attachment ensure high dumping. The technical output of power shovels fitted with a longer working attachment is by 20–40% lower than that of a conventional power shovel, having a bucket of the same capacity. High dumping is efficient when, due to much more efficient operation of transport facilities, the effective output of excavator's increases, transport costs decrease, and other general mining indices improve (the rate of advance of the mining front, rate of penetration). Stripping power shovels with a bucket capacity E up to 100 m^3 and more are used for overcast operations (dumping rock in the mined-out space) and not as often (at $E = 15\ m^3$) for upper loading (dumping). They are suited for benches 10–50 m high. The main technological parameters of power shovels: bucket capacity, operating parameters, over all dimensions, negotiated grade, mass, and ground pressure. The operating parameters include the radius and height of digging and dumping that depend on the length of the handle, or stick, and boom, angle of boom inclination, and also on the position of the digging and dumping spots, Fig. 7.10. The digging radius R_{dg} is defined as the horizontal distance from the axis of swing (revolving) of the excavator to the bucket lip in the course of digging. The following digging radii are distinguished: maximum digging radius R_{dgmax} at a maximum advanced horizontal handle; minimum digging radius R_{dgmin}

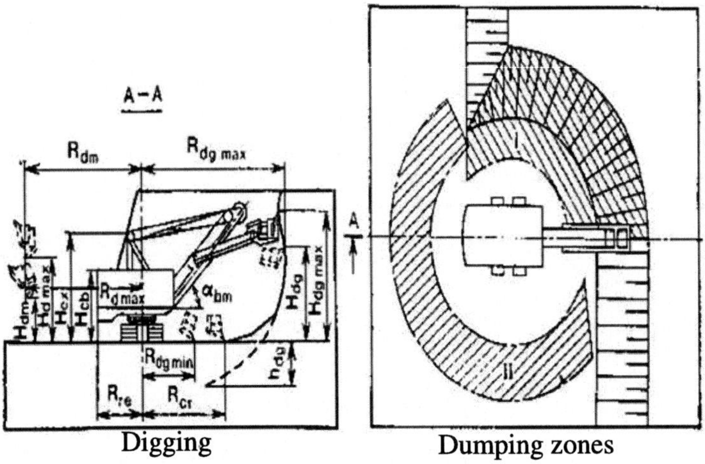

Fig. 7.10 Mining face and power shovel's operating parameters

with the handle and bucket drawn to the crawler tracks at the level where the excavator is positioned; excavator digging radius on the level of crawlers R_{cr}-maximum radius of digging at the excavator positioning level; and also digging radius at a maximum digging height $R_{dg}H$.

The digging height H_{dg} is the vertical distance from the excavator crawlers to the bucket lip during digging; the maximum digging height H_{dgmax} corresponds to a maximum lifted handle. Also distinguished are the following digging heights: digging height at a maximum digging radius H_{dm}, and maximum digging depth below the excavator crawler's h_{dg}. The dumping radius R_{dm} is the horizontal distance from the excavator swing axis to the bucket lip during dumping; the maximum dumping R_{dmax} radius corresponds to a maximum advanced horizontal excavator handle. The dumping height H_{dm} is defined as the vertical distance from the excavator crawlers to the lower edge of the tipped bottom; the maximum dumping height H_{dmax} corresponds to a maximum lifted handle. The minimum values of the digging and dumping radii do not coincide with the corresponding maximum values of the digging height.

The operating zone of a power shovel is predetermined by the radii and heights of digging and dumping. The boom angle $\alpha_{bm} = 45°$. The boom angle of some power shovels may vary within 30–50°. The dumping height increases and the radii of shovel operation decrease with the boom angle. The overall dimensions of a power shovel are determined by the rear-end clearance radius R_{re}, the excavator height H_{ex} (the vertical distance from the crawlers level to the upper edge of the most protruding non-detachable part of the shovel) and the height of excavator cabin H_{cb}.

The travelling speed of crawler-mounted power shovels is 0.9–3.7 km/h. The upgrade negotiated is up to 12° at a shovel mass of up to 100 t and to 7° for heavier models. The shovel ground pressure at the instant the bucket, or dipper is being filled, the shovel is swung and the bucket emptied varies within a wide range and it may exceed the average (certified) ground pressure 15–20 times. The maximum ground pressure should not exceed the carrying capacity of the base that varies from 2 kgf/cm^2 for soft sand to 5–6 kgf/cm^2 for compact clay and 8–10 kgf/cm^2 for marl.

7.7 Working Parameters of Draglines

The main working parameters of draglines, Fig. 7.11, are:

- The digging radius R_{dg} and the dumping radius R_{dm} that are determined by the length of the dragline boom and its angle.
- The digging radius with throwing of the bucket R_{dgth} (the length of throwing depends on the dragline model and skill of the operator and reaches 1/4 of boom length (2.5–15 in.); when the bucket is thrown the hoist rope makes an angle of 15–20° with the vertical).
- The radius of digging at the dragline level R_{dr}.
- The digging depth of penetration H_{dp}, depending on the boom length and angle, location of the dragline in the face, excavation ability of the mined rock, the length of ropes (lines) and skill of the operator.
- The height H_{dm} determined by the boom length and angle.
- Dragline body swing radius R_b.
- Height wise dimensions of the upper pillar and body of the dragline H_p and H_b.

The dragline boom angle is within 20–35°. Similar to a power shovel the boom angle may vary due to a respective change in other dragline parameters. Heavy-duty draglines are of the walking type, light-duty ones are crawler-mounted. The negotiated upgrade is up to 7–12°. The ground pressure is 1–2 kgf/cm^2. Draglines are usually operated for excavating soft rock. Draglines are used with buckets of

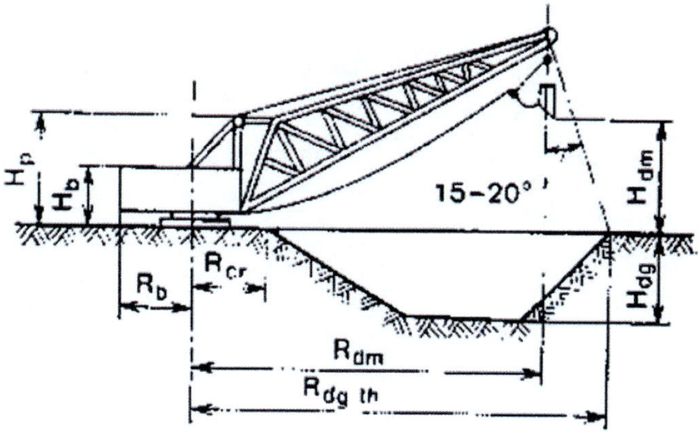

Fig. 7.11 Main working parameters of a dragline

8–10-m^3 capacity and over deal successfully with fine- and medium-blasted rocks. When used to excavate rock blasted into large lumps, the bucket is filled insufficiently and the bucket and the drag rope wear rapidly. In the USSR, draglines constitute about 20% of all the working attachments of single-bucket excavators and they are responsible for about 20% of the total volume of mining and earth-moving work.

7.7.1 Dragline Faces

The main type of dragline face is the end face. The possible parameters of the face are dependent on dragline location and the associated method of digging.

When rock is extracted by digging, the dragline is arranged on the roof of a. bench (muck pile) outside the limits of the heap of rock due to possible caving, Fig. 7.12a. Then the certified depth of penetration and the slope angle determine the possible height of the face. The dragline face is curvilinear in the massif in accordance with the path of the dragline bucket, and the face in a muck pile is close to straight line due to the periodic sloughing of the extracted rock. When the dragline is located on a bench toe and rock is extracted by deep digging, Fig. 7.12b, to prevent slippage and ensure filling of the bucket, the face slope angle αf must not exceed 20–25°. The height of the face (bench) fbh must not be over 0.8 H_{dm} so as to prevent digging under bucket-stop conditions. Usually, H_{bh} = (0.5–0.7) H_{dm}. High digging is efficiently done only by draglines with buckets of a capacity of 10–15 m^3 and over. However at H_{bh} > 0.4 H_{dm}, the output of these draglines is lower than that ensured by deep digging. A heavy-duty dragline may be located on an intermediate berm carrying two cuts in sub-benches by high and deep digging, Fig. 7.12c. Equation 7.27 gives

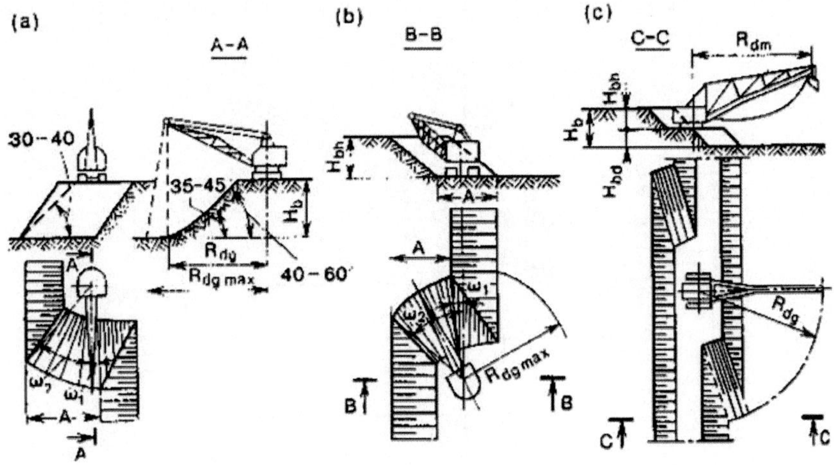

Fig. 7.12 End faces of a dragline in a through cut with deep, high and combination digging

the maximum width of a cut carried with a dragline, Fig. 7.12a:

$$A_{max.} = R_{dg}(\sin \omega_1 + \sin \omega_2),\ m \tag{7.27}$$

where:

ω_1 and ω_2 are the excavator swing angles during digging, each of the angles not exceeding 30–45°.

Usually $\omega_1 = 0$; the excavator swing angle during digging $\omega_1 = \omega_2 = 30$–45°, and the total swing angle of the dragline for dumping falls within 45–90°. Then the design width of cut $A = R_{dg}\sin\omega$ and for various type of draglines the width of cut is:

Dragline	I	II	III	IV	V	VI
A, m	23	23	29	33	36	42

Draglines with 4–10-m^3 buckets are sometimes used for loading rock into haulage facilities, Fig. 7.13a. Soft rocks may be excavated with heavy-duty draglines ($E = 8$–$15\ m^3$), practicing deep digging and bulk dumping onto the roof of a bench, wherefrom the rock is loaded into transport means with the aid of power shovels, Fig. 7.13b. Then the following relationship must be valid, Eq. 7.28:

$$R_{dm} \geq A - 0.5C_g + lb_s,\ m \tag{7.28}$$

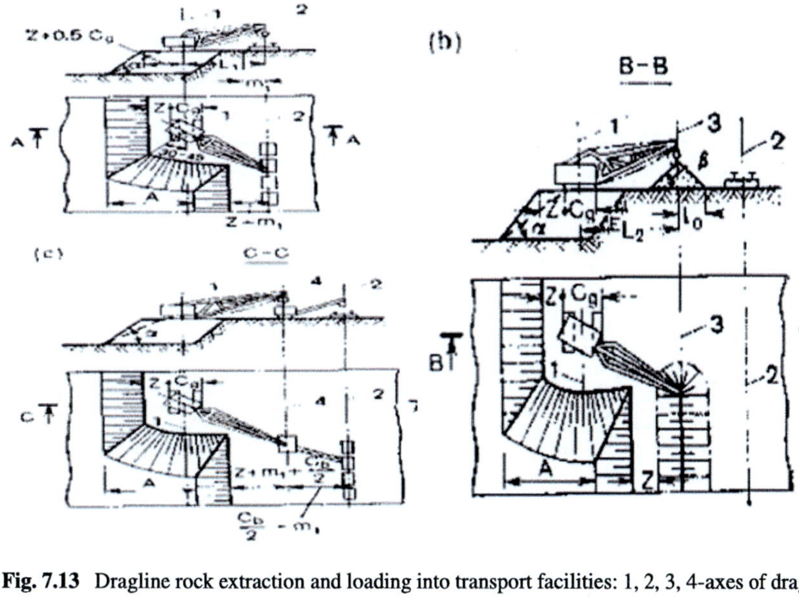

Fig. 7.13 Dragline rock extraction and loading into transport facilities: 1, 2, 3, 4-axes of dragline, center line of transport facilities, axes of heap and re-loader

where:

C_g is the dragline gauge, m;
lb_s is the horizontal equivalent of bulk slope, m.

$$lb_s = \sqrt{H_b A K_1 \cot \beta},\ m \tag{7.29}$$

where:

β is the bulk slope angle, deg.

If the dumping radius of the dragline, R_{dm}, is insufficient for condition to be valid, the volume of the stripping cut must be decreased, usually at the expense of its width. In the presence of a hopper-re-loader, heavy-duty draglines can load rock onto a conveyer, into railway cars or other transport facilities, Fig. 7.13c.

In the examples of extraction-loading work considered, the dragline carries normal or narrow through longitudinal cut-panels. It is possible to increase the width of a panel and diminish the number of times transport communications (tracks, roads, etc.) have to be shifted along the front of the bench, just as when power shovels are operated, by installing intermediate hopper-feeders fitted with discharge conveyers and individual conveyer-type re-loaders with sufficient operating parameters. Depending on their installation and the mode in which draglines are operated, panels may be worked carrying one or several normal or wide through longitudinal cuts, through transverse cuts or dead-end longitudinal wide cuts. The parameters of panels are determined in almost the same way as when rock is extracted with power shovels.

When deep digging is practiced face conveyers, hopper-feeders and re-loaders may be placed both on the upper and lower berm of the bench. In the former case the width of the excavated panel may be increased. When draglines are used for extracting sandy and sandy-gravel material from underwater faces the face angle of slope αf diminishes and, consequently, its possible height; the underwater part of the face caves intensively and this causes often shiftings of the excavator; $\alpha f = 22 + 0.16\,G$, deg, where G is the content of gravel over 5 mm in size in the mined material. Bucket filling is increased by using perforated buckets with openings 25–35 mm in diameter (Fig. 7.14).

7.7.2 Road Width

This is dependent on the location of the road and the size of equipment. It is necessary to consider that if a road is elevated then it should be wider than a similar road in a cutting. Other factors are:

- Where drainage is required this may need an additional 1–2 m. Provision for a safety shoulder for marker guides etc.

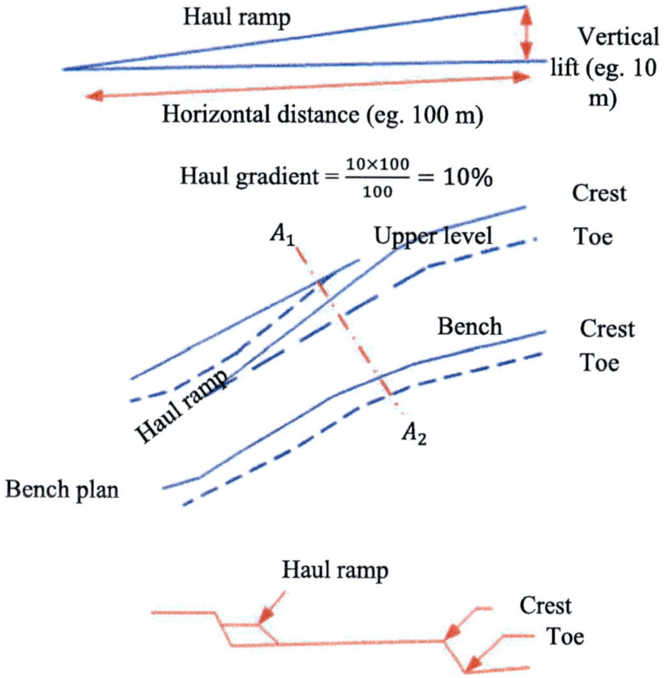

Fig. 7.14 Haul ramps

- Whether the road is either single of double haul.

As a general rule, road width should be twice the track width for each vehicle traveling simultaneously on the road, e.g. truck clearance +100. An example applicable to off-highway haulers of 80 tonne capacity and being 5 m wide passing each other would then require:

$$(5 + 5) + (5 + 5) + \text{Drainage} + \text{shoulder}\ 25\ \text{m}$$

This would be adequate to provide for road hogs (windrows) or whopper stoppers as a precaution against runaway trucks. This applies particular where down the hill or mountain loads are hauled. The road width can then be transferred onto the plan and the road: to be drawn from crest to toe as mentioned earlier, Fig. 7.15.

7.7.3 Services

Other factors in roadway design include flats on benches such that flat turning areas are available before trucks enter the next ramp. Safety cuts on roadways carrying

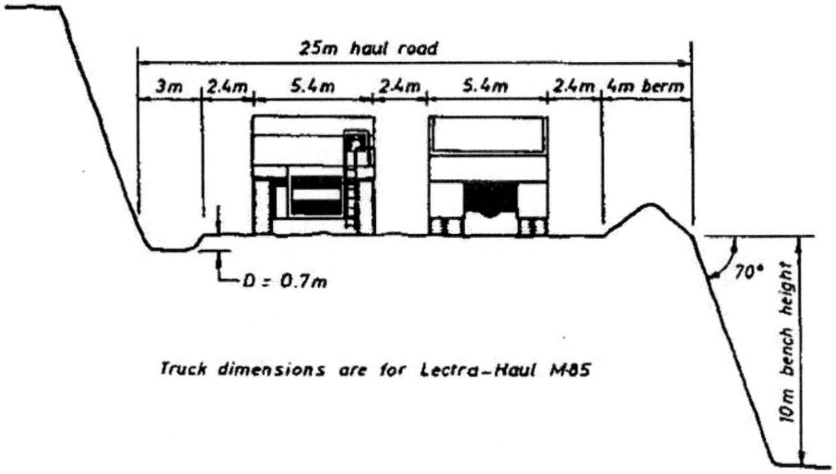

Fig. 7.15 Typical design haul-road width for two-way traffic using 77.11t (85St.) trucks road width for double width haul road

down traffic need to be provided with a 1–1.5 m high pile of fines some 20 m long to catch runaway vehicles. Alternative escape roads may be built turning back into the hillside. These include provision of such items as electricity, especially where large electrically-powered drilling rigs and excavators are used. Here, ring mains supplying power at 33 kV and possibly 3.3 kV for supplying the units by tracking cable. Cable sleds and cable reeling devices may be necessary to facilitate cable handling. A water supply may need to be maintained to working benches, haul roads: stockpiles and dumps for dust—suppression purposes. Other services may include provision for pumping and mine drainage facilities, material supply including fuel, explosives, and consumable stores.

7.7.4 Stockpiles

Allowance must be made for the storage of material such as in-pit stocks. Where blending is necessary, sufficient broken ore stocks must be maintained to meet blending and production requirements. Orebody variability and production variability have to be adequately considered. In an operating, bench operation it is desirable to open up two or more benches as soon as possible. This ensures:

- Adequate areas for working, i.e. broken stocks.
- Flexibility in grade control, if necessary.

7.7.5 Mine Layout

A classification of mineral deposits is illustrated in Fig. 7.16. This indicates the influence of the shape of a deposit on the design work. In general, stratified deposits are in softer, younger sedimentary rocks and non-stratified are in harder, older, igneous rocks. The shape of a mineral deposit fundamentally determines the shape of the

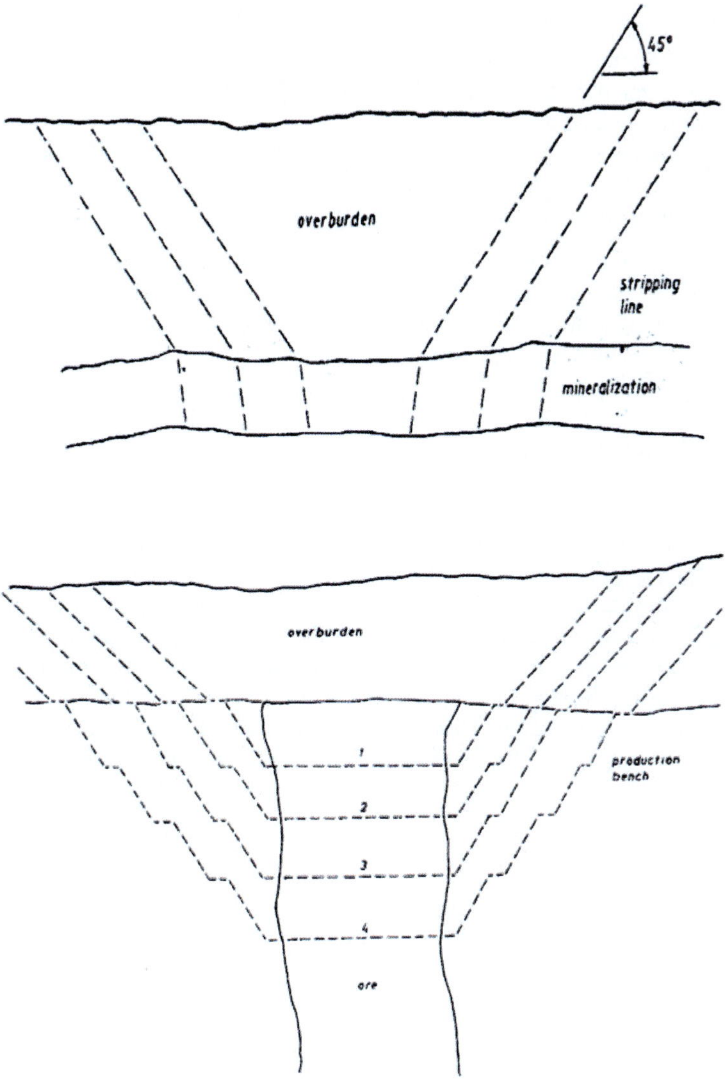

Fig. 7.16 Pit profile for a horizontal deposit

pit; the physical characteristics of the mineral and waste determine the choice of machines. In summary, the factors affecting layout are:

7.7.5.1 The Shape and Depth of the Deposit

This is of particular importance since the selection of the transport system is restricted by the distance and vertical height through which overburden and ore must be moved, that is, routes and gradients. The properties of the ore and overburden:

- Slope angle limitations, both of working faces, sidewalls and dumps.
- Degree of hardness and abrasiveness, which influence the selection of excavation and transport equipment.
- Grade variations which may require selective mining to be practiced and affect the shape of the pit due to the cut-off grade limits.
- Competence of the various rocks to support the equipment.

The drainage requirements, particularly the maintenance of drainage slopes to stamps so as to safeguard faces, benches, dumps and access roads at all times during the life of the mine from the influence of ground water. Saturated slopes are more prone to collapse than well-drained slopes.

7.7.5.2 Dimensions of Open-Pit

These are determined by the dimensions of the deposit itself. The objective is to plan for optional dimensions of the pit so the total capital outlays per 1 tonne of ore mined should be minimal, account being taken of the period over which the initial investments are made. In general the extent across the strike is smaller than the length. However, circular type pits occur where length and width approach one another. Pit types range from quarry-type operations of relatively limited life and yearly output (these may be hill-side type operations) to large pit operations (in general below-ground pits) with long life and multi-million tonne annual production rates. Figures 7.16 and 7.17 are provided to give a concept of pit profiles and layouts.

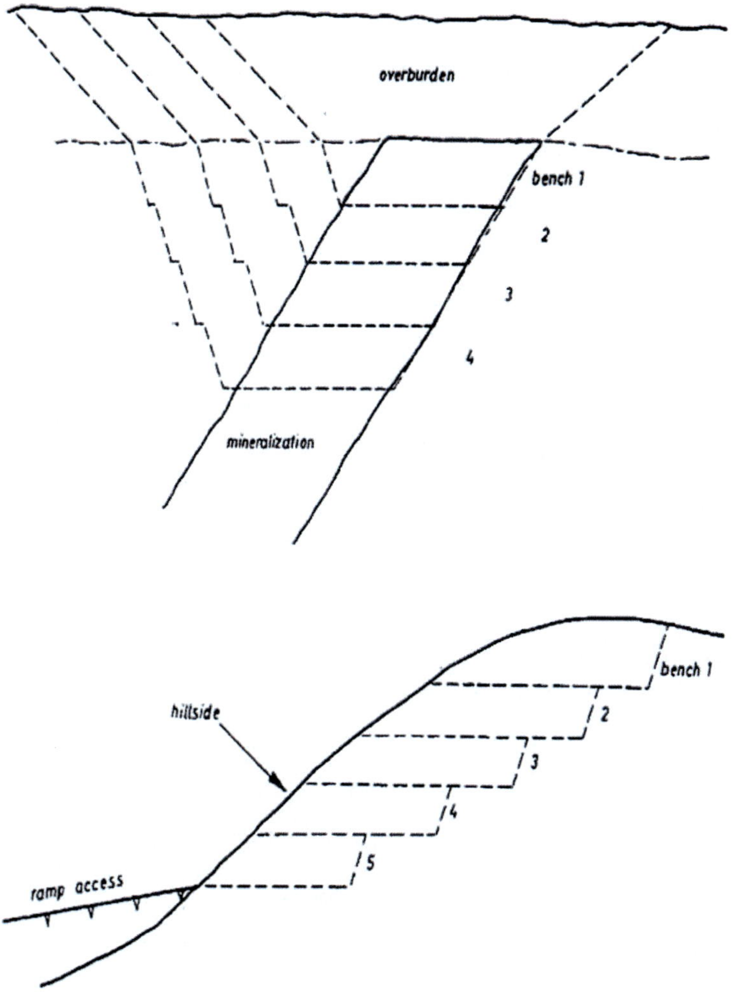

Fig. 7.17 Pit profile for an inclined deposit

References

1. Alshibani A, Moselhi O (2012) Least cost optimization of scraper–pusher fleet operations. Can J Civ Eng 39
2. Montaser A, Moselhi O (2013) Tracking scraper-pusher fleet operations using wireless technologies. In: 4th construction specialty conference. Montréal, Québec, May 29 to June 1
3. Alkass S, Harris F (1988) An expert system for earthmoving equipment selection in road construction. J Constr Eng Manag 114(3):426–440. https://doi.org/10.1061/(ASCE)0733-9364(1988)114:3(426)
4. Anon GD (1997) Caterpillar performance handbook. Caterpillar Inc., Illinois, IL, USA

5. Brinkschulte L, Geimer M (2017) Real-time estimation of the remaining lifetime of components. ATZoffhighway worldwide 03:54–60
6. Burmistrov KV, Osintsev NA (2020) Principles of sustainable development of mining and technical systems in transitional periods. Bull Tomsk Polytec Univ Geo Assets Eng 331:179–195
7. Bulldozer blades capacities. The society of automotive engineers. http://www.cwsindustries.com/images/pdf/techdata/rating_dozer_blades.pdf
8. Caterpillar performance handbook, 29th edn. Catterpillar Inc.
9. Cerny PA (2001) Data mining and neural networks from a commercial perspective. In: ORSNZ (Operational Research Society of New Zealand) conference twenty naught one, Christchurch, New Zealand
10. Peurifoy RL, Schexnayder CJ, Shapira A. Construction planning, equipment and method, 7th edn. McGraw Hill Publishers, New York
11. Gransberg, Popescu, Ryan (2006) Construction equipment management for engineers, estimators, and owners. Taylor & Francis Group, Routledge
12. Demidenko E (2004) Mixed models: theory and applications. Wiley, New York, NY, USA
13. Edwards DJ, Holt GD (2000) ESTIVATE: a model for calculating excavator productivity and output costs. Eng Constr Archit Manag 7(1):52–62
14. Eldin N, Mayfield J (2005) Determination of most economical scrapers fleet. J Constr Eng Manag 131(10):1109–1114. https://doi.org/10.1061/(ASCE)0733-9364(2005)131:10(1109)
15. Eldrandaly K, Eldin N (2006) A knowledge-based decision support system for scraper selection and cost estimation. Int Arab J Inf Technol 3(4):339–343
16. Farid F, Husaini H (1990) Optimum multi-pusher-scraper fleets for earthmoving. In: Proceedings of the first international symposium on uncertainty modelling and analysis, Maryland, 3–5 Dec 1990, pp 628–634
17. Han SW, Hong TH, Lee SY (2008) Production prediction of conventional and global positioning system-based earthmoving systems using simulation and multiple regression analysis. Can J Civ Eng 35(6):574–587
18. Ingersoll-Rand (1939) Modern methods for scraper mucking and loading. Form No. 2412
19. Karshenas S, Xin F (1992) Application of neural networks in earthmoving equipment production estimating. In: Proceedings of 8th conference in computing in civil engineering and geographic information systems, ASCE, New York, NY, pp 841–847
20. Komatsu S (2003) Specifications and application handbook, 24th edn. Komatsu Publications and Training Group, Tokyo, Japan
21. Komatsu performance handbook, 27th edn. Komatsu Inc.
22. Brinkschulte L, Hafner J (2019) Real-time load determination of wheel loader components. ATZheavy duty worldwide 12(4):62–68. https://doi.org/10.1007/s41321-019-0043-5
23. Klanfar M, Kujundžić T, Vrkljan D (2014) Calculation analysis of bulldozer's productivity in gravitational transport on open pits. Tech Gaz 21(3):517–523
24. Mayfield JC (2004) An advisory system for scraper selection. M.A. Sc. thesis, Texas A&M University, USA
25. McGill M, Freadrich J (1994) The effect of fragmentation on loader productivity. In: Proceedings of 5th state of the art seminar on blasting technology, instrumentation and explosives application, New Orleans/Louisiana/USA
26. Morris AA (1991) Continuous scrapers set to revolutionize gold mine gully rock handling. Chamber of Mines Research Organization
27. Montaser A, Bakry I, Alshibani A, Moselhi O (2011) Estimating productivity of earthmoving operations using spatial technologies. In: 3rd international/9th construction specialty conference, Ottawa, Ontario, Canada, June 14–17, vol 3, pp 2318–2327
28. Nielson K (1987) Model studies of loading capacity as a function of fragmentation from blasting. In: Proceedings of 3rd mini-symposium, explosives and blasting research, Miami/Florida/USA
29. Rashidi A, Rashidi Nejad H, Maghiar M (2014) Productivity estimation of bulldozers using generalized linear mixed models. KSCE J Civ Eng 18(6):1580–1589

30. Rupprecht SM (2003) Guidelines for scraper cleaning in narrow vein mining layouts. In: Kuruppu MD, Lilly PA (eds) 12th international symposium on mine planning and equipment selection, Kalgoolie, Western Australia
31. Rupprecht SM (2013) Loading and hauling of broken rock in a narrow tabular orebody utilizing scraper winches. The Southern African Institute of Mining and Metallurgy
32. Rupprecht SM (2019) The potential of automatic scraper winches in underground narrow vein mining in South African
33. Seung CO, Sinha SK (2006) Construction equipment productivity estimation using artificial neural network model. J Constr Manag Econ 24:1029–1044
34. Singh SP, Narendrula R (2006) Factors affecting productivity of loaders in surface mines. Int J Min Reclam Environ 20–32
35. Smith SD (1999) Earthmoving productivity estimation using linear regression techniques. J Constr Eng Manag ASCE 125(3):133–141
36. Son J, Mattila K, Myers D (2005) Determination of haul distance and direction in mass excavation. J Constr Eng Manag 131(3):302–309. https://doi.org/10.1061/(ASCE)0733-9364(2005)131:3(302)
37. Shmulevich I, Asaf Z, Rubinstein D (2007) Interaction between soil and a wide cutting blade using the discrete element method. Soil Tillage Res 97:37–50
38. Tam CM, Tong TKL, Tse SL (2002) Artificial neural networks model for predicting excavator productivity. J Eng Constr Architectural Manag 9(5–6):446–452
39. Tosuna A, Konak G (2013) Estimation of loader capacity based on the heap size distribution calculated by using numerical models. J Min Sci 49(3)
40. Tsuji T, Nakagawa Y, Matsumoto N, Kadono Y, Takayama T, Tanaka T (2012) 3-D DEM simulation of cohesive soil-pushing behavior by bulldozer blade. J Terrramech 49:37–47

Chapter 8
Surface Mining Methods and Systems

8.1 Surface Mining Methods

Surface mining deposits can be primarily divided as follows:

- Relatively horizontally stratified deposits with thick or thin overburden.
- Stratified and vein deposits dipping at angles; greater than the angle of repose of the spoil so that overburden cannot be deposited within the pit.
- Massive deposits, e.g. pipes, stockwork, etc., of considerable depth and lateral extent where the provision of spoil space within the mine is not possible. For further reading: Bohnet [9] and Hartman [19, 20].

The main mining methods may be briefly classified as Strip Mining, Terrace Mining, and the Conical Pit Mining (next section).

8.1.1 Strip Mining

There are two main categories.

8.1.1.1 Contour Mining

Usually in hilly terrain where a steep slope limits the width that can be economically mined from the mineral out crop. The operation is often followed by angering the exposed mineral bed (particularly coal). Small crawler-mounted draglines and shovels are used for stripping and in recent years mobile equipment, e.g. bulldozers, rippers, rubber-tired front-end loaders and tractor scrapers have been used. For environmental reasons the "haul-back stripping method has become common, overburden being hauled along the contour and dumped in the mined-out area. For further reading: Anderson and Kirk [2] and Andrew Cole [3].

M. M. Ali Elbeblawi et al., *Surface Mining Technology*, Topics in Mining, Metallurgy and Materials Engineering, https://doi.org/10.1007/978-981-16-3568-7_8

8.1.1.2 Area Mining

Where, the surface and the deposit are relatively horizontal. So that a wide area may be is mined in a succession of strips. Usually large tonnage reserves (especially coal) are mined in a continuous progression across the mining area with gradual changes in overburden thickness. The trend is to larger stripping machines and a walking dragline with a 170 m (220 yd.) bucket has been in service for ten years. The conditions favorable for strip mining are, Fig. 8.1a:

- Relatively thin overburden.

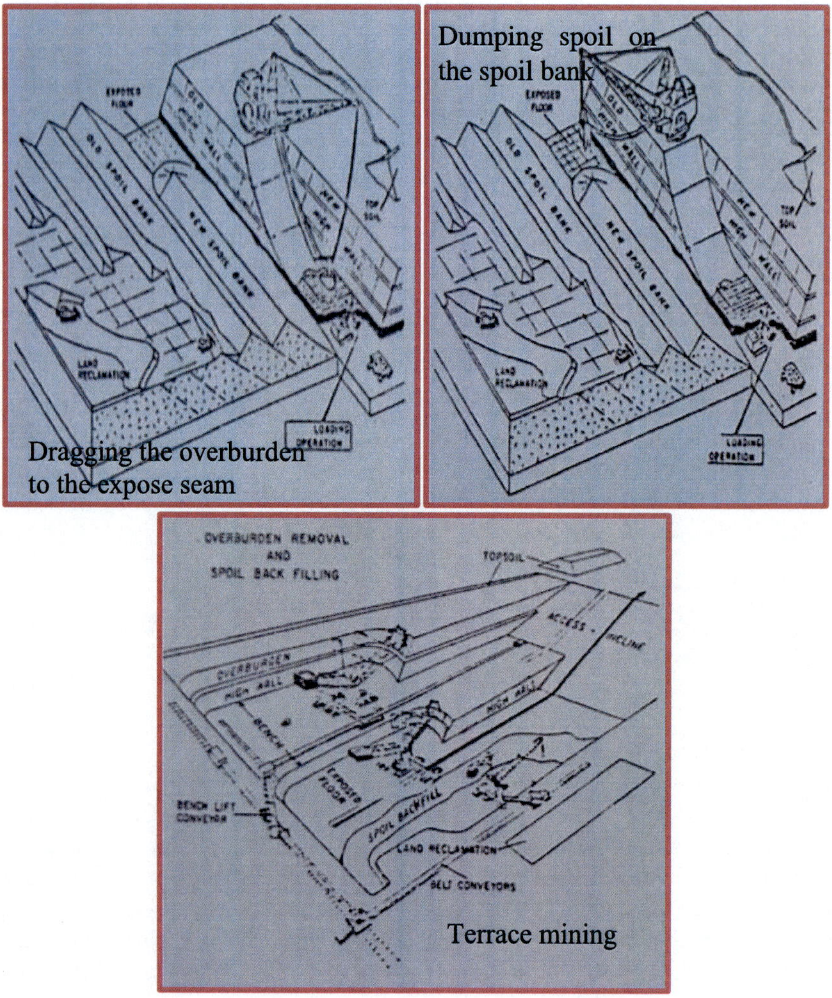

Fig. 8.1 Strip mining

- Relatively thin mineral.
- Horizontal, regular topography and seam.

For thicker overburden beyond the limits of the stripping machine, rehandling of the spoil, e.g. by use of the "extended bench" method, becomes necessary with Increased stripping costs. Strip mining of deep multi-seam deposits is becoming increasingly necessary as near-surface, single seam deposits become depleted. A number of methods have been developed to deal with multi-seam settings, but, because of widely varying deposit geometry and characteristics, it is difficult to classify the methods and adopt a logical selection procedure as for single seam deposits. In general planning for multi-seam operations is more complex and deposit-specific than for single and two seam operations.

8.2 Introduction

The walking dragline is the preferred machine for stripping overburden in strip mining operations. Figure 8.1a illustrates the method of operation. The dragline bucket is dragged up the face of the highwall until it is filled with overburden. The dragline upper-structure is rotated through approximately 90° and the overburden dumped into the void left by mining the previous strip, to form the spoil heap. The mineral seam is exposed and mined by a shovel/truck operation. For further reading: Atkinson [7]; Hrebar [22] and McDonald [33].

The walking dragline is the preferred machine for strip mining because the dump radius is greater, therefore it digs in thicker overburden or competent overburden with flatter slope angles it can span a wider pit pressure can be reduced by increasing the tub diameter, so that it may weaker overburden or on the spoil heap. Strip mining starts with the excavation of a "box cut". It is not always possible to locate the box cut entirely on economic or technical grounds. Where possible it should be:

- Located in thin overburden, i.e. near the outcrop (or sub-crop).
- Orientation should avoid unfavorable structural features, e.g. bedding planes dipping into the cut, joint planes creating instability.
- Avoidance of unfavorable geological features, e.g. faults wash-outs.
- Public services, i.e. avoidance of power lines, road or rail services, gas, oil, water or sewage pipelines—or it will be necessary to arrange prior diversion of such services.
- Bodes of surface water, e.g. rivers, canals, reservoirs, drainage patterns.
- Buildings and other structures.
- Old mine workings, underground or surface.
  It is not always possible to select the optimum location and a compromise must he made, e.g. it may be necessary to start at the property boundary. Since the problem is primarily an economic one, all feasible locations should be investigated.

8.3 Opening up the Deposit

The box cut may be excavated in thinner overburden by:

- End cut, Fig. 8.2.
- Side cut, Fig. 8.3.

End cut is preferred for very thin overburden, since the dragline swing angle is 90° or less. The Side cut is used where the overburden thickness is too great for the

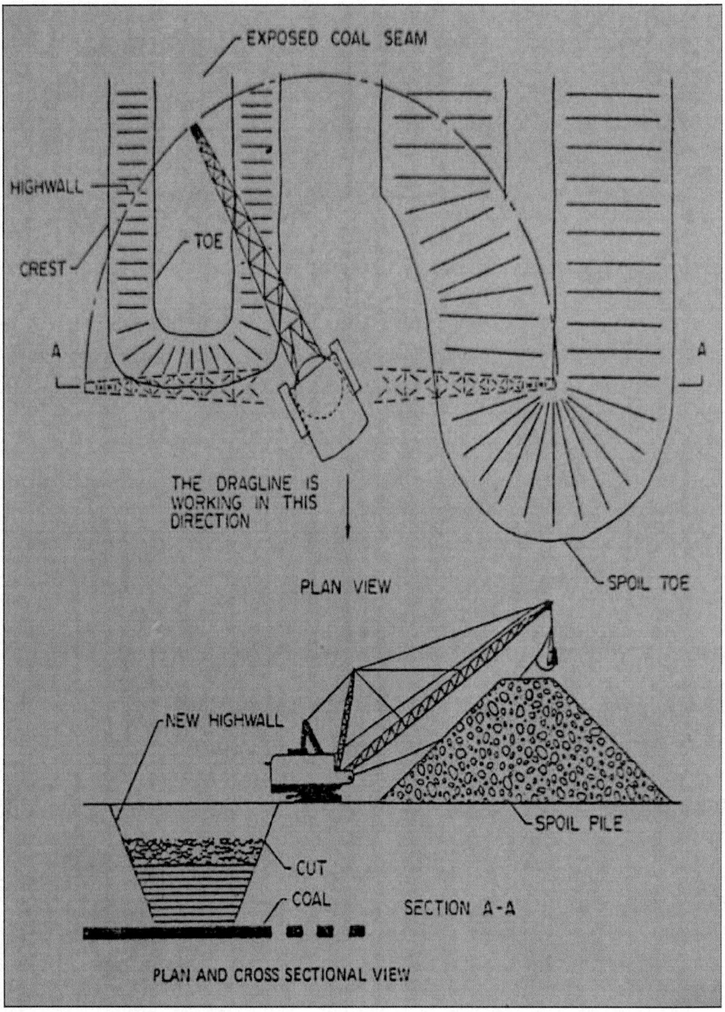

Fig. 8.2 End-cut mining method

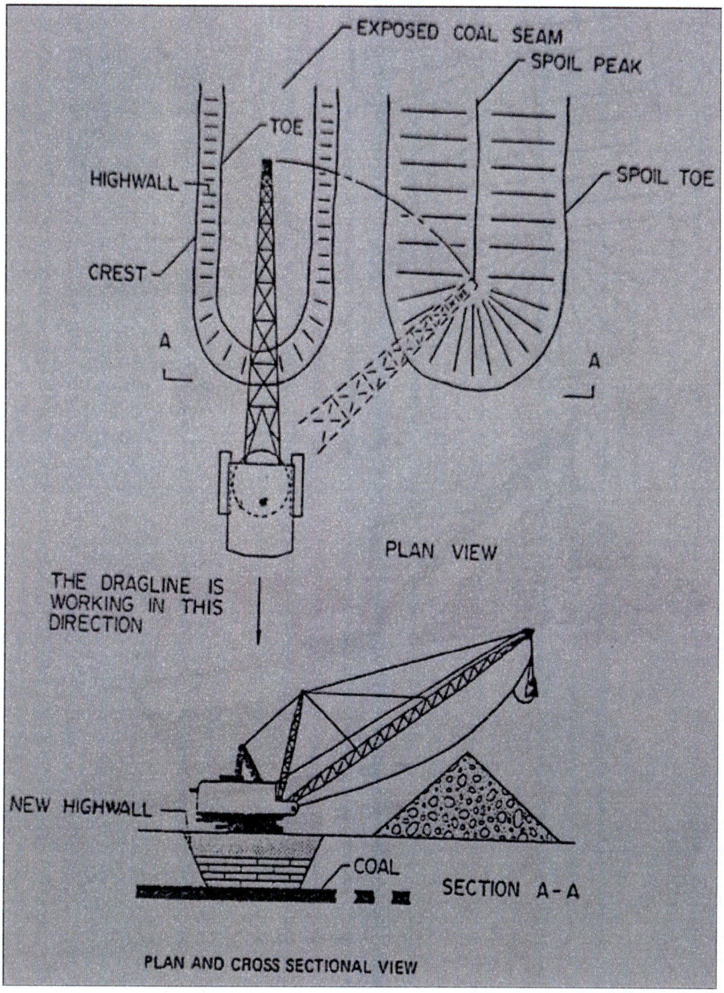

Fig. 8.3 Side-cut mining method

End cut and the overburden spoil must be dumped at a greater distance from the edge of the box cut. The extra reach needed is gained by using a dragline swing angle of up to 180°. If the overburden is thick a rehandle method is necessary. The methods are available: For further reading: Anon [5, 6] and Rand [38].

- Re-handle-End cut, Fig. 8.4.
- Borrow Pit, Fig. 8.5.

In the Re-handle-End cut method spoil dumped on both sides of the box-cut. The spoil on the side in the direction of advance must be subsequently re-handled, usually by spreading it out with a bulldozer and stripping it with the dragline in subsequent

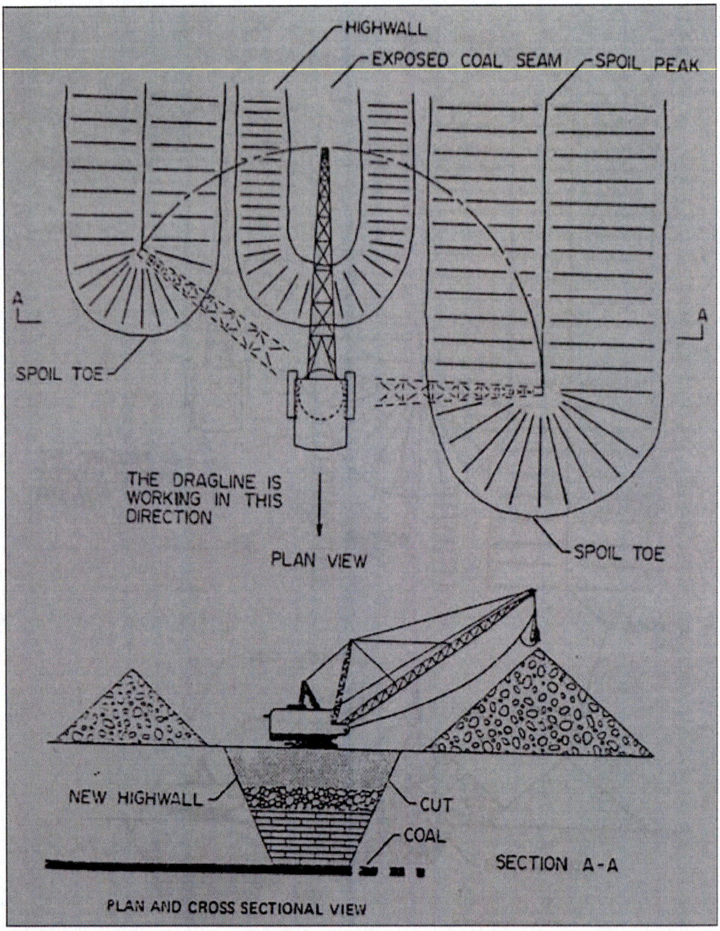

Fig. 8.4 Re-handle-end-cut mining method

cuts. The borrow Pit method does not require a bu1ldozing operations to assist in rehandle. The initial cut forms the 'borrow pit', parallel to the box cut. Excavating the borrow pit is the rehandling operation. The box cut is formed by dumping the box cut spoil into the borrow pit. Normally only one box cut is necessary and represents a relatively small cost. The geometry of the cut may be determined from Fig. 8.6 as follows:

- Draw the highwall at angle β, showing the cut width C, the overburden thickness D and the seam thickness T.
- Draw the two spoil heap faces at the angle of repose, θ, distance C apart.
- Draw the chain dotted line, x–y, at distance = Dx, swell factor, above the base (pavement) of the seam.

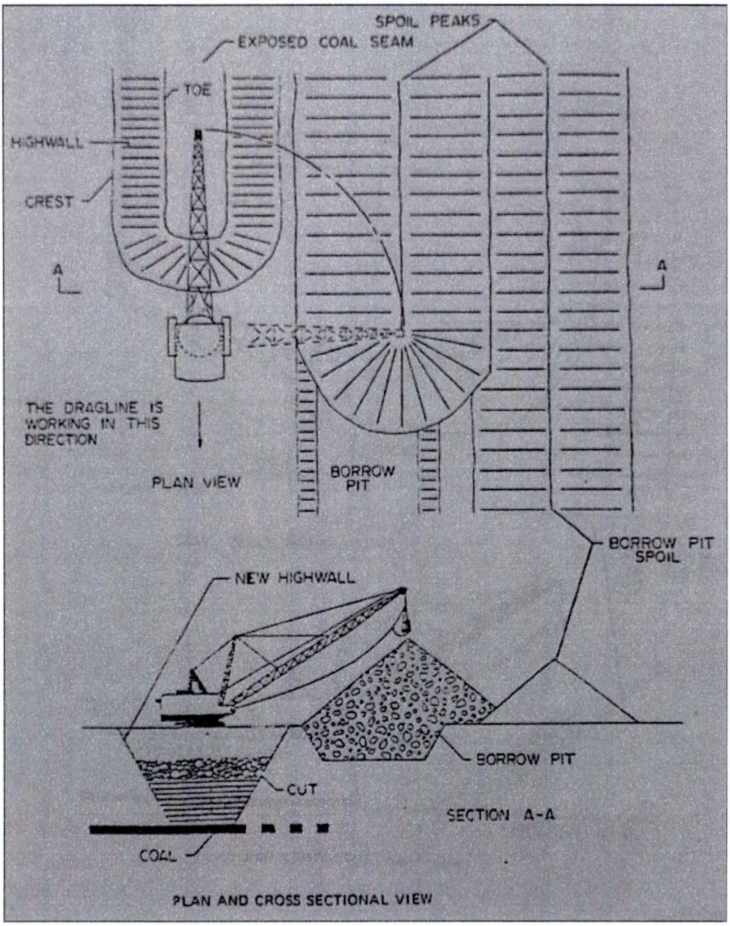

Fig. 8.5 Borrow pit mining method

- Bisect this line as shown and draw line, a–b at angle θ to the base.

From this diagram the Dumping Radius R, and the Dumping Height Ha, of the dragline can be determined. Note that this is one section only. In practice it is uneconomic to fix the dragline geometry from the extreme pit sections. A dragline is selected that will be suitable for the majority of the deposit. For thicker overburden the effective reach of the dragline may be increased by the use of extended bench method; or a fore cut using a shovel/ truck operation may be used to reduce the overburden thickness. For further reading: Stubblefield and Fish [41].

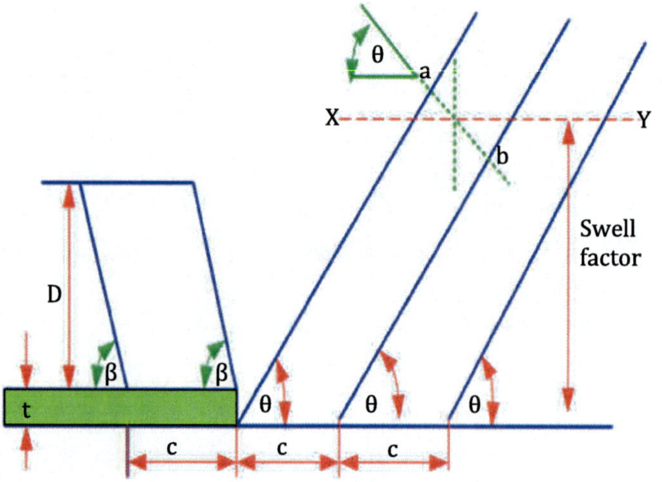

Fig. 8.6 Geometry of the cut

8.4 Advance Benching (Side Benching or Chop-Down)

Often it is necessary to form a working bench for the dragline at a lower elevation than the original ground surface. It is used when the surface is incompetent and a horizon of suitable bearing strength must be selected if the ground surface is irregular. Advance Benching may be adopted to provide a suitable bench for the dragline. In extremely undulating terrain, advance benching may be used to reduce the ridges and fill the valleys, thereby providing a regular dragline working bench and reducing fill requirements in the depressions. Usually the advance bench is excavated one cut width in advance of the working cut. Chop down can be used as shown in Fig. 8.7 in weak materials. Note that in this method weak materials form an unfavorable potential slip plane in the spoil heap. If this could result in slope instability, then selective spoil placement is essential. For further reading: Rupprecht [40].

To avoid this problem a key-cut operation (advance benching) is adopted and more competent material is placed at the toe of the spoil heap, Fig. 8.8.

The Final Cut exposes the mineral, Fig. 8.9. The Final Cut spoil is placed as shown, enclosing the Advance Bench spoil. This is satisfactory for mainly frictional materials. If however, the Advance Material is cohesive and weak, selective placement using the "lead" principle is needed. The dragline geometry can be determined in the same way as previously described.

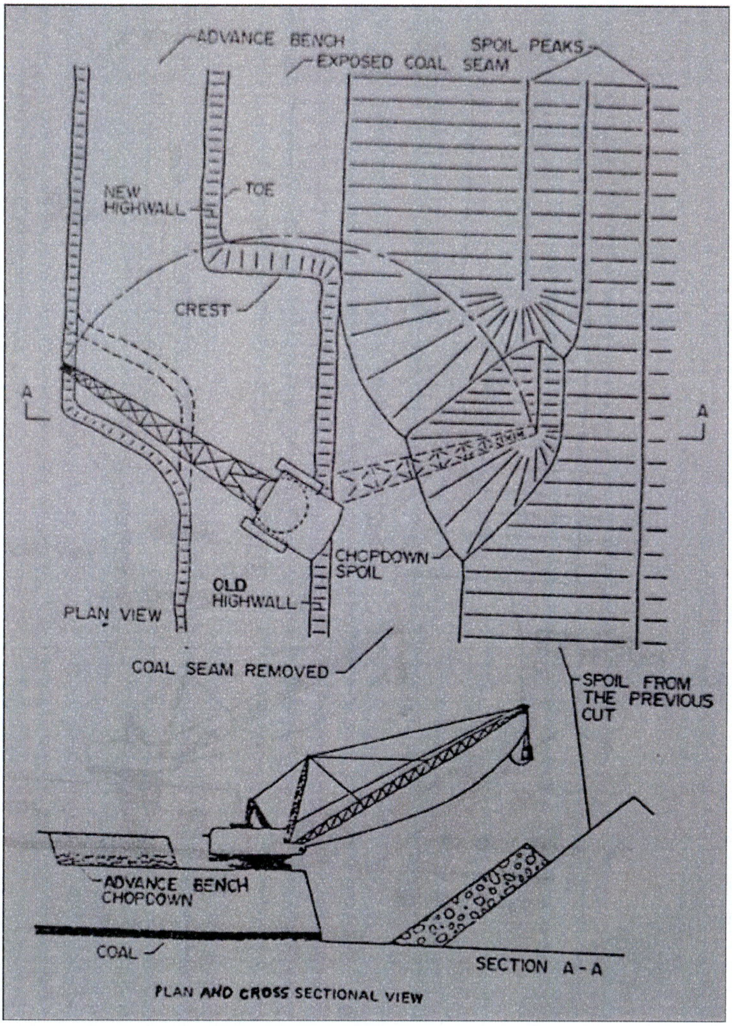

Fig. 8.7 Advance benching (chop down operation)

8.5 Dragline Bucket Size

Having determined the dragline geometry it is necessary to determine the bucket size. For further reading: McLean and McDowell [34]. The main parameter is the overburden to be stripped per annum, but the following also must be accounted for:

- Dragline cycle time (typically 55–70 s).
- Material swell factor (1.2–1.5).
- Bucket fill factor (0.8–0.95).

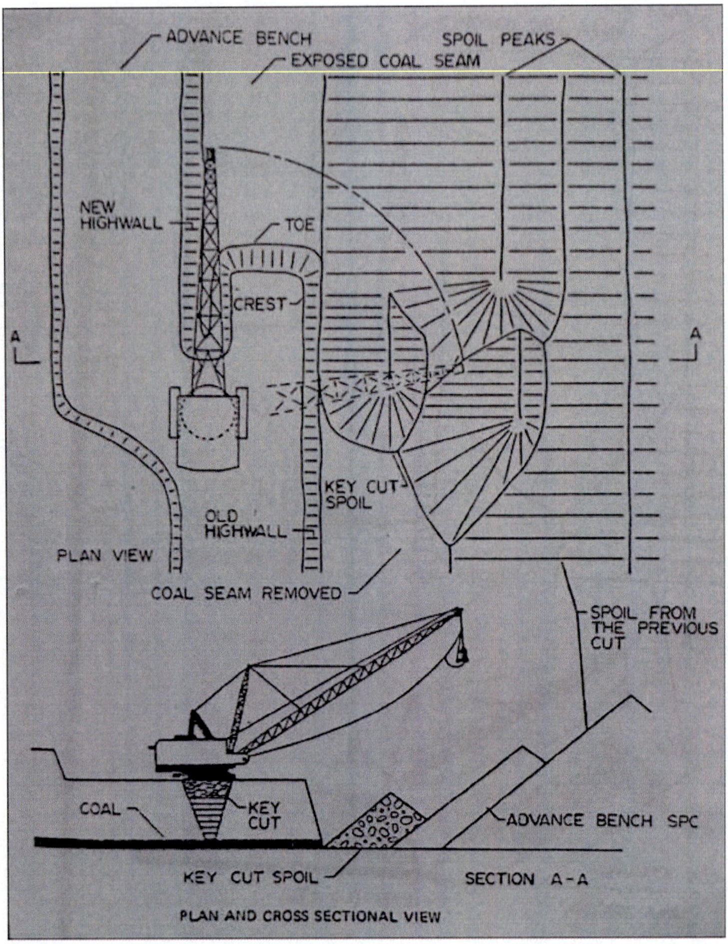

Fig. 8.8 Advance benching (cut operation)

- Dragline scheduled operating time (>90%).
- Dragline availability (80–90%).
- Dragline utilization, i.e. that part of the time the dragline is being used as a fraction of the time available (Circa 0.9).

The dragline bucket capacity can be calculated from Eq. 8.1:

$$B_C = \frac{Q}{C \times S \times A \times O \times B_f \times P} \tag{8.1}$$

where:

B_C bucket capacity, m^3.

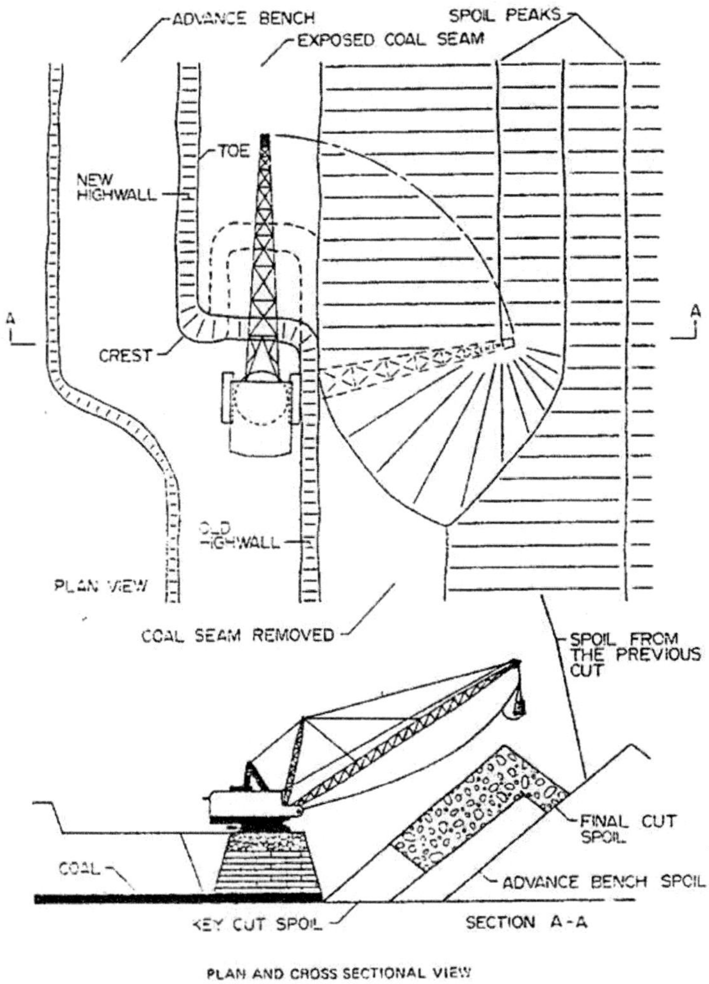

Fig. 8.9 Advance benching (Final cut operation)

Q production required (bank volume per hour).
C theoretical cycles per hour for a 90° swing.
t dragline cycle time for a 90° swing in minutes.
A mechanical availability during the scheduled hours of work.
O job operational factor.
B_f bucket factor = Filling factor/Swell Factor.
S swing factor (for angles differing from 90°).
P propel time factor.

8.5.1 Dragline Selection

The following factors must be reviewed during machine selection:

- Highwall and spoil heap angle may change with varying materials.
- Swell factors (difficult to determine) may change with material type changes.
- Check the maximum digging depth requirement. Greater than 40 m depth can create problems even with large draglines.
- If an Advance Bench is used in thick overburden check the maximum dumping height (as the dragline is lower down in the overburden).
- Note draglines are rated by maximum suspended load not bucket size. Check this is satisfactory.
- Check if there are areas in the deposit that require extended periods of stripping at depths greater than the designed depth. Can mineral production be maintained while stripping the extra overburden in these areas?
- Are any blending processes needed, i.e. will the mineral quality vary outside the specified quality? If so will stockpiling, or working of separate areas simultaneously, is required?
- Check the dragline bearing pressure against the overburden dragline bench surface materials.
- Note that the dragline must rehandle spoil at the junctions of the "access ramp roads" with the cut. The dragline must have adequate reach for this purpose.

8.5.2 Dragline Geometry

Dragline selection is normally based on geological columns for the deposit, for the direction of highwall advance being considered. The extreme situations must be considered as well as typical pit cross sections. Where the maximum capacity of a dragline, that is otherwise adequate for the rest of the deposit, is exceeded in a few isolated locations, these may be dealt with by temporary localized auxiliary operations. The limitations of dumping height and dumping radius can be conveniently super-imposed on the pit cross-sections using transparent overlays and templates. In order to decide where the peaks of spoil dumps can be placed dragline profiles are prepared using the manufacturer's machine dimensions. A typical dimensional profile is shown in Fig. 8.10. The profile includes the "throw" of the bucket beyond the dumping radius (when spare dumping height is available to achieve this). A dragline bucket can be thrown about (1/3 – 1/2) × dumping height. Profiles are drawn to the same scale as the pit Sections on transparent overlays so they can be reversed for spoil heap side investigations. Although Fig. 8.10 shows all possible bucket-boom combinations, this can lead to error and it is safer to prepare one profile for the preferred combination. For further reading: McLean and McDowell [34].

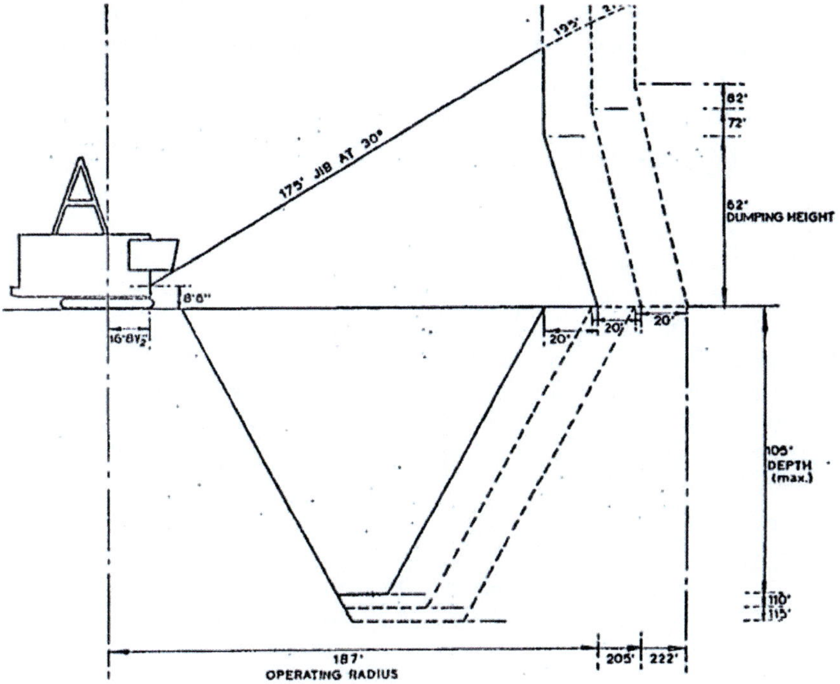

Fig. 8.10 All possible bucket-boom combinations

8.6 Introduction to Strip Mine Design

What are the objectives of Strip Mining? For further reading: Tussey [46] and Woodruff [48].

- To recover useful minerals at economic cost.
- To cause minimum disturbance to the population, wild life, of the area.
- To avoid stream and air pollution.
- To restore the mined out land in accordance with a pre-determined plan.

8.6.1 Major Factors

- Available reserves.
- Market demand.
- Quality of the mineral.
- Stripping Ratio.
- Environmental constraints.

Strip Mining is normally chosen where the overburden thickness is such that excavation can be carried out by a single stripping machine such as a dragline.

8.6.2 *Stripping Ratio*

- Overburden Ration = d/t.
- This is a depth ratio often used by geologists.
- Mining engineers use a volumetric ratio.
- Overburden (m^3)/mineral (m^3)—dimensionless.
- Or overburden (m^3)/mineral (tonnes) = m^3/t.
- The mining engineer's Stripping Ratio includes slopes, access ramps, and "dead work".
- The stripping ratio can be for the deposit as a whole, or for part of the deposit (a "block"):

$$S_B = \frac{\text{Volume of overburden removed, including slopes, etc.}}{\text{Tonnes of mineral recovered per block}}$$

where:

S_B Block stripping ratio.

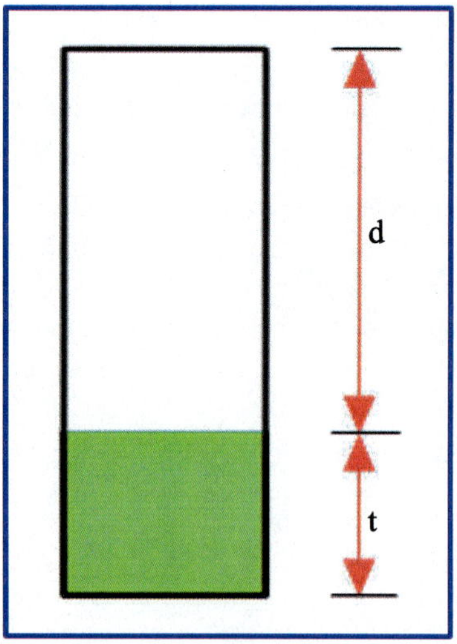

The economic stripping ratio above which it is unprofitable to mine depends on the price obtained for the mineral, the mine location (which determines transport costs), nature of the overburden (which determines stripping costs), etc. In the UK where the selling price is high, the market always near the mine and the overburden relatively weak, very high ratios can be worked, e.g. 25 m^3/t. Most mine planning offices carry out reserves and stripping ratio calculations with a desk top computer or use charts relating mineral seam thickness, percentage recovery, overburden thickness and tonnes per square kilometer (or hectare).For further reading: Anna Gogolewska [4]; Ogedengbe [37] and Superstein [42].

Example The quantity of mineral required fixes the production target.

Consider a deposit with a single flat seam, 4.22 m thick. Assume 20%, mineral losses due to mining. Production requirement is 5,000 tonnes per day rom (run-of-mine). Overburden is 15 m thick. Mineral density = 1.28. Calculate the volume and area of overburden to be stripped per day. This type of deposit can be stripped using a simple dragline operation. There will be no additional volume to be stripped except for "dead-work", e.g. re-handling of spoil at the access road junctions.

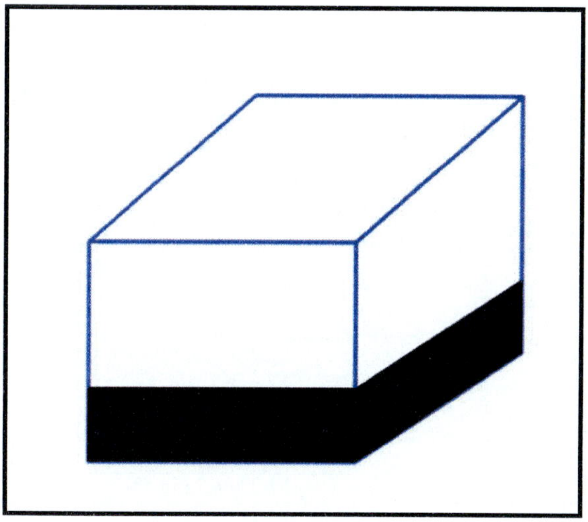

At 80% recovery and 10% dead work, the stripping ratio will be, Eq. 8.2:

$$1.15\frac{\times 1.10}{1.28 \times 0.80 \times 4.22} = 3.82\ m^3/t \quad (8.2)$$

- Therefore, the daily stripping rate = 5000 × 3.82 = 19,100 m.
- The area to be stripped daily = 19,100 / 15 = 1273 m^3.

These charts, Fig. 8.11, relate mineral seam thickness, percentage recovery, overburden thickness and tonnes per acre. For this example determine the dragline geometry (Dumping Height and Dumping Radius) and bucket capacity. Assume a single flat seam with flat topography. The following data apply:

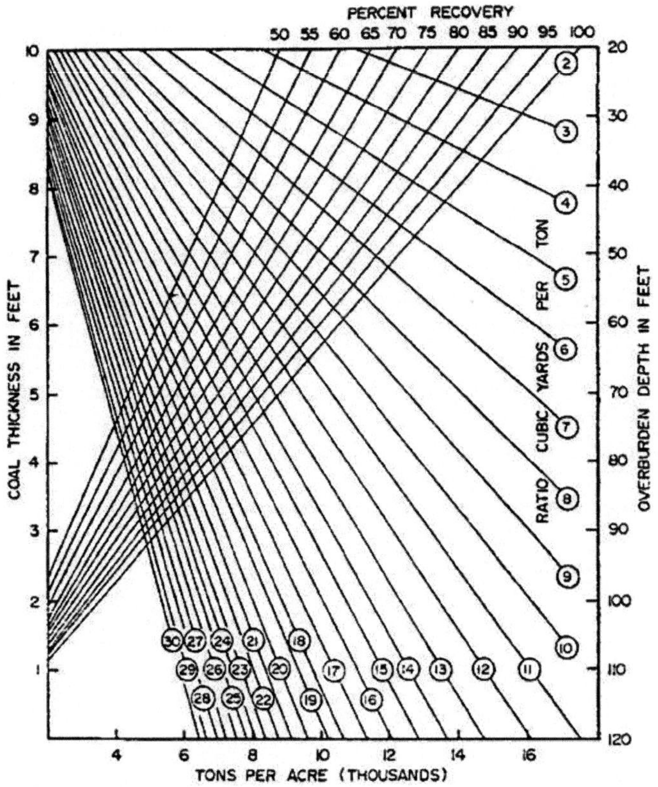

Fig. 8.11 Stripping ratio calculations using charts

Highwall slope angle	1 on 2 (63.5°)
Spoil heap slope angle	1.25 on 1 (38.5°)
Swell factor	1.25
Dragline cycle time	65 s
Bucket fill factor	0.9
Dragline operation	24 h/day
Management	Excellent
Operating conditions	Fair
Tub diameter	20 m
Positioning	0.75 of tub diameter

Note that the actual, not theoretical cycle time is given. What box cutting method should be used for this deposit?

8.7 Terrace Mining (Multi-bench, Lateral Advance)

Where the overburden and/or mineral is too thick to permit direct across-the-pit casting of overburden, even by the largest stripping machines, the spoil must be transported round the pit by conveyor, rail or truck transport and dumped in the void left by previous mining. Figure 8.1 (bottom) shows a basic layout for a Terrace Mine. Such mines invariably must achieve large economies of scale and individual mine outputs currently approach 45×10^6 tonnes/year and involve total depths of 300 m, although new projects of greater output and depth are being developed. BWE's of 240 m $\times 10^3$ output per day are in service.

There is a marked trend in Terrace Mining to the use of large, "beefed-up" Bucker Wheel Excavators (BWE) in conjunction with high-speed, side-slewable, belt conveyors, operating at speeds up to 7 m/s, with widths up to 3 m. It is only by using such equipment on a massive scale that low grade minerals, such as lignite can be mined economically. This has led to the development of BWE's capable of digging much stronger rocks than previously considered possible, e.g. Cuddalore Sandstone, Neyveli; Athabasca Tar Sands, Saskatchewan. For further reading: Bricker [10]; Hood and Roxborough [21] and McCarter [32].

8.7.1 Terrace Mining

Terrace mining, i.e. lateral advance, multi-bench mining, is adopted where the deposit is relatively horizontal, either or both the overburden and mineral are too thick to permit direct casting of overburden even by very large draglines and the spoil must be transported round the pit and dumped in the void left by previous mining. (Some dragline operations in deep overburden involving rehandling of the spoil are occasionally referred to as "Terrace Mining" (Sect. 8.1) but these methods are really extensions of strip mining techniques, e.g. the intermediate extended bench method and the dragline "terrace" method.) Terrace mining is best typified by the lignite mines of Germany and Australia. Individual mine outputs are currently of the order of 45 million tonnes per year involving total depths of 300 m, and new projects of greater output and depth are being developed. Because of the low calorific value of the fuel it is essential to achieve large economies of scale to ensure their viable operation. Such mines are characterized by:

- Thick seams.
- Flat dips; but very thick seams of steep dip may also be mined by this method if the areal extent of the "in-ratio" coal justifies establishing a long life mine.
- Thick overburden.
- Geologically young, with poorly consolidated formations that can be dug by continuous excavators, e.g. bucket wheel excavators (BWE) with high speed belt conveyor transport; but shovel/truck operations are equally applicable. (Note that

a BWE requires a minimum cut depth of 0.5–0.7 × wheel diameter, so that a thick seam is essential for mineral extraction by EWE).

- Severe groundwater problems (e.g. the German Rhine; Korewelt, Australia; and Neyveti, India).
- Severe slope stability problems, requiring very flat overall excavated and spoil slope profiles.
- Extensive land surface reclamation and the construction of very large external spoil dumps.

8.8 Reclamation

Reclamation is greatly facilitated because:

- Spoil deposit operations are not directly linked to coal production as in direct casting or other cross-pit operations. Spoil bank can be independently designed.
- Heights and elevations of spoil benches and overall slopes can 1e selected to avoid instability and to obtain a preplanned final surface, i.e. Reclamation is part of the mine plan.
- Minimum of final leveling is required.
- Spoil can be dumped in any desired interchange.
- Spoil from neighboring developing fly ash can be introduced into the dumping.
- Mineral of commercial value can be conveyor system, e.g. kaolin, sand sale or stockpiling against future sales.

For further reading: Banerjee and Gupta [8]; Gitt and Dollhopf [16]; Gupta [17]; Hao et al. [18]; Hu Caudle and Chong [23]; Jordaan [25]; Jha and Singh [28]; Jim [29], Kundu and Ghose [30] and Maiti [31].

8.9 Conveyor Advancement

Side-slewable, high speed conveyors are best advanced by a side boom tractor with a track shifter head. In soft ground with wet conditions the ties (sleepers or pontoons) on which the conveyors are mounted, tend to sink in, causing delays. If this is likely to occur: use several conveyors in series, advance each in turn in good weather, advance final conveyor (can be short length) quickly—this avoids delays. Also use a bandwagon to increase the block width-this reduces frequency of conveyor shifting. If upward and downward cuts required or if high and tow spoil banks required—then BWE's and spreaders must cross the bench conveyor. This is facilitated by using several conveyors. (Note: If the electric trailing cable is terminated at the center point of the bench, there should be a transfer point there.). Electrical Supplies to BWE, Spreader, Conveyor drive-heads, etc. are most complex and should be carefully planned. For further reading: Anath and Rakesh [1]; Brown [11]; CEMA [12] and Daniyan et al. [14].

Leave bights of power and control cables between each conveyor trestle:

$$\text{Cable length} = 1.3 \times \text{conveyor length} \tag{8.3}$$

- To allow for slewing when advancing.
- Fit plug-couplers to cables between conveyors.
- In very bad (sticky) conditions:
  - Grade off conveyor track.
  - Blind-off with sand and gravel, fly ash or similar material.
  - Reclaim blinding material, after advancing, for further use.
  - This procedure involves extra costs (Identify the need to do so in advance).

8.9.1 Bench Conveyors

Use sheet piles and "Tirfor" or "Sylvester" devices for anchoring at drive-head and tail stations. Equip a light Front-End loader with a pile hammer (hydraulic) (allow in the CAPEX).

- At pivot points install permanent spillage reclaim devices.
- Large drive-heads—up to 300 tonnes—use walking mechanisms or crawlers for advancement.
- Electrical Controls—Emergency stop—belt slip—broken belt -sequence starting and stopping—belt mis-alignment—Klaxon start-up alarms.

For further reading: Fenner Dunlop [15]; Jacek K. Radlowski [24]; Joy [26, 27] and Nekoufar [36].

8.9.2 Bench Lift Conveyors

- It can be rail track or crawler mounted.
- Much easier to install on level ground then propel into inclined position.

Select a design that operates with level receiving and discharge booms regardless of inclination of main structure. Fit electric motors (synchronized) for traction. Fit thruster brakes and run-back preventers on the conveyor. Electrical Controls—Emergency stop—belt slip—broken belt—belt mis-alignment Klaxon start-up alarms—sequence starting and stopping. Note the Bench Lift Conveyor stops much more quickly than the Bench Conveyor feeding it—use a time delay for sequence stopping, Fig. 8.12. This avoids flooding of the bench lift conveyor receiving hopper.

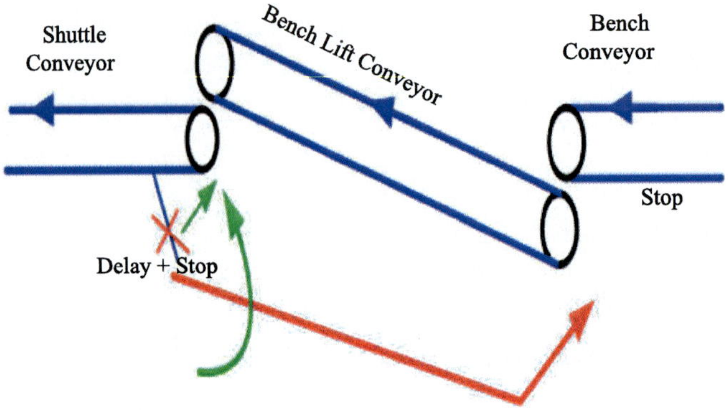

Fig. 8.12 Handling materials with different type conveyors

Emergency Stop—No delay! The Bench Lift Conveyor provides an excellent structure for carrying electric power supply cables to a lower beach—but weight can be significant allow for in design of the structure—use plug couplers at each end, but at position that allows easy coupling and uncoupling. Install at 45° to Bench Conveyor and Shuttle Conveyor—then BWE can excavate farther into the corner. For further reading: Rulmeca [39]; Vanamane et al. [47] and Wheeler [50, 51].

8.9.3 Shuttle Conveyors

It is mounted on rail track. Reclaim rail track from behind after advancement then re-install in front for future advancement. Standard gauge suitable for most conveyors—but additional wider gauge track may be required at drive-heads for stability due to height of structure. For further reading: Zdzislaw Murzynski [53].

- Use sheet piles and tensioning devices plus anti-creep devices on the rails.
- Electrical Controls—Emergency Stop—belt slip—broken belt–belt misalignment (especially because of inherently small clearances)—sequence starting and stopping—Klaxon start-up alarms.

8.10 The Conical Pit Mining

This method is used for mining irregular deposits, pipes, stockwork and steeply inclined stratified deposits. Despite the description the pit shape may deviate considerably from an inverted cone to accommodate irregularities due to deposit shape and mineral grade. Usually the pit is excavated by shovel/truck operations. In deep pits inclined or vertical skip haulage may be used from selected levels. This type of operation is usually carried out in competent metalliferous ores and their country rocks, so that extensive ground preparation is necessary. The superior efficiency of rotary blast-hole drilling has led to its almost universal use over the past 15 years, in most conical pits; holes up to 400 mm diameter being common, ANFO (ammonium nitrate/fuel oil) boosted with aluminum granules or metallized slurries, both more economic than traditional explosives, have become the preferred explosives. Conical pits are typified by the large number of boreholes. Bingham Canyon, Utah, USA is probably best known with 75 benches, being over 4 km long, 2.5 km wide and 0.3 km deep. For further reading: Meade [35].

8.10.1 Introduction

The design and planning of conical pit mines (often referred to as "open-pit") is a complex, inter-related process. The design process is shown in Fig. 8.13. A plan of a typical conical pit is shown in Fig. 8.14 and a cross-section in Fig. 8.15. Ore bodies mined by conical pits vary considerably (Figs. 8.16, 8.17 and 8.18) and the geometrical configuration has a strong influence on mine design. Figure 8.19 shows a typical topographical plan of a deposit with base and section lines. A series of geological sections are prepared from core drilling information as shown in Fig. 8.20. From the geological cross sections a series of bench plans are prepared showing the geology of the sine area for each mining bench, Fig. 8.21. "Note the units for Figs. 8.19, 8.20, and 8.21 are feet, not meters".

Usually because of the large amount of information to be stored computer processing of the geological data is essential and the deposit is divided into blocks. The block height is usually conveniently chosen as the bench height. Average mineral values are determined for each block. Figure 8.22 shows this concept.

8.10.2 Design Considerations

8.10.2.1 Bench Height

Selection of bench height is governed by the reach of the excavating equipment and the depth capability of the blast-hole drilling equipment, e maximum digging height of the excavator determining the maximum bench height. Generally an increase in bench height is desirable because:

- A greater bench height reduces the "set-up" time per meter drilled; and the sub-grade drilling arid explosives are used over a greater length of blast hole (and hence a greater tonnage). The greater the bench height the greater the saving.
- There is a trend to larger blast hole diameter. As blast hole diameter increases the bench height should be increased to maintain good blast geometry. For further reading: Sweigard [43].

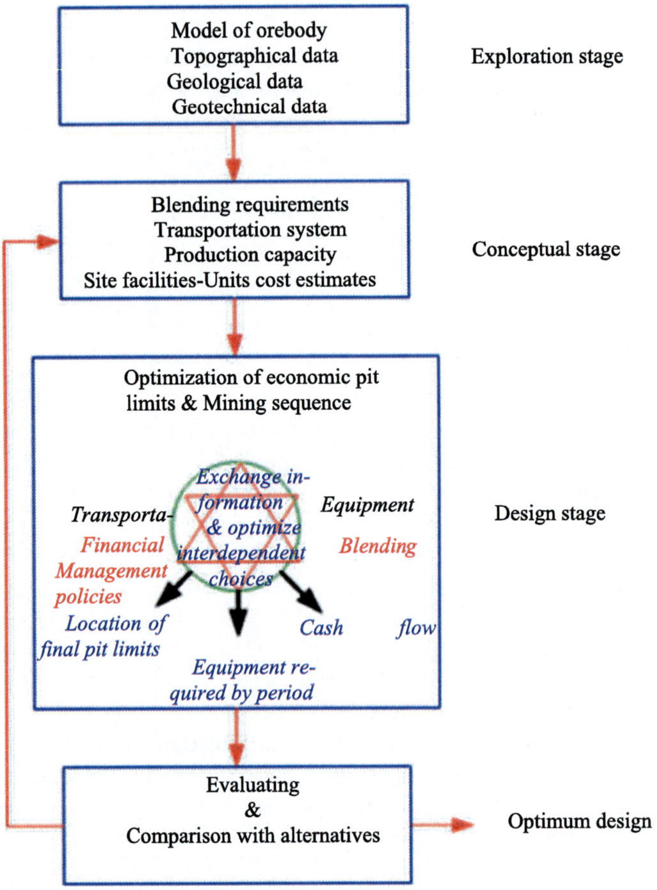

Fig. 8.13 Design and planning process of conical pit mines

- The number of rows of blast holes per blast that can be drilled, while still maintaining good horizontal displacement, is governed by the blast hole diameter rather than the bench height. Thus for a given blast hole diameter and explosive type, the broken ore that can be generated is directly proportional to the bench height. An increase in broken ore per blast reduces the frequency of blasting, thereby reducing shovel moves and increasing shovel efficiency.

Typical crowd shovels can effectively excavate benches of 11.5–18.5 m in height. Special long boom shovels can excavate 23 m high benches.

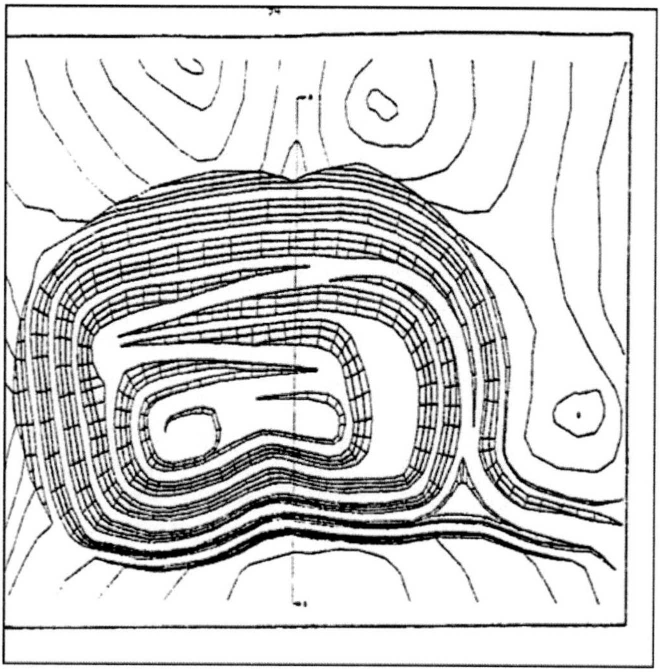

Fig. 8.14 A plan of a typical conical pit

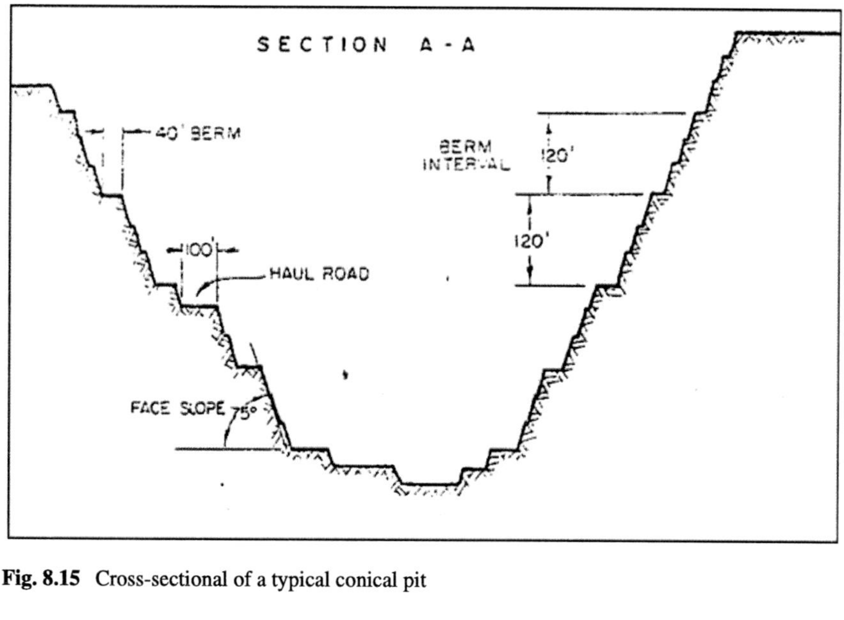

Fig. 8.15 Cross-sectional of a typical conical pit

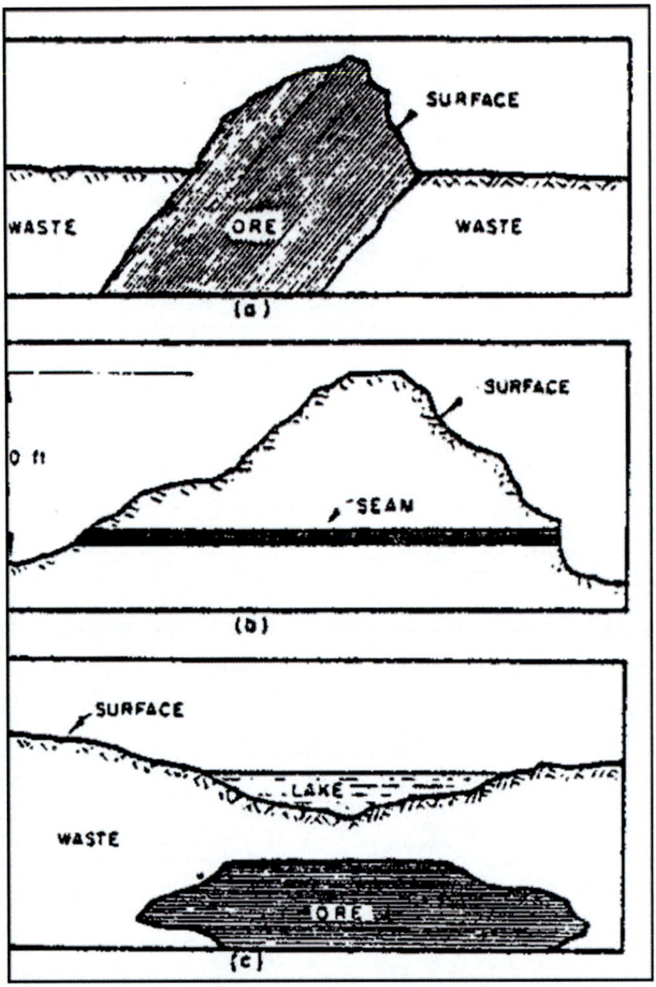

Fig. 8.16 Orebodies mined by conical pits at various topographical settings

8.10.2.2 Haul Roads

This is dealt with in detail elsewhere. The location of the final haul road requires considerable investigation. Ideally this road should be established as soon as possible to avoid construction of temporary roads and where possible is located in the footwall of the deposit, Fig. 8.23. It is generally not economic, for cash flow purposes to mine the total bench into the hanging wall to the ultimate pit limits. Figure 8.24 shows a typical configuration after several benches of ore have been mined. For further reading: Tannant and Regensburg [44] and Thompson [45].

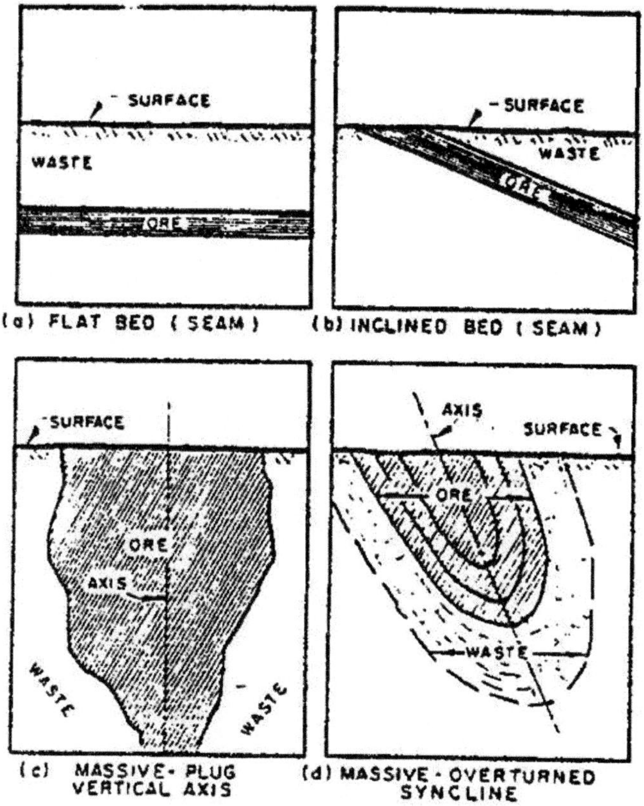

Fig. 8.17 Orebodies mined by conical pits at various geological settings

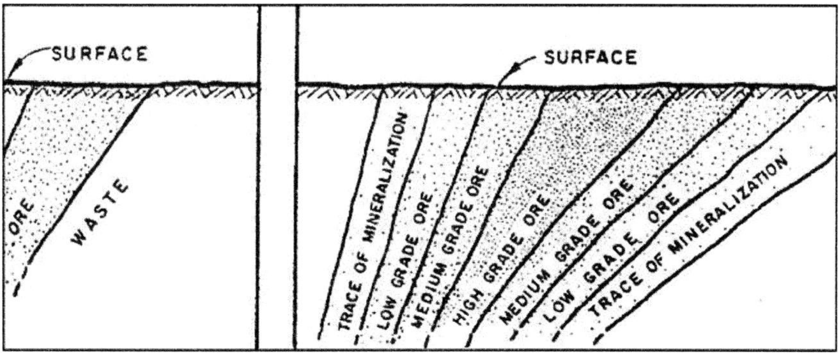

Fig. 8.18 Orebodies mined by conical pits at various inclined settings

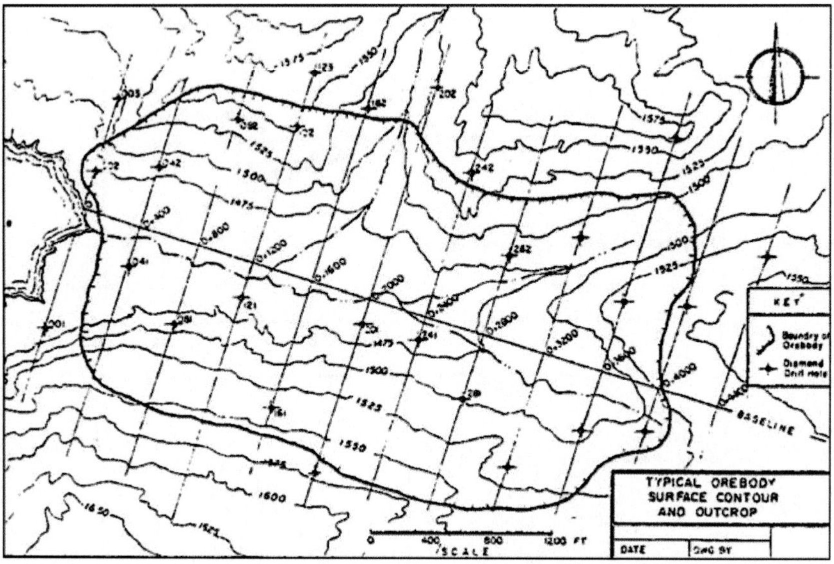

Fig. 8.19 Typical topographical plan of a deposit with base and section lines

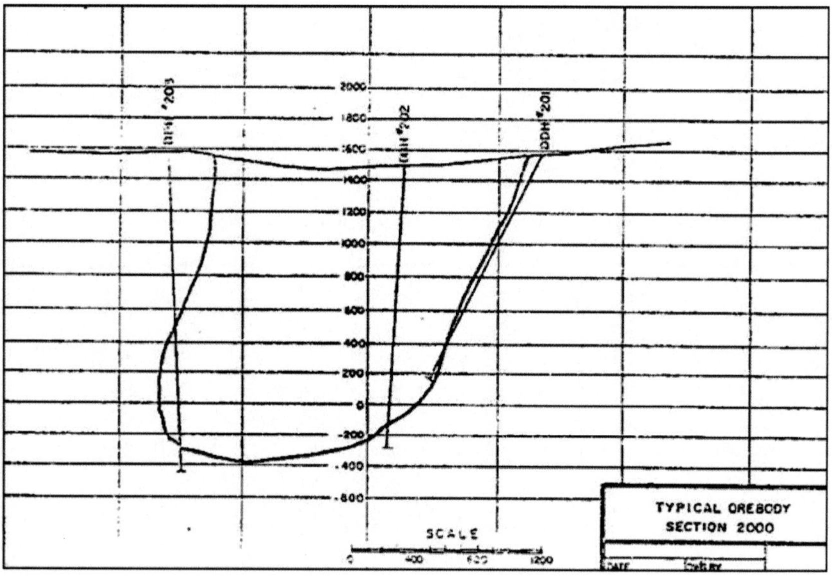

Fig. 8.20 Series of prepared geological sections from core drilling information

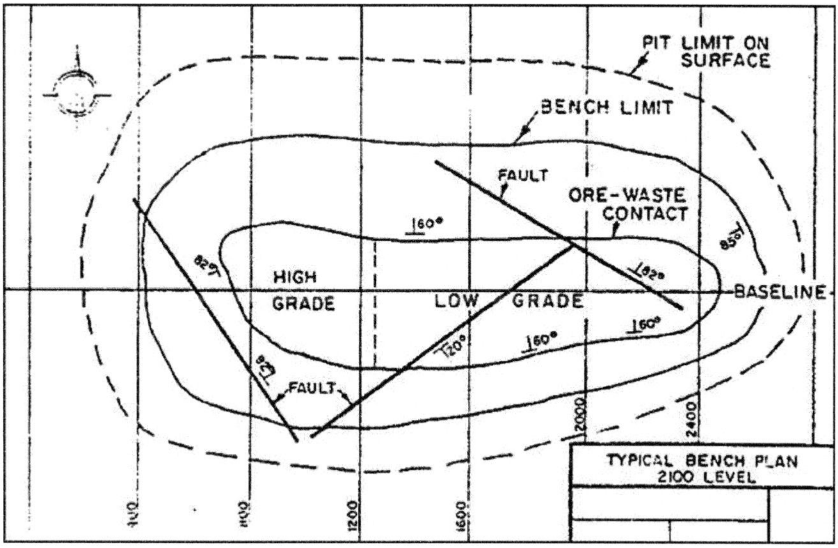

Fig. 8.21 Series of prepared bench plans showing the geology of the sine area for each mining bench

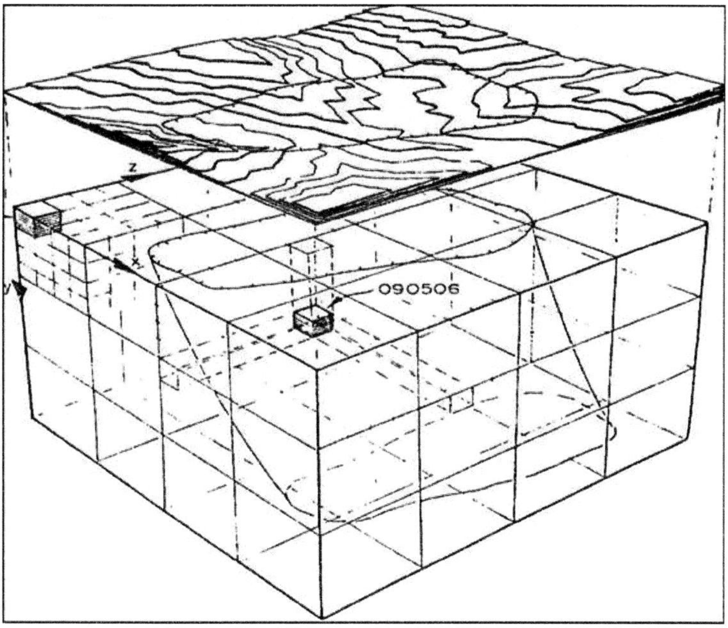

Fig. 8.22 Average mineral values determined for each block

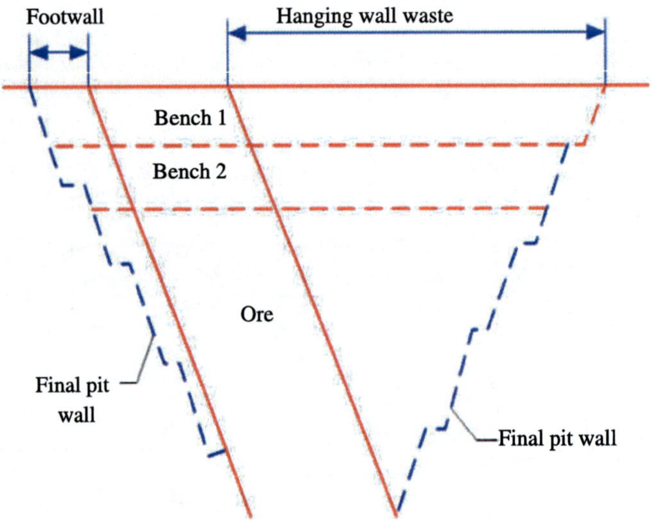

Fig. 8.23 Distribution of footwall and hanging wall waste on an inclined orebody

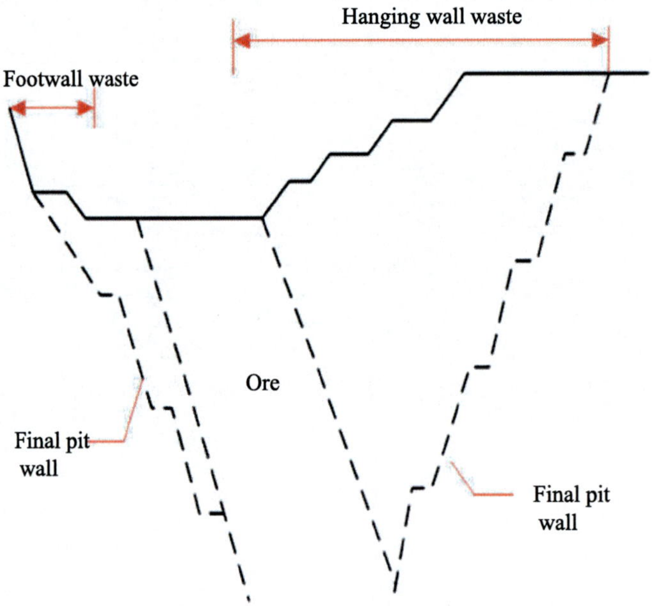

Fig. 8.24 Typical configuration after several benches of ore have been mined

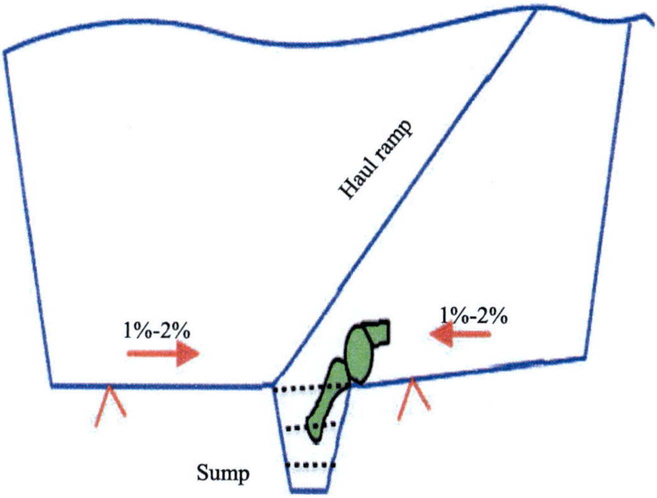

Fig. 8.25 Pit floor sloping to facilitate drainage

8.10.2.3 Pit Floor

The pit floor slope during periods of high rainfall should be traded as shown in Fig. 8.25.

8.10.2.4 Overall Pit Slope Angle

The ultimate pit slopes must be fixed on haul road as well as to technical considerations, Fig. 8.26. In this case the bench height is fixed by the shovel size selected. Because of the jointing system it is decided the maximum safe berm interval is 100 feet. Because of the blast hole size selected and the bench wall control planned the minimum berm width is set at 45 feet. The footwall dips at 75°, so the face slopes between berms cannot exceed 75°. If the haul road is included the overall slope becomes 52°.

A geotechnical analysis is then required to determine whether this overall slope is stable to the planned mine depth. In some cases this viii indicate a much steeper overall slope is stable, but no advantage can be gained as the other parameters are fixed by other considerations.

8.10.2.5 Planning Considerations

The following are considered:

- Annual production requirements.

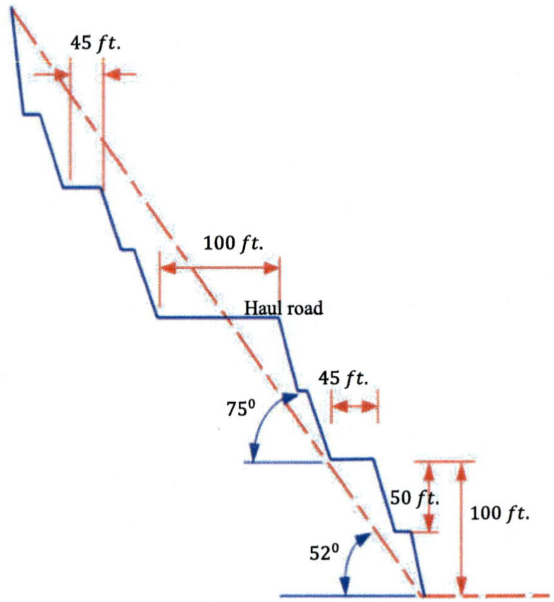

Fig. 8.26 Diagram of an open pit using computer codes

- CAPEX, OPEX, unit costs.
- Mineral grade values for processing.

The general basis for planning is to maximize the Net Present Value of the deposit, but this may be difficult to achieve. A simplistic method of analysis is to determine the pit limits the "cut-off" stripping ratio, Fig. 8.27, i.e. cost of mining ore + waste = revenue earned from the ore recovered. In practice incremental analysis is used, each increment representing one year of mining. The profits are discounted at an appropriate rate, over a period until the discounted h flow no longer increases.

For short term planning a limit is not determined but the value production is calculated. This method is used for comparing alternatives. These plans will consider:

- Shovel mining sequence (moves, ore and waste schedule), production, and number of operating units required.
- Drill sequence, meterage requirements by blast hole size, blast hole site access development, and number of operating units required.
- Haulage routes, haul distances, road construction capability, and number of units required.
- Stripping requirements to uncover the ore.
- Explosive requirements based on rock types, the type and quantity of explosive required.
- Dewatering requirements.
- Power distribution requirements.
- Waste dumps location, capacity and geotechnical design.

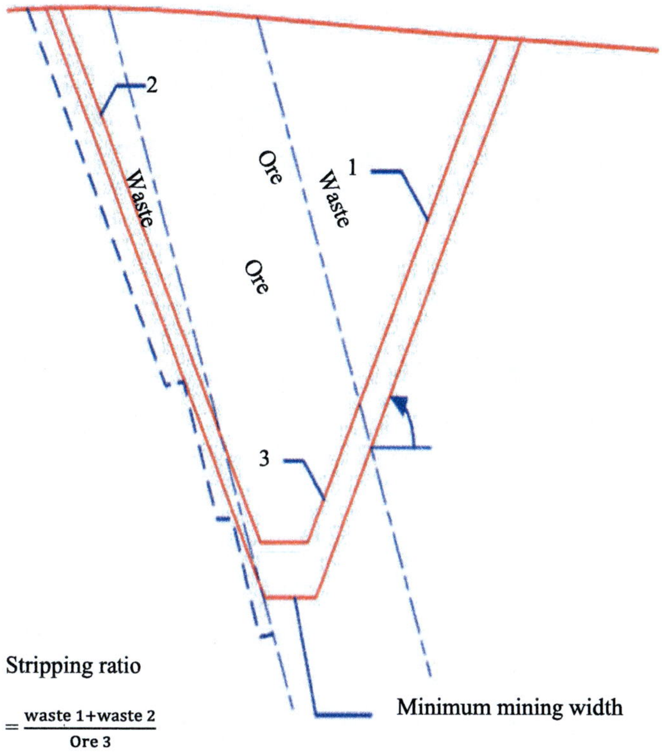

Fig. 8.27 Three-dimensional computer generated diagrams of an open pit

- Co-ordination of items 1–8 to ensure adequate lead time to bring into service.
- Schedule ore sequence to meet the process plants requirements.
- Define the cut-off grade.
- Surface facilities required for maintenance, engineering and administration.

Normally production receives first priority since usually long term contracts must be fulfilled. Some degree of variation may be allowable in grade, but contaminants and liberation properties roust be satisfactory cost, although vitally important may not be of first priority because the long term plan should be such that the ore will be profitable, unless there has been an unprecedented drop in the metal price. For further reading: Whitman [49] and Whyte and Cumming [52].

8.10.2.6 Ramp Roads

Because of the flat aspects of strip mines the Bottom Dump Truck is the preferred haulage unit for strip mines. For gradients greater than 6% (3.5°) however Rear Dump (alternatively unitized) trucks may be essential. Careful attention must therefore be

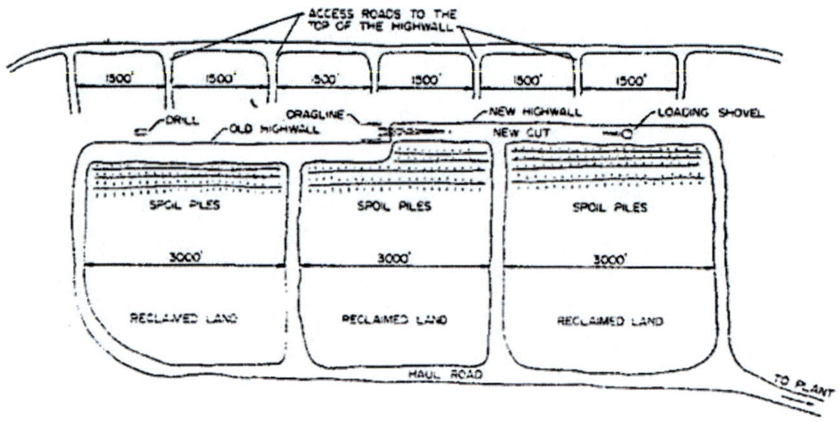

Fig. 8.28 Diagrammatic plan of strip mine showing ramp road access (dimensions in feet)

paid to dips, pavement contours, etc., during layout. Strip mine access is via 'ramps' roads through the spoil, Fig. 8.28. The layout is characterized by the "T" junctions formed by the intersections of the "ramps" through the spoil with the cut. The ramps eliminate spoil space, resulting in higher spoil heaps at corners of the "T" junctions; this can be a cause of spoil heap failure.

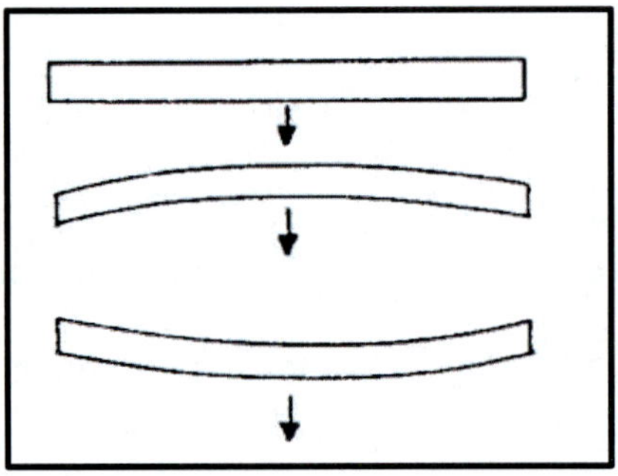

- Straight—best for transport. Reduced truck maintenance costs.
- "Outside" Curve—more spoil space created, easier spoil disposal.
- "Inside" Curve—more stable slopes on highwall.

8.10.2.7 Initial Cut

The line of the cut can be adjusted from the original box cut on the crop line by a succession of short cuts. This can reduce initial dragline productivity, and must be justified by future gains.

8.10.2.8 Strip Mining Operations

Relatively horizontal stratiform deposits are usually most economically exploited by strip mining. This simple operation is:

- Easily supervised.
- Has the shortest possible transport distance, and therefore low stripping costs.
- It permits economic land surface reclamation of the mined out area without delay.

Over the past 30 years the walking dragline has become the preferred machine for stripping regular, relatively-horizontal, stratified deposits under relatively thin cover, e.g. < 50 m (165 ft.). The main reasons for this preference are:

- The stripping shovel is usually considered the more productive as it casts approximately an average of 42 m (bank) per cubic meter of dipper capacity per hour, whereas a walking dragline casts approximately 35 m (bank) per cubic meter of bucket capacity per hour. On the other hand, for the same width and bucket or dipper capacity the dragline usually has the lower mass (service weight) and involves a smaller capital investment. The inherently greater reach of the dragline is a distinct advantage in thick overburden.
- Stripping shovels must stand on top of the mineral bed and subject it to average bearing pressures of 350–420 kN/m^2 (50–60 lbf/in^2). If the mineral cannot withstand this load, a dragline must he used, A competent working horizon can be selected for a dragline anywhere within the overburden.
- If the mineral bed surface is irregular, subjected to faulting, etc., and the surface reasonably regular, a dragline is more suitable.
- If the mineral lies in tenses, or if substantial lateral variations in grade occur, the economically recoverable mineral lying in patches, the stripping machine must 'deadhead' over the surface relatively often. These conditions favor the dragline: because the tub has a low bearing pressure, more easily prepared routes are available and a shovel must dig itself out of the pit.
- Draglines are more suited to the excavation of the initial box-cut.
- Where the spoil has a relatively flat angle of repose, a dragline is more suited to provide the necessary reach.
- Large stripping machines represent substantial capital investments, they are relatively immobile and in wet conditions, or where rock falls can occur, there are obvious advantages in that the dragline does not stand on the pit floor.
- The shovel has more wear points, e.g. hydraulic leveling jacks, crawlers and their drives, etc., and incurs greater maintenance costs that draglines in hot, dusty conditions.
- Where the overburden is strong and dense, the stripping shovel may be preferred.
- If bad fragmentation occurs owing to hard beds, fissures, open joints and other discontinuities, a shovel is more suitable.

Compared with shovel/truck or scraper operations the dragline (or stripping shovel) is relatively quiet and may often be used 24 h per day in populated areas. The choice should be made on the basis of economic factors, but some considerations are difficult to quantify and prejudice may also be involved. The advent of inexpensive explosives, i.e. ANFO and metallized slurries, has resulted in considerable reductions in blasting costs, and this has tended to favor the dragline; which requires better fragmentation of the overburden. Large modem shovels have relatively light dippers and also need well, fragmented overburden; hence, some of their advantages are reduced. In general, overall costs are reduced by good fragmentation because of increased availability, reduced maintenance costs, reduced bucket or dipper filling time and improved fillability.

8.10.2.9 Dragline Terminology

There is no universally accepted dragline terminology. For convenience B.E. terminology is used here. Other terms are used but for the purpose of this seminar this terminology is adequate. The operating radius readily identifiable from Fig. 8.30, but the reach is more suitable for describing the maximum distance that the dragline can deposit spoil from the cut (Fig. 8.29).

A	Clearance radius	H	Boom point height
B	Operating radius	J	Digging depth
C	Boom foot radius	K	Point sheave diameter
D	Clearance height	L	Tub diameter
E	Boom foot height	M	Boom angle
F	Dumping clearance	P	Positioning
G	Dumping height	R	Reach

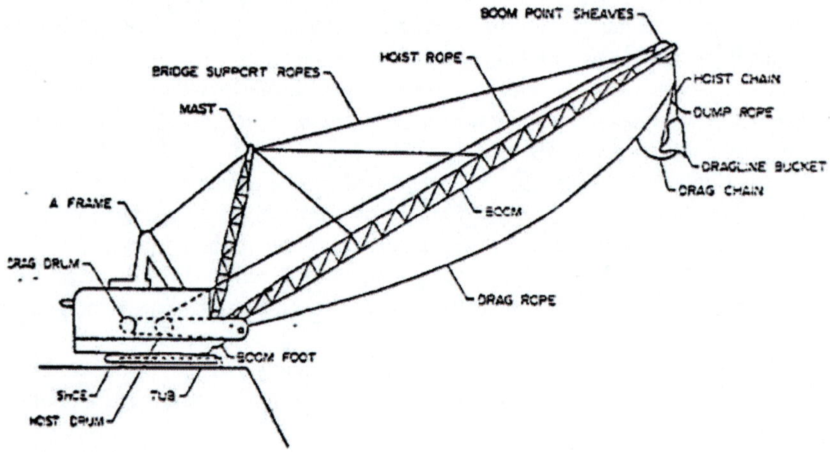

Fig. 8.29 Schematic diagram of dragline

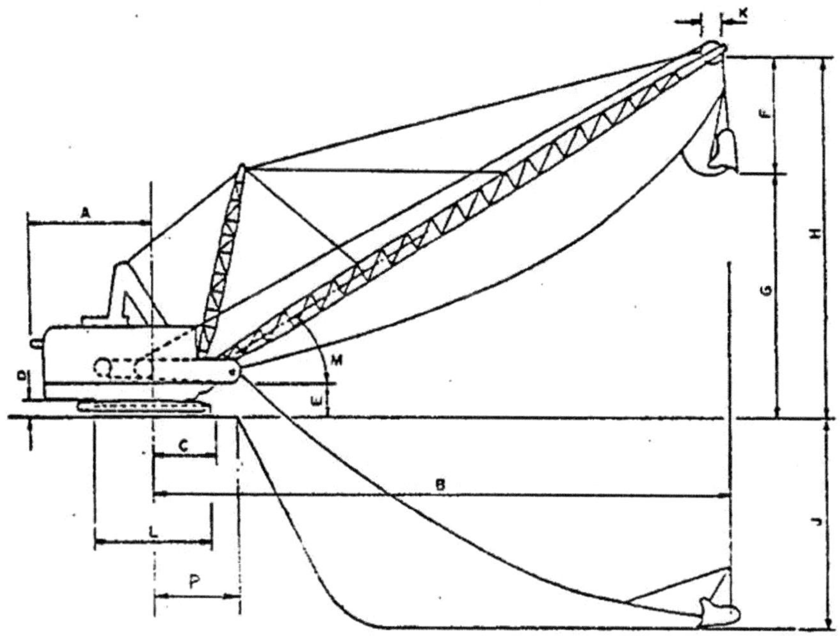

Fig. 8.30 Dragline operating radius

Reach is the horizontal distance from the crest of the dragline bench to the center line of the hoist ropes drooping vertically from the boom point sheaves. Positioning is the horizontal distance from the center of machine rotation to the crest of the dragline bench. Often positioning is expressed as a decimal or a percentage of the tub diameter L, i.e. 0.5 L or 0.5 L. Positioning is not a fixed parameter; it can be varied by the operator according to conditions. Where the highwall is stable it may be possible to safely operate at 0.5 L but in most operations is nearer 0.75 L. It is not a fixed factor. Slope in strip mines may be referred to, in North America, as 1 on 2, 1 on 3, etc., the horizontal length of run on the height of rise.

8.11 Classification of Opencast Mining Systems

In opencast mining, a mining system is understood as the order and sequence of accomplishing opencast mining work within the limits of an open-pit field or of its section. A mining system must ensure safe, planned and efficient extraction of all minerals, the specified production capacity of the enterprise, full extraction of the mineral reserves, and proper conservation of the earth and environment. In the general case, opencast mining consists of development work, stripping work, and mining work (extraction) proper. When mining low-valuable minerals outcropping immediately onto the surface, the stage of stripping work may be absent or inessential. In such cases, the mining (or working) system is the order and sequence of performing of extraction work and of opening and preparation of working levels within the limits of an open-pit field.

In some cases of mining of horizontal deposits, there is no need for opening of new levels upon completion of the mining-construction work; in such cases, the working system is characterized by only the order and sequence of accomplishing the stripping and mining work. A working system is naturally associated with the integrated sets (complexes) of mining and hauling equipment employed in a quarry. Whereas the working system determines the order and sequence of mining work, the complexes of equipment determine the types, capacity and arrangement of equipment units to ensure the accomplishment of mining work in the specified volume and order. Classifications of mining systems and equipment complexes will be considered later, since certain mining systems can be characterized by a definite sequence and regularity of performance of stripping, mining and development work, irrespective of the types of integrated mechanization means which are employed in the system.

Depending on interrelations between stripping, mining and development work, the following working systems may be distinguished:

- Dependent (or heavily dependent) systems, i.e. those in which the stripping, mining and development operations are closely interrelated as regards their sequence in time and space; in that case, the planned stripped reserves of the mineral are rather limited (usually sufficient for operation of a system for 15–45 days) and the order of mining work is strictly specified by the calendar schedule.
- Semi-dependent systems in which the stripping, mining and development work can be carried out without strict coordination in time; the planned stripped reserves of the mineral may he substantial (for a period of operation of 3–6 months); the order of various operations is specified by the annual calendar schedule which allows substantial time reserves between various operations and these may be carried out with different intensity.
- Independent systems in which the stripping, mining and development work can be carried out practically independently from one another; the stripped reserves of the mineral are not practically restricted by work organization and the time reserves for performing various mining operations are rather large.

When working horizontal and gently dipping deposits, the initial spread of work for stripping and mining work in a quarry is formed on completion of the development work; development work can only be renewed for reconstruction of the quarry. Thus, the working systems of horizontal and gently dipping deposits during the period of exploitation are characterized only by the order and sequence of stripping and mining work and variations in the length of the working line or the height of individual benches and the dimensions of working hems. Such systems are called continuous (or bulk) systems.

When working dipping and steep deposits, development work can be carried out both during the construction and exploitation of a quarry, in order to form a new line of stripping and mining work. The composition of the development work in the exploitation period includes stripping and cutting of new working levels. Thus, the working systems of dipping and steep deposits are characterized by the order of accomplishment of stripping and mining work and regular development work. Such systems may be called deepening (or sinking) systems. Hillside deposits are worked out by using the systems of the first group (continuous systems). With steep slopes of the surface and steeply dipping deposits, the systems of the second group (deepening systems) are employed. When working deposits with complicated topographic and geological conditions of bedding, systems of both groups can be used in a quarry. By the direction of advance of the working line in plan, the following working systems can be distinguished:

- Longitudinal systems in which a one-sided or two-sided line of stripping and mining work is advanced parallel to the longer axis of the open-pit field.
- Transverse systems in which a one-sided or two-sided line of stripping and mining work is advanced parallel to the shorter axis of the open-pit field.
- Fan-like systems in which the line of stripping and mining work is advanced in a fan-like mode with one central or a number of dis-concentrated turning points.
- Circular systems in which the working zone includes all flanks of the quarry around its perimeter and working is carried out in concentric circular strips from the center to the boundaries of the open-pit field or from the boundaries to the center.

With all versions of working systems, of prime importance is the position of waste dumps (external, internal or combined waste dumps), which determine the direction of movement of overburden rock. The classification of working systems by the features indicated above is given in Table 8.1 and shown graphically in Fig. 8.31. This classification, which is based on geological and geometrical characteristics of working systems, characterizes the essence of the technology of opencast mining and can help in calculations of working systems. The choice of a particular working system is substantiated by establishing the quantitative relationships between the main dimensions of the deposit and open-pit field, the parameters of the elements of a working system, the parameters and arrangement of mining equipment, and the production capacity of the quarry for mining, stripping and development work. The choice of a working system is based on the following positions:

- The highest production capacity of the quarry for extraction of the mineral is established for a particular case as permitted by the natural and engineering conditions. The highest production capacity of a quarry depends on the kind of integrated mechanization of mining work chosen for the quarry and used in calculations of the working system. They are done in the design of new and reconstruction of existing quarries.
- It must be ensured that the operating quarry will have the specialized planned production capacity for extraction of the mineral. In calculations, a number of

Table 8.1 Classification of open cast mining system

Group index	Group of systems	Subgroup index	Subgroup	System index	Working system
C	Continuous	CL	Continuous longitudinal	CLO	Continuous longitudinal one sided
				CLT	Continuous longitudinal two sided
		CT	Continuous transverse	CTO	Continuous transverse one sided
				CTT	Continuous transverse two sided
		CF	Continuous fan like	CFC	Continuous fan like central
				CFD	Continuous Fan like dis-central
		CC	Continuous circular	CCC	Continuous Circular Central
				CCP	Continuous circular peripheral
D	Depending	DL	Depending longitudinal	DLO	Depending longitudinal one sided
				DLT	Depending Longitudinal two sided
		DT	Depending transverse	DTO	Depending transverse one sided
				DTT	Depending transverse two sided
		DF	Depending fan-like	DFD	Depending fan-like dis-concentrated
		DC	Depending circular	DCC	Depending circular central
DC	Combined		Various combinations of systems indicated above		

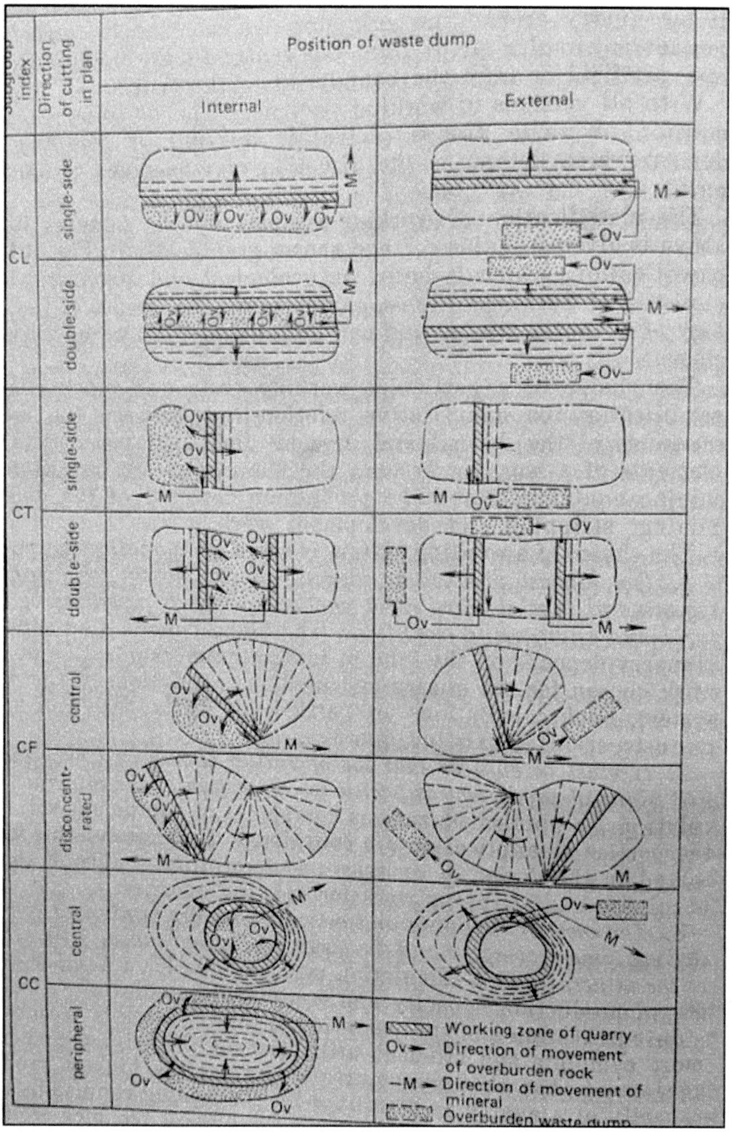

Fig. 8.31 Classification of opencast mining systems

various complexes of equipment suitable for the particular application are considered. Problems of this kind are solved in the design of quarries and feasibility studies of the plans of mining and stripping work for existing quarries.

- The costs of stripping and mining work are minimized (for a selected equipment complex and the given production capacity for extraction of the mineral). This

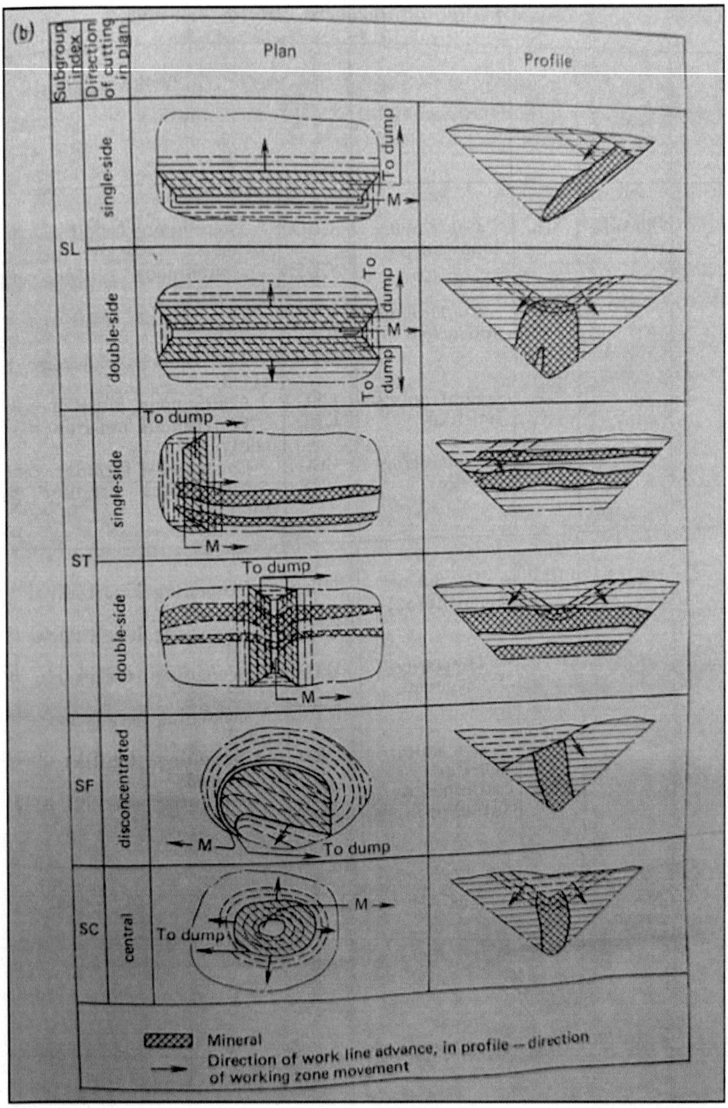

Fig. 8.31 (continued)

calculation may be done for a number of versions of production capacity and equipment complex.

Intensification and concentration of mining work is favorable for more complete and efficient utilization of the mining and hauling equipment. For that reason, realization of the highest production capacity of a quarry as permitted by the mining conditions is associated in most cases with attaining the highest technico-eonomical results of working.

8.12 Classification of Mining Systems

8.12.1 Based on the Direction of Transfer of Overburden and the Method of Stripping Work

In 1947, Sheshko proposed a classification of mining systems by the direction of displacement (transfer) of the overburden rock to waste dumps. By this characteristic, all mining systems can be divided as follows, Fig. 8.32:

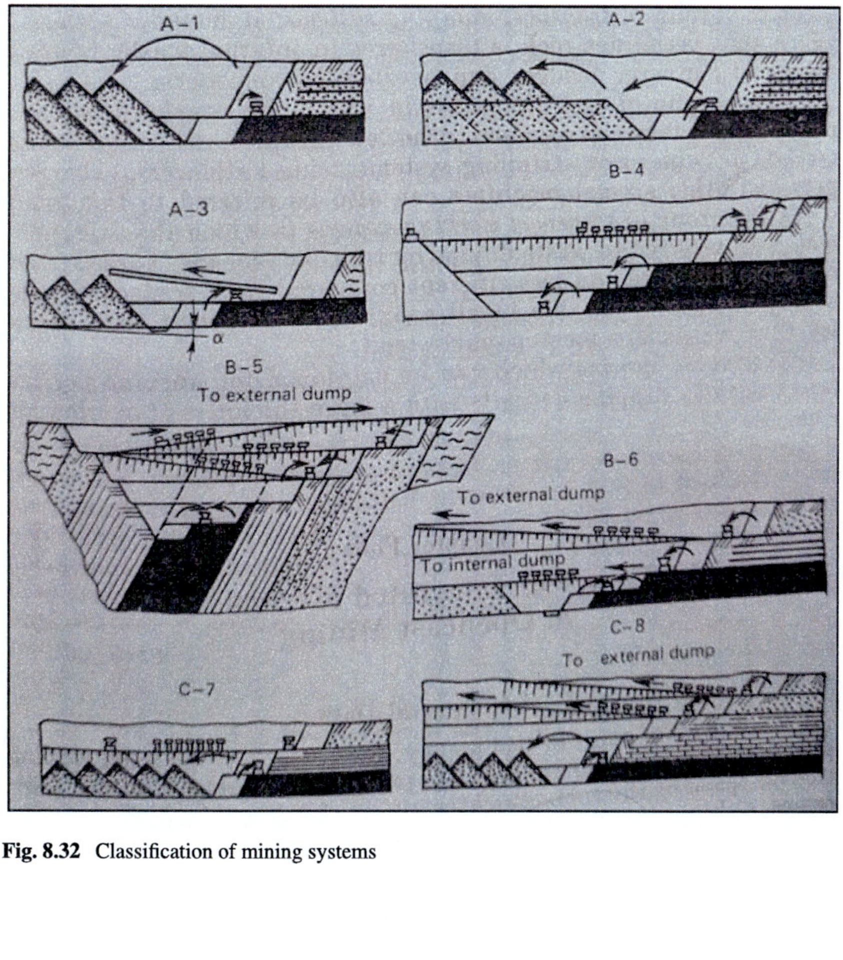

Fig. 8.32 Classification of mining systems

- Systems with transverse displacement of the overburden rock to waste dumps without the use of transport means; these systems may be called transport- less.
- Systems with longitudinal (frontal) transfer of the overburden rock to waste dumps with the application of transport means; they may be called transport mining systems.
- Combined systems with transverse and longitudinal transfer of the overburden rock to waste dumps; they have the characteristic features of both transport and transport-less systems.

Systems in these groups are further subdivided by the method and difficulties associated with accomplishment of hauling and dumping work. Group A is divided by the method of accomplishment of hauling and dumping work into systems A-1, A-2 and A-3. A system A-0 is separated specially for the cases when the volume of stripping work is insignificant. Group B is divided by the relative difficulty of rock transportation into systems B-4, B-5, and B-6. Group C includes two working systems C-7 and C-8, which are classified by the relative prevalence of transport-less or transport displacement of the overburden rock. In 1952, Melnikov proposed a classification of mining systems based on the method of stripping work. All mining systems are divided into five groups, Fig. 8.33:

1. The group of transport-less mining systems, where the transfer of the rock from a face to an internal waste (lump is performed by stripping excavating machines (power shovels and draglines).
2. The group of transport-dumping systems; it includes systems in which the overburden rock is transferred to internal waste dumps by means of conveyer bridges and overburden spreaders.
3. The group of special systems in which the overburden rock is removed by tower excavators, wheeled scrapers, hydro mechanized method or cableways; stripping systems using bulldozers, cable scrapers and other special machines can also be referred to this group.
4. The group of transport working systems in which the overburden rock is transferred to waste dumps by transport means. These systems are more intricate amid less efficient compared with transport-less systems, but are applicable in all conditions of deposit bedding and for that reason are most popular.
5. Combined systems which can be employed for working horizontal and gently dipping deposits with a large thickness of overburden.

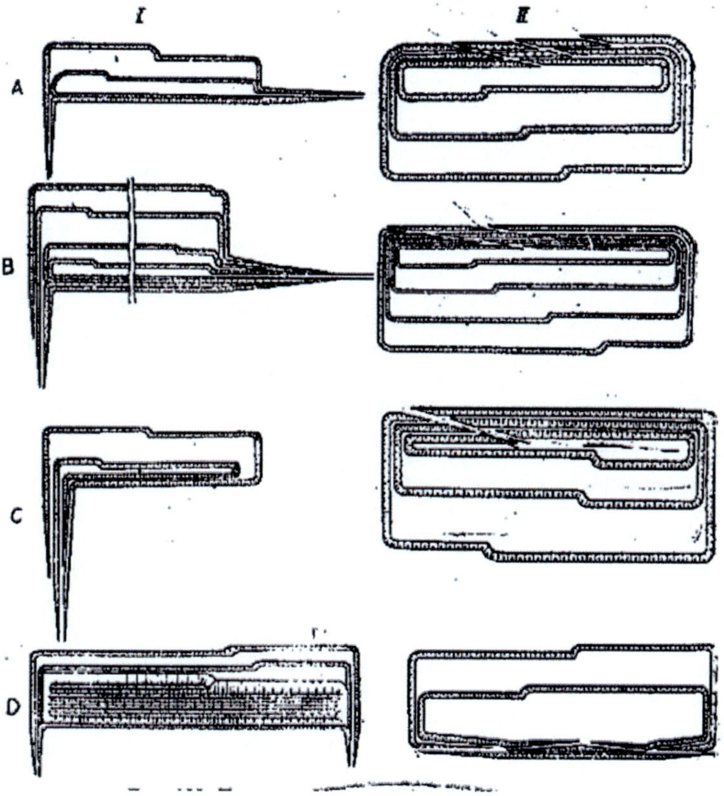

Fig. 8.33 Schemes of surface mining development

References

1. Anath KN, Rakesh V (2013) Design and selecting proper conveyor belt. Int J Adv Technol 4(2):43–49
2. Anderson GP, Kirk SJ (1992) Open cast (strip) mining. In: Hartman HL (eds) SME mining engineering handbook. Society for Mining, Metallurgy, and Exploration, Inc., Littleton, CO
3. Cole A (1999) Environmental geology. Chapter: Surface mining, strip mining, quarries. https://doi.org/10.1007/1-4020-4494-1_318
4. Gogolewska A (2011) Surface and underground mining technology: Mining and Power Engineering. Wroclaw University of Technology
5. Anon (1979) Illustrated surface mining methods. Skelly and Loy, McGraw-Hill, New York
6. Anon (1983) Anatomy of a mine from Prospect in Production. General Technical Report INT-35 (Rev.), US Department of Agriculture, Forest Service, Ogden, UT
7. Atkinson T (1992) Strip mine planning and design. In: Hartman HL (eds) SME mining engineering handbook.Society for Mining, Metallurgy, and Exploration, Inc., Littleton, CO
8. Banerjee SK, Gupta BN (1996) Biological reclamation of mined out land. TFRI Series IV. Tropical Forest Research Institute, Jabalpur (Madhya Pradesh)
9. Bohnet E Jr (1992) Surface mining: comparison of methods. In: Hartman HL (eds) SME mining engineering handbook. Society for Mining, Metallurgy, and Exploration, Inc., Littleton, CO

10. Bricker ML (1992) Surface mining: mechanical extraction methods. In: Hartman HL (eds) SME mining engineering handbook. Society for Mining, Metallurgy, and Exploration, Inc., Littleton, CO
11. Brown S (2004) Conveyor noise specification and control. In: in Acoustics, Gold Coast, Australia
12. CEMA (2014) Belt conveyor of bulk materials
13. Chironis NP (ed) (1984) improving mining methods and larger equipment reach deeper seams. Coal Age 89(7)
14. Daniyan IA, Adeodu AO, Dada OM (2014) Design of material handling equipment: belt conveyor system for crushed limestone using 3roll Idlers. J Adv Eng Technol 1(1)
15. Dunlop F (2009) Conveyor handbook. Conveyor Belting Australia, pp 1–70
16. Gitt MJ, Dollhopf DJ (1991) Coal waste reclamation using automated weathering to predict lime requirement. J Environ Qual 20:285–288
17. Gupta RK (1979) Reclamation and use of coal-mined ecosystem in India: a review. In: Wali MD (ed) Ecology and coal resource development. Pergamon Press, New York, NY, pp 510–515
18. Hao XZ, Zhou DM, Wang YJ, Chen HM (2004) Study of rye grass in copper mine tailing treated with peat and chemical fertilizer. Acta Pedol Sin 41(4):645–648
19. Hartman HL (1987) Surface mining: method comparison and summary. In: Introductory mining engineering. Wiley-Interscience, New York
20. Hartman HL (ed) (1992) SME mining engineering handbook, 2nd edn. Society for Mining, Metallurgy, and Exploration, Baltimore, MD
21. Hood MC, Roxborough FF (1992) Rock breakage: mechanical. In: Hartman (ed) SME mining engineering handbook. Society for Mining, Metallurgy, and Exploration, Inc., Littleton, CO
22. Hrebar MJ (1992) Strip mine planning and design. In: Hartman (ed) SME mining engineering handbook. Society for Mining, Metallurgy, and Exploration, Inc., Littleton, CO
23. Hu Z, Caudle RD, Chong SK (1992) Evaluation of firm land reclamation effectiveness based on reclaimed mine properties. Int J Surf Min Reclam Environ 6:129–135
24. Radlowski JK (1988) In-Pit crushing and conveying as an alternative to an all truck system in open pit mines. M.Sc. thesis. Faculty of Graduate Studies. University of British Columbia. Department of Mining and Mineral Process Engineering. Canada\
25. Jordaan JT (2011) Determining waste mining capacities for open pit mines. In: Potvin Y (ed) Proceedings of the fourth international seminar on strategic versus tactical approaches in mining, Australian Centre for Geomechanics, Perth, pp. 339–346. Downloadable from: https://papers.acg.uwa.edu.au/p/1108_27_Jordaan/
26. Joy J (1919) Loading machine. United States of America Patent 1306064, 10 06 1919.
27. Joy J (1959) Mobile conveyor apparatus for underground mines. United States of America Patent 2879884, 31 03 1959.
28. Jha AK, Singh JS (1991) Spoil characteristics and vegetation development of an age series of mine spoils in a dry tropical environment. Vegetatio 97:67–76
29. Jim CY (2001) Ecological and landscape rehabilitation of a quarry site in HongKong. Restor Ecol 9:85–94
30. Kundu NK, Ghose MK (1998) Studies on the existing plant communities in Eastern coalfield areas with a view to reclamation of mined out lands. J Environ Biol 19(1):83–89
31. Maiti SK (2003) MoEF Report, an assessment of overburden dump rehabilitation technologies adopted in CCL, NCL, MCL, and SECL mines (Grant no. J-15012/38/98-IA IIM)
32. McCarter MK (1992) surface mining: mechanical extraction methods. In: Hartman (ed) SME mining engineering handbook. Society for Mining, Metallurgy, and Exploration, Inc., Littleton, CO
33. McDonald DB (1992) Open cast (strip) mining. In: Hartman (ed) SME mining engineering handbook. Society for Mining, Metallurgy, and Exploration, Inc., Littleton, CO
34. McLean CA, McDowell AWK (1992) Dredging and placer mining. In: Hartman (ed) SME mining engineering handbook. Society for Mining, Metallurgy, and Exploration, Inc., Littleton, CO

35. Meade LP (1992) Surface mining: mechanical extraction methods. In: Hartman (ed) SME mining engineering handbook. Society for Mining, Metallurgy, and Exploration, Inc., Littleton, CO
36. Nekoufar S (2014) Trucks vs. pocket belt conveyors. Australian Bulk Handling Review, no. January/February, p 6
37. Ogedengbe TI (2010) Lecture note on applied techniques of production management, pp 16–20
38. Rand G (1992) Open cast (strip) mining. In: Hartman (ed) SME mining engineering handbook. Society for Mining, Metallurgy, and Exploration, Inc., Littleton, CO
39. Rulmeca, Technical information. Project and Design Criteria for Belt conveyors, pp 1–50
40. Rupprecht S (2012) Mine development—access to deposit. University of Johannesburg, Johannesburg
41. Stubblefield GM, Fish RW (1992) Open cast (strip) mining. In: Hartman (ed) SME Mining Engineering Handbook. Littleton, CO: Society for Mining, Metallurgy, and Exploration, Inc.
42. Superstein LW (1992) Production operations. In: Hartman HL (eds) SME mining engineering handbook. Society for Mining, Metallurgy, and Exploration, Inc., Littleton, CO
43. Sweigard RJ (1992) Materials Handling: Loading and Haulage. In: Hartman HL (ed) SME mining engineering handbook. Society for Mining, Metallurgy, and Exploration, Inc., Littleton, CO
44. Tannant DD, Regensburg B (2001) Guidelines for mine haul road design. (University of British Columbia: Kelowna, B.C. Canada). 115 pp. Downloadable from: https://www.researchgate.net/profile/Dwayne_Tannant/publication/277759950_Guidelines_for_Mine_Haul_Road_Design/links/5584333f08aeb0cdaddbb03d/Guidelines-forMine-Haul-Road-Design.pdf
45. Thompson (2015) Principles of mine haul road design and construction. Course Notes, 156 pp. Downloadable from: http://mineravia.com/yahoo_site_admin/assets/docs/Principles_of_mine_haul_road_design_and_construction_v5_Sep_2015_RJTs.28192929.pdf
46. Tussey IJ (1992) Open cast (strip) mining. In: Hartman HL (ed) SME mining engineering handbook. Society for Mining, Metallurgy, and Exploration, Inc., Littleton, CO
47. Vanamane SS, Mane PA, Inamder KH (2011) Design and its verification of belt conveyor system used for mould using belt comp software. Int J Appl Res Mech Eng 1(1):48–52
48. Woodruff SD (1966) Methods of working coal and metal mines: Chapter 1—Strip mining (open-cast mining) of coal. Pergamon, pp 393–468
49. Whitman GB (1992) Surface mining: mechanical extraction methods. In: Hartman HL (ed) SME mining engineering handbook. Society for Mining, Metallurgy, and Exploration, Inc., Littleton, CO
50. Wheeler CA (2013) Rail conveyor system. Australia NSW Patent US20130118371 A1, 16 May 2013
51. Wheeler CA (2015) An Overview of specialized belt conveyors. TUNRA Bulk Solids ICCHPS, Newcastle, 2015
52. Whyte J, Cumming J (eds) (2007) Mining explained, discovery, extraction, refining, marketing, investing. The Northern Miner, Toronto, ON
53. Zdzislaw Murzynski, Belt conveyors and mechanization of storage process in polish surface mining. M.Sc. Eng., Central Research and Design Ins. for Opencast Mining. Peltegor

Glossary of Terms

Acidic rocks Igneous rock carrying a high (greater than 65%) proportion of silica.

Adit An opening driven horizontally into the side of a mountain or hill for providing access to a mineral deposit.

Advance stripping The removal of barren or sub-ore-grade earthy or rock materials required to expose and permit the minable grade of ore to be mined: The removal of these non-ore materials is known as stripping:

Alluvial Of or pertaining to alluvium: Adjectively used to identify particular types of, or minerals found associated with, deposits made by flowing water as, alluvial fan, alluvial terrace, alluvial gold, alluvial tin.

Alluvial deposit placer deposit: Earth, sand, gravel, or other rock or mineral materials transported by and laid down by flowing water: Compare alluvium:

Alluvium Clay silt, sand, gravel, or other rock materials transported by flowing water and deposited in comparatively recent geologic time as sorted or semi-sorted sediments in river beds, estuaries, and floor plains, on lakes, shores and in fans at the base of mountain slopes, and estuaries: The term is not applied to subaqueous sediments deposited in seas or lakes or to non-sorted sediments carried or deposited by glaciers: Compare till:

Angle of rest; angle of repose The maximum slope at which a heap of any loose or fragmented solid material will stand without sliding or come to rest when poured or dumped in a pile or on a slope:

Angle of slide The slope, measured in degrees of deviation from the horizontal, on which loose or fragmented solid materials will start to slide; it is a slightly greater angle than the angle of rest:

Apex The top or terminal edge of a vein on surface or its nearest point to the surface.

Assay A chemical test performed on a sample of ores or minerals to determine the amount of valuable metals contained.

Bank; bench face Specifically, a usually steeply sloping mass of any earthy or rock material rising above the digging level from which the soil or rock is to be dug from its natural or blasted position in an open pit mine or quarry:

M. M. Ali Elbeblawi et al., *Surface Mining Technology*, Topics in Mining, Metallurgy and Materials Engineering, https://doi.org/10.1007/978-981-16-3568-7

Bank height; bench height; digging height The vertical height of a bank as measured between its highest point or crest and its toe at the digging level or bench: Compare berm:

Bank slope; bench slope The angle, measured in degrees of deviation from the horizontal, at which the earthy or rock material will stand in an excavated, terrace-like cut in an open-pit mine or quarry:

Bench A ledge, which, in open-pit mines and quarries, forms a single level of operation above which mineral or waste materials are excavated from a contiguous bank or bench face: The mineral or waste is removed in successive layers, each of which is a bench, several of which may be in operation simultaneously in different parts of, and at different elevations in an open-pit mine or quarry: Compare berm:

Berm A horizontal shelf or ledge built into an embankment or sloping wall of an open pit or quarry to break the continuity of an otherwise long slope for the purpose of strengthening and increasing the stability of the slope or to catch or arrest slope slough material. A berm may also be used as a haulage road or serve as a bench above which material is excavated from a bank or bench face: This berm is sometimes used as a synonym for bench: Compare bench.

Berm interval Vertical distance from crest of berm to its underlying toe, as in a bank or bench:

Blasthole A drill hole in a mine that is filled with explosives in order to blast loose a quantity of rock.

Boom A telescoping, hydraulically powered steel arm on which drifters, man-baskets and hydraulic hammers are mounted.

Breast A working face in a mine usually restricted to a stope.

Broken reserves The ore in a mine which has been broken by blasting but which has not yet been transported to surface.

Bucket-ladder dredge; bucket-line dredge; ladder-bucket dredge A dredge whose digging mechanism consists of a ladder like truss on the periphery of which is attached an endless chain which rides on sprocket wheels and on which buckets are attached.

Bucket Wheel Excavator, BWE A continuous digging machine originally designed and used in large-scale

Stripping and mining of East German brown coal deposits Its digging mechanism is essentially a boom on which is mounted a rotating vertical wheel having buckets on its periphery: As the rotating wheel is pressed into the material to be dug, the buckets cut, gather, and discharge the material onto a conveyor belt where it is moved to the mined-materials transport system.

Bulk mining Any large-scale, mechanized method of mining involving many thousands of tonnes of ore being brought to surface per day.

Bulldozer; dozer A tractor on the front end of which is mounted a vertically curved steel blade held at a fixed distance by arms secured on a pivot or shaft near the horizontal center of the tractor: The blade can be lowered or tilted vertically by cables or hydraulic rams: It is a highly versatile piece of earth excavating and moving equipment especially useful in land clearing and leveling work, in stripping topsoil, in road and ramp building and in floor or bench cleanup and gathering operations.

Burden The distance between the explosive charge and the free face of the material to be blasted: b: Barren or non-ore material that overlies and must be removed to gain access to minable grade material: Frequently called overburden or cover.

Capitalization A financial term used to describe the value financial markets put on a company. It is determined by multiplying the number of outstanding shares of a company by the current stock price.

Carryall; carryall scraper; scraper A self—loading carrier device with a scraper-like, retractable bottom usually self -propelled and used especially for excavating and hauling unconsolidated or crushed rock and earthy materials.

Carryall Trade name for Le Tourneau Westinghouse scrapers:

Claim A portion of land held either by a prospector or a mining company. In Canada, the common size is 1,320 ft. (about 400 m) square, or 40 acres (about 16 ha).

Clamshell A twin-jawed bucket without teeth, usually hung from the boom of a crane that can be either crawler or wheel mounted: The bucket is dropped in the open position onto the material to be excavated or handled: It is then closed, encompassing material between the hinged two halves:

Chute An opening usually constructed of timber and equipped with a gate, through which ore is drawn from a stope into mine cars.

Continuous miner A piece of mining equipment which produces a continuous flow of ore from the working face.

Country rock Loosely used to describe the general mass of rock adjacent to an orebody. Also it is known as the host rock.

Coyote blasting; coyote-hole blasting; gopher-hole blasting A term applied to the method of blasting in which large charges are fired in small adits or tunnels driven, at the level of the floor, in the face of a quarry or slope of an open-pit mine.

Crawler One of a pair of an endless chain of plates driven by sprockets and used instead of wheels, by certain power shovels, tractors, bulldozers, drilling machines, etc.:, as a means of propulsion: Also any machine mounted on such tracks.

Crosscut A horizontal opening driven from a shaft and (or near) right angles to the strike of a vein or other orebody.

Deck The area around the shaft collar where men and materials enter the cage to be lowered underground.

Decline A sloping underground opening for machine access from level to level or from surface; also called a ramp.

Development The preparation of a mineral deposit for commercial production, including installation of plant and machinery and the construction of all related facilities Underground work carried out for the purpose of opening up a mineral deposit. Includes shaft sinking, crosscutting, drifting and raising.

Dip The angle at which a vein, structure or rock bed is inclined from the horizontal as measured at right angles to the strike.

Dipper A digging bucket rigidly attached to a stick or arm on an excavating machine; also the machine itself.

Dipper stick The straight shaft which connects the digging bucket with the boom on an excavating machine or power shovel.

Down-the-hole drill A percussive or hammer drill in which the bit-driven mechanism is located immediately behind the drill bit and is small enough in diameter to permit it to enter and follow the bit down into the hole drilled.

Dragline A type of excavating equipment which casts a rope-hung bucket a considerable distance, collects the dug material by pulling the bucket toward itself on the ground with a second rope, elevates the bucket, and dumps the material on a spoil bank, in a hopper, or on a pile.

Dredge Any apparatus used for excavating under water: Usually a large raft, barge, or boat on which are mounted a chain of buckets, suction pumps, or other devices to dig and elevate the materials from a subaqueous deposit.

Dredging The act of using a dredge to excavate underwater deposits.

Drift A horizontal underground opening that follows along the length of a vein or rock formation as opposed to a crosscut which crosses the rock formation.

Drop cut The initial cut made in the floor of an open pit or quarry for the purpose of developing a bench at a level below the floor.

Dump A pile of broken rock or ore on surface.

Dump leaching Term applied to dissolving and recovering minerals from sub-ore-grade materials from a mine dump. The dump is irrigated with water, sometimes acidified, which percolates into and through the dump and run-off from the bottom of the dump is collected and mineral in solution is recovered by a chemical reaction.

Earth slide A term applied to the down-slope movement of a part of an earth embankment when the distance moved is sufficient to break up the blocks and pulverize the earth enough so that the major part of the moving mass moves in a somewhat fluid manner: Compare earth slump.

Earth slump A term applied to down-slope movement of part of an earth embankment in block-like masses without other apparent deformation than the change in level: Compare earth slide.

Environmental impact study A written report, compiled prior to a production decision that examines the effects proposed mining activities will have on the natural surroundings.

Exploration Prospecting, sampling, mapping, diamond drilling and other work involved in searching for ore.

Face The end of a drift, crosscut or stope in which work is taking place.

Footwall The rock on the underside of a vein or ore structure

Front end loader A tractor loader with a digging bucket mounted and operated at the front end of the tractor.

Gangue The worthless minerals in an ore deposit.

Geology The science concerned with the study of the rocks which compose the Earth.

Geophysics The study of the physical properties of rocks and minerals.

Geophysical survey A scientific method of prospecting that measures the physical properties of rock formations.

Giant See hydraulic monitor.

Glory hole A pit from which or in which material dumped into it is fed by gravity to an underground haulage system.

Grade The incline of a road, railway, skip way, etc.: in terms of degrees from the horizontal, percent of rise to the horizontal distance, or inches per foot of horizontal projection. b: The percentage of each valuable mineral in an ore.

Ground sluicing To strip ground down-slope by means of a directed stream of water to excavate placer material and transport it to a riffled trough in which the valuable mineral is recovered: Compare hydraulic king.

Hanging wall The rock on the upper side of a vein or ore deposit.

Haul road A road built to carry heavily loaded trucks at a good speed: The grade is limited on this type of road and usually kept to less than 17 percent of climb in direction of load movement.

Heap leaching A process whereby valuable metals, usually gold and silver, are leached from a heap, or pad, of crushed ore by leaching solutions percolating down through the heap and collected from a sloping, impermeable liner below the pad.

Highwall The unexcavated face of exposed overburden and coal or ore in an opencast mine or the face or bank on the uphill side of a contour strip mine excavation.

Host rock The rock surrounding an ore deposit.

Hydraulic monitor; giant; monitor A device for directing a high-pressure jet of water in hydraulic king. It is essentially a swivel-mounted, counter-weighted nozzle attached to a tripod or other type of stand and so designed that one man can easily control and direct the vertical and lateral movements of the nozzle.

Hydraulic-king Excavating alluvial or other mineral deposits by means of high-pressure water jets.

Industrial minerals Non-metallic, non-fuel minerals used in the chemical and manufacturing industries. Examples are asbestos, gypsum, salt, graphite, mica, gravel, building stone and talc.

Intermediate rock An igneous rock containing 52–66% quartz.

Jet piercing The use of high-velocity jet flames to drill holes in hard rocks, as taconite, and to cut channels in granite quarries: It involves combustion of oxygen and a fuel oil fed under pressure through a nozzle to produce a jet flame generating a temperature of over 5,000o F: A stream of water joins the flame, and the combined effect is a thermodynamic spalling and disintegration of the rock into fragments which are blown from the hole or cut:

Ladder-bucket dredge See bucket-ladder dredge.

Leach material Material sufficiently mineralized to be economically recoverable by selectively dissolving the wanted mineral in a suitable solvent. See also leaching.

Leach pile Mineralized materials stacked so as to permit wanted minerals to be effectively and selectively dissolved by application of a suitable solute.

Leaching The extraction of a mineral from an ore by selectively dissolving it in a suitable solvent, as water or water somewhat acidified by addition of sulfuric acid, hydrochloric acid, etc.

Level The horizontal openings on a working horizon in a mine; it is customary to work mines from a shaft, establishing levels at regular intervals, generally about 50 m or more apart.

Loading equipment Mechanical shovels or other machines singly or in combination used to load excavated or stockpiled materials into trucks, mine cars, conveyors, or other materials transportation or haulage units:

Lode A mineral deposit in solid rock.

Long ton 2240 lbs. avoirdupois (compared with a short ton, which is 2000 lbs.).

Margin Cash deposited with a broker as partial payment of the purchase price for any type of listed stock. The stock is held by the broker as security for the loan.

Marginal deposit An orebody of minimal profitability.

Metalized slurry blasting The breaking of rocks, etc.: using a slurries explosive medium containing a powdered metal, as powdered aluminum.

Mineral A naturally occurring homogeneous substance having definite physical properties and chemical composition and, if formed under favorable conditions, a definite crystal form.

Monitor See hydraulic monitor.

Muck Ore or rock that has been broken by blasting.

Muck sample A representative piece of ore that is taken from a muck pile and then assayed to determine the grade of the pile.

Monitor See hydraulic monitor.

Native metal A metal occurring in nature in pure form, uncombined with other elements.

Net profit interest A portion of the profit remaining after all charges, including taxes and bookkeeping charges, such as depreciation, have been deducted.

Nugget A small mass of precious metal, found free in nature.

Open-pit mine; opencast mine; Open cut mine; strip mine A mine working or excavations open to the surface.

Orebody A natural concentration of valuable material that can be extracted and sold at a profit.

Ore Reserves The calculated tonnage and grade of mineralization which can be extracted profitably; classified as possible, probable and proven according to the level of confidence that can be placed in the data.

Overbreak; back break The rock which is broken (and sometimes must be excavated) by blasting outside the intended area or line of break. Over break may occur as the result of misalignment or unintentional overcharging of blasthole or intentionally over breaking as in the toe of a bench face to facilitate digging to grade.

Outcrop An exposure of rock or mineral deposit that can be seen on surface that is, not covered by soil or water.

Percolation leaching The selective removal of a mineral by causing a suitable solvent to seep into and through a mass or pile of material containing the desired soluble mineral.

Pillar A block of solid ore or other rock left in place to structurally support the shaft, walls or roof of a mine.

Pit limits The vertical and lateral extent to which the mining of a mineral deposit by open pitting may be economically carried: Cost of removing overburden or waste material versus the minable value of the ore so exposed is usually the factor controlling the limits of a pit.

Pit slope The angle at which the wall of an open pit or cut stands as measured along an imaginary plane extended along the crests of the berms or from the slope crest to its toe: Compare slope.

Placer deposit See alluvial deposit: For purposes of this volume, unconsolidated deposits of detrital material containing valuable mineral.

Power shovel An excavating and loading machine consisting of a digging bucket at the end of an arm suspended from a boom, which extends crane-like from that part of the machine which houses the power plant: When digging the bucket moves forward and upward so that the machine does not usually excavate below the level at which it stands.

Profit and loss statement The income statement of a company detailing revenues minus total costs to give total profit.

Prospect A mining property, the value of which has not been determined by exploration.

Quarry An open or surface working usually for the extraction of building stone, slate, aggregate, limestone, etc.

Rail haulage system A materials transportation system consisting of gondola cars, and the steel rails on which the cars are moved about which a suitably powered traction unit as a locomotive.

Raise A vertical or inclined underground working that has been excavated from the bottom upward.

Reclamation The restoration of a site after mining or exploration activity is completed.

Recovery The percentage of valuable metal in the ore that is recovered by metallurgical treatment.

Resource The calculated amount of material in a mineral deposit, based on limited drill information.

Rippability A measure of the ease or difficulty with which a rock or earth material can be broken by tractor drawn rippers or rigid steel tines into pieces that can be economically moved by other equipment, usually scrapers.

Ripping The act of breaking, with a tractor-drawn ripper or long angled steel tooth compacted soils or rock into pieces small enough to be economically excavated or moved by other equipment as a scraper or dozer.

Rockbolting The act of supporting openings in rock with steel bolts anchored in holes drilled especially for this purpose

Rockburst A violent release of energy resulting in the sudden failure of walls or pillars in a mine, caused by the weight or pressure of the surrounding rocks.

Rock Any natural combination of minerals; part of the earth's crust.

Rocker A small bucket or box mounted on two rocker arms in which auriferous alluvial sands are agitated by oscillation, in water, to collect gold: A shortened term for rocker shovel; rocker arm shovel.

Rocker shovel; rocker arm shovel A digging and loading machine consisting of a bucket attached to a pair of semicircular runners which when rolled lifts and dumps the bucket load into a car or other materials transport unit behind the machine.

Rock factor The number of cubic meters of a particular rock type required to make up one tonne of the material. One tonne of a highly siliceous ore may occupy 0.40 cubic meters, while a tonne of dense sulphide ore may occupy only 0.25 cubic meters.

Rotary drill Broadly, various types of drill machines that rotate a rigid, tubular string of rods to which is attached a bit for cutting rock to produce bore-holes: The bit may be a roller cone bit, a toothed or fishtail drag bit, an auger bit, or a diamond bit.

Run-of-mine A term used loosely to describe ore of average grade.

Sawing The cutting of blocks of stone using a continuous one or three-strand steel wire cable running over pulleys as a bolt, which when fed by slurry of sand and water and held against the rock by tension cuts a narrow uniform channel by abrasion.

Secondary blasting Reblasting oversize pieces of rock to reduce them to a size suitable for handling by the excavating, materials transporting, and crushing equipment available.

Shaft A vertical or inclined excavation in rock for the purpose of providing access to an orebody. It is usually equipped with a hoist at the top, which lowers and raises a conveyance for handling workers and materials.

Short ton 2000 lbs. avoirdupois.

Shovel Any bucket-equipped machine used for digging and loading earthy or fragmented rock materials.

Skip A self-dumping bucket used in a shaft for hoisting ore or rock.

Slack-line cableway A cableway having one low and one high tower and a track cable with adjustable tension suspended between them: One end of the track cable is attached to a hoist drum by means of which the tension on the cable can be rapidly changed so as to position, lower, or raise the digging skip.

Slope The inclined surface of a hill, or any part of the surface of the earth; the inclination of a roadway: An in dined tunnel: Compare pit slope.

Slope stability The resistance of any inclined surface, as the wall of an open pit or cut, to failure by sliding or collapsing.

Sluice box A long inclined trough or launder containing riffles in the bottom that provide a lodging place for heavy minerals in ore concentration: The material to be concentrated is carried down through the trough on a current of water:

Sluice boxes are widely used in placer operations for separating minerals, as gold, platinum, or other heavy minerals, from alluviums.

Sluicing A separation of minerals in a flowing stream of water: See also sluice box: Moving earth, sand, gravel, or other rock or mineral materials by flowing water.

Spiral system In open-pit mining, a haul road arranged spirally along the perimeter walls of the pit so that gradient of road is more or less uniform from bottom to the top of the pit.

Spoil; waste The overburden or non-ore material removed in gaining access to the ore or mineral material in surface mining.

Spoil pile See waste dump.

Stockpile Broken ore heaped on surface, pending treatment or shipment.

Stope An excavation in a mine from which ore is, or has been, extracted.

Strike The direction, or bearing from true north, of a vein or rock formation measure on a horizontal surface.

Stripping The removal of earth or non-ore rock materials as required gaining access to the ore or mineral or materials wanted; the process of removing overburden or waste material in a surface mining operation.

Stripping ratio The ratio of tonnes removed as waste relative to the number of tonnes of ore removed from an open-pit mine.

Sublevel A level or working horizon in a mine between main working levels.

Suction dredge A dredge in which the material excavated is lifted by pumping through a suction pipe.

Sustainable development Industrial development that does not detract from the potential of the natural environment to provide benefits to future generations.

Switchback A zigzag arrangement of a roadway (or rail tracks) for surmounting the grade of a steep hill or the slope wall of a surface or open-pit mine.

Till Non-sorted, non-stratified earth or rock materials carried or deposited by glaciers: Equivalent to moraine of some European geologists and to boulder clay of some British geologists: Compare alluvium.

Toe The base of a bank, bench, or slope in a quarry or open-pit mine.

Toe-hole A blasting hole, usually drilled horizontally or at a slight inclination into the base of a bank ~ bench, or slope of a quarry or open-pit mine.

Tower excavator A cableway excavator designed specifically for levee work but which is used occasionally in the stripping of overburden, spoils, or waste in surface mining: The unit is basically a Sauer man-type excavator with towers either fixed or movable: With the head-tower located on the spoil pile and the tail tower on the unexcavated wall, it is possible to dig pits of almost unlimited width.

Track haulage Movement or transportation of excavated or mined materials in cars or trucks running on rails.

Track shifter A machine or appliance used in shifting a railway track laterally: Also, the operator of such a device or machine.

Tram To haul cars of ore or waste in a mine.

Trench A long, narrow excavation dug through overburden, or blasted out of rock, to expose a vein or ore structure.

Truck Any wheeled vehicle, usually self-propelled, used to transport heavy articles or materials: In mining, term is usually applied to dump and/or bottom-dump semi-trailers used to transport mined waste and ore materials.

Umpire sample or assay An assay made by a third party to provide a basis for settling disputes between buyers and sellers of ore.

Vein A fissure, fault or crack in a rock filled by minerals that have traveled upwards from some deep source.

Wall rocks Rock units on either side of an orebody such as the hanging wall and footwall rocks of an orebody.

Waste dump; spoil pile The area where mine waste or spoil materials are disposed of or piled.

Winze is a shaft in a mine which joins two levels, by definition mined downwards (winzed).

Zone An area of distinct mineralization.